COMPUTATIONAL

PHYSICS

Steven E. Koonin
Professor of Theoretical Physics
California Institute of Technology

The Benjamin/Cummings Publishing Company, Inc.

Menlo Park, California • Reading, Massachusetts
Don Mills, Ontario • Wokingham, U.K. • Amsterdam • Sydney
Singapore • Tokyo • Mexico City • Bogota • Santiago • San Juan

Sponsoring Editor: Richard W. Mixter
Production Editor: Karen Gulliver
Cover Designer: Victoria Ann Philp

Library of Congress Cataloging-in-Publication Data

Koonin, Steven E.
  Computational physics.

  Bibliography: p.
  1. Mathematical physics. 2. Numerical analysis.
3. Physics--Computer programs. 4. Differential
equations--Numerical solutions. I. Title.
QC20.K665 1986   530.1'5'0285   85-15052
ISBN 0-8053-5430-1
BCDEFGHIJKL-HA-89876

The Benjamin/Cummings Publishing Company, Inc.
2727 Sand Hill Road
Menlo Park, CA 94025

This book was typeset by the author using the UNIX-based TROFF
text-processing system running on a Digital Equipment Corpora-
tion VAX-11/750 computer. Camera-ready output from a Versatec
V-80 printer-plotter was reduced 15% to produce the final copy.

For my parents

Preface

Computation is an integral part of modern science and the ability to exploit effectively the power offered by computers is therefore essential to a working physicist. The proper application of a computer to modeling physical systems is far more than blind "number crunching", and the successful computational physicist draws on a balanced mix of analytically soluble examples, physical intuition, and numerical work to solve problems which are otherwise intractable.

Unfortunately, the ability "to compute" is seldom cultivated by the standard university-level physics curriculum, as it requires an integration of three disciplines (physics, numerical analysis, and computer programming) covered in disjoint courses. Few physics students finish their undergraduate education knowing how to compute; those that do usually learn a limited set of techniques in the course of independent work, such as a research project or a senior thesis.

The material in this book is aimed at refining computational skills in advanced undergraduate or beginning graduate students by providing direct experience in using a computer to model physical systems. Its scope includes the minimum set of numerical techniques needed to "do physics" on a computer. Each of these is developed in the text, often heuristically, and is then applied to solve non-trivial problems in classical, quantum, and statistical physics. These latter have been chosen to enrich or extend the standard undergraduate physics curriculum, and so have considerable intrinsic interest, quite independent of the computational principles they illustrate.

This book should not be thought of as setting out a rigid or definitive curriculum. I have restricted its scope to calculations which satisfy simultaneously the criteria of illustrating a widely applicable numerical technique, of being tractable on a microcomputer, and of having some particular physics interest. Several important numerical techniques have therefore been omitted, spline interpolation and the Fast Fourier Transform among them. *Computational Physics* is perhaps best thought of as establishing an environment offering opportunities for further exploration. There are many possible extensions and embellishments of the material presented; using one's imagination along these lines is one of the more rewarding parts of working through the book.

Computational Physics is primarily a physics text. For maximum benefit, the student should have taken, or be taking, undergraduate courses in classical mechanics, quantum mechanics, statistical mechanics, and advanced calculus or the mathematical methods of physics. This is *not* a text on numerical analysis, as there has been no attempt at rigor or completeness in any of the expositions of numerical techniques. However, a prior course in that subject is probably not essential; the discussions of numerical techniques should be accessible to a student with the physics background outlined above, perhaps with some reference to any one of the excellent texts on numerical analysis (for example, [Ac70], [Bu81], or [Sh84]). This is also *not* a text on computer programming. Although I have tried to follow the principles of good programming throughout (see Appendix B), there has been no attempt to teach programming *per se*. Indeed, techniques for organizing and writing code are somewhat peripheral to the main goals of the book. Some familiarity with programming, at least to the extent of a one-semester introductory course in any of the standard high-level languages (BASIC, FORTRAN, PASCAL, C), is therefore essential.

The choice of language invariably invokes strong feelings among scientists who use computers. Any language is, after all, only a means of expressing the concepts underlying a program. The contents of this book are therefore relevant no matter what language one works in. However, *some* language had to be chosen to implement the programs, and I have selected the Microsoft dialect of BASIC standard on the IBM PC/XT/AT computers for this purpose. The BASIC language has many well-known deficiencies, foremost among them being a lack of local subroutine variables and an awkwardness in expressing structured code. Nevertheless, I believe that these are more than balanced by the simplicity of the language and the widespread fluency in it, BASIC's almost universal availability on the microcomputers most likely to be used with this book, the existence of both BASIC interpreters convenient for writing and debugging programs and of compilers for producing rapidly executing finished programs, and the powerful graphics and I/O statements in this language. I expect that readers familiar with some other high-level language can learn enough BASIC "on the fly" to be able to use this book. A synopsis of the language is contained in Appendix A to help in this regard, and further information can be found in readily available manuals. The reader may, of course, elect to write the programs suggested in the text in any convenient language.

This book arose out of the Advanced Computational Physics Laboratory taught to third- and fourth-year undergraduate Physics majors at Caltech during the Winter and Spring of 1984. The content and presentation have benefitted greatly from the many inspired suggestions of M.-C. Chu, V. Pönisch, R. Williams, and D.

Meredith. Mrs. Meredith was also of great assistance in producing the final form of the manuscript and programs. I also wish to thank my wife, Laurie, for her extraordinary patience, understanding, and support during my two-year involvement in this project.

Steven E. Koonin
Pasadena
May, 1985

How to use this book

This book is organized into chapters, each containing a text section, an example, and a project. Each text section is a brief discussion of one or several related numerical techniques, often illustrated with simple mathematical examples. Throughout the text are a number of exercises, in which the student's understanding of the material is solidified or extended by an analytical derivation or through the writing and running of a simple program. These exercises are indicated by the symbol ▯ ▯ ▯ in the outer margin.

The example and project in each chapter are applications of the numerical techniques to particular physical problems. Each includes a brief exposition of the physics, followed by a discussion of how the numerical techniques are to be applied. The examples and projects differ only in that the student is expected to use (and perhaps modify) the program which is given for the former in Appendix B, while the book provides guidance in writing programs to treat the latter through a series of steps, also indicated by the symbol ▯ ▯ ▯ in the outer margin. However, programs for the projects have also been included in Appendix C; these can serve as models for the student's own program or as a means of investigating the physics without having to write a major program "from scratch". A number of suggested studies accompany each example and project; these guide the student in exploiting the programs and understanding the physical principles and numerical techniques involved.

The diskette included with this book (360 kB, double-sided, double-density format) also contains the BASIC source codes for the examples and projects; it is suitable for use on any microcomputer system operating under MS-DOS Version 2.0 or higher. Further information about these programs can be found at the beginning of Appendix B and in the file README on the diskette, which can be read by inserting the diskette into the default disk drive and entering the DOS command "TYPE README". Note that it is wise to back up this write-protected diskette before beginning to use the programs.

An attempt has been made to use only the most primitive BASIC statements, so that the codes for the projects and examples should be appropriate for most BASIC dialects. All of the programs will run under a BASIC interpreter, but most require enough computation to make execution speed an important consideration. For serious study, it is therefore recommended that the codes be compiled through the IBM or Microsoft BASIC compiler, after which they will run between five and ten times faster. The programs have also been written in such as way as to make relatively straightforward their transcription into another high-level language, such as FORTRAN.

A "laboratory" format has proved to be one effective mode of presenting this material in a university setting. Students are quite able to work through the text on their own, with the instructor being available for consultation and to monitor progress through brief personal interviews on each chapter. Three chapters in ten weeks (60 hours) of instruction has proved to be a reasonable pace, with students typically writing two of the projects during this time, and using the "canned" codes to work through the physics of the remaining project and the examples. The eight chapters in this book should therefore be more than sufficient for a one-semester course. Alternatively, this book can be used to provide supplementary material for the usual courses in classical, quantum, and statistical mechanics. Many of the examples and projects are vivid illustrations of basic concepts in these subjects and are therefore suitable for classroom demonstrations or independent study.

Contents

Contents

Contents

The problem with computers is that they only give answers
-attributed to P. Picasso

COMPUTATIONAL PHYSICS

Chapter 1

Basic Mathematical Operations

Three numerical operations - differentiation, quadrature, and the finding of roots - are central to most computer modeling of physical systems. Suppose that we have the ability to calculate the value of a function, $f(x)$, at any value of the independent variable x. In differentiation, we seek one of the derivatives of f at a given value of x. Quadrature, roughly the inverse of differentiation, requires us to calculate the definite integral of f between two specified limits (we reserve the term "integration" for the process of solving ordinary differential equations, as discussed in Chapter 2), while in root finding we seek the values of x (there may be several) at which f vanishes.

If f is known analytically, it is almost always possible, with enough fortitude, to derive explicit formulas for the derivatives of f, and it is often possible to do so for its definite integral as well. However, it is often the case that an analytical method cannot be used, even though we can evaluate $f(x)$ itself. This might be either because some very complicated numerical procedure is required to evaluate f and we have no suitable analytical formula upon which to apply the rules of differentiation and quadrature, or, even worse, because the way we can generate f provides us with its values at only a set of discrete abscissae. In these situations, we must employ approximate formulas expressing the derivatives and integral in terms of the values of f we can compute. Moreover, the roots of all but the simplest functions cannot be found analytically, and numerical methods are therefore essential.

This chapter deals with the computer realization of these three basic operations. The central technique is to approximate f by a simple function (such as first- or second-degree polynomial) upon which these operations can be performed easily. We will derive only the simplest and most commonly used formulas; fuller treatments can be found in many textbooks on numerical analysis.

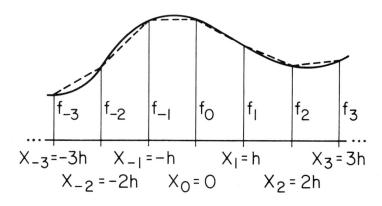

Figure 1.1 Values of f on an equally-spaced lattice. Dashed lines show the linear interpolation.

1.1 Numerical differentiation

Let us suppose that we are interested in the derivative at $x=0$, $f'(0)$. (The formulas we will derive can be generalized simply to arbitrary x by translation.) Let us also suppose that we know f on an equally-spaced lattice of x values,

$$f_n = f(x_n);\ x_n = nh\ (n = 0, \pm 1, \pm 2, \cdots),$$

and that our goal is to compute an approximate value of $f'(0)$ in terms of the f_n (see Figure 1.1).

We begin by using a Taylor series to expand f in the neighborhood of $x=0$:

$$f(x) = f_0 + x f' + \frac{x^2}{2!} f'' + \frac{x^3}{3!} f''' + \cdots ,\tag{1.1}$$

where all derivatives are evaluated at $x=0$. It is then simple to verify that

$$f_{\pm 1} \equiv f(x = \pm h) = f_0 \pm h f' + \frac{h^2}{2} f'' \pm \frac{h^3}{6} f''' + O(h^4),\tag{1.2a}$$

$$f_{\pm 2} \equiv f(x = \pm 2h) = f_0 \pm 2h f' + 2h^2 f'' \pm \frac{4h^3}{3} f''' + O(h^4),\tag{1.2b}$$

where $O(h^4)$ means terms of order h^4 or higher. To estimate the size of such terms, we can assume that f and its derivatives are all of the same order of magnitude, as is the case for many functions of physical relevance.

Upon subtracting f_{-1} from f_1 as given by (1.2a), we find, after a slight rearrangement,

$$f' = \frac{f_1 - f_{-1}}{2h} - \frac{h^2}{6} f''' + O(h^4).\tag{1.3a}$$

The term involving f''' vanishes as h becomes small and is the dominant error associated with the finite difference approximation that retains only the first term:

$$f' \approx \frac{f_1 - f_{-1}}{2h}. \qquad (1.3b)$$

This "3-point" formula would be exact if f were a second-degree polynomial in the 3-point interval $[-h, +h]$, because the third- and all higher-order derivatives would then vanish. Hence, the essence of Eq. (1.3b) is the assumption that a quadratic polynomial interpolation of f through the three points $x = \pm h, 0$ is valid.

Equation (1.3b) is a very natural result, reminiscent of the formulas used to define the derivative in elementary calculus. The error term (of order h^2) can, in principle, be made as small as is desired by using smaller and smaller values of h. Note also that the symmetric difference about $x = 0$ is used, as it is more accurate (by one order in h) than the forward or backward difference formulas:

$$f' \approx \frac{f_1 - f_0}{h} + O(h); \qquad (1.4a)$$

$$f' \approx \frac{f_0 - f_{-1}}{h} + O(h). \qquad (1.4b)$$

These "2-point" formulas are based on the assumption that f is well approximated by a linear function over the intervals between $x = 0$ and $x = \pm h$.

As a concrete example, consider evaluating $f'(x=1)$ when $f(x) = \sin x$. The exact answer is, of course, $\cos 1 = 0.540302$. The following BASIC program evaluates Eq. (1.3b) in this case for the value of h input:

```
10 X=1: EXACT=COS(X)
20 INPUT "enter value of h (<=0 to stop)";H
30 IF H<=0 THEN STOP
40 FPRIME = (SIN(X+H)-SIN(X-H))/(2*H)
50 DIFF=EXACT-FPRIME
60 PRINT USING "h=#.######, ERROR=+#.######";H,DIFF
70 GOTO 20
```

(If you are a beginner in BASIC, note the way the value of H is requested from the keyboard, the fact that the code will stop if a non-positive value of H is entered, the natural way in which variable names are chosen and the mathematical formula (1.3b) is transcribed using the SIN function in line 40, the way in which the number of significant digits is specified when the result is to be output to the screen in line 60, and the jump in program control in

Table 1.1 Error in evaluating $d\sin x/dx\,|_{x=1}=0.540302$

h	Symmetric 3-point Eq. (1.3b)	Forward 2-point Eq. (1.4a)	Backward 2-point Eq. (1.4b)	Symmetric 5-point Eq. (1.5)
0.50000	0.022233	0.228254	-0.183789	0.001092
0.20000	0.003595	0.087461	-0.080272	0.000028
0.10000	0.000899	0.042938	-0.041139	0.000001
0.05000	0.000225	0.021258	-0.020808	0.000000
0.02000	0.000037	0.008453	-0.008380	0.000001
0.01000	0.000010	0.004224	-0.004204	0.000002
0.00500	0.000010	0.002108	-0.002088	0.000006
0.00200	-0.000014	0.000820	-0.000848	-0.000017
0.00100	-0.000014	0.000403	-0.000431	-0.000019
0.00050	0.000105	0.000403	-0.000193	0.000115
0.00020	-0.000163	-0.000014	-0.000312	-0.000188
0.00010	-0.000312	-0.000312	-0.000312	-0.000411
0.00005	0.000284	0.001476	-0.000908	0.000681
0.00002	0.000880	0.000880	0.000880	0.000873
0.00001	0.000880	0.003860	-0.002100	0.000880

line 70.)

Results generated with this program, as well as with similar ones evaluating the forward and backward difference formulas Eqs. (1.4a,b), are shown in Table 1.1. Note that the result improves as we decrease h, but only up to a point, after which it becomes worse. This is because arithmetic in the computer is performed with only a limited precision (5-6 decimal digits for a single precision BASIC variable), so that when the difference in the numerator of the approximations is formed, it is subject to large "round-off" errors if h is small and f_1 and f_{-1} differ very little. For example, if $h=10^{-6}$, then

$$f_1=\sin(1.000001)=0.841472;\ f_{-1}=\sin(0.999999)=0.841470,$$

so that $f_1-f_{-1}=0.000002$ to six significant digits. When substituted into (1.3b) we find $f'\approx1.000000$, a very poor result. However, if we do the arithmetic with 10 significant digits, then

$$f_1=0.8414715251;\ f_{-1}=0.8414704445,$$

which gives a respectable $f'\approx0.540300$ in Eq. (1.3b). In this sense, numerical differentiation is an intrinsically unstable process (no well-defined limit as $h\to0$), and so must be carried out with caution.

It is possible to improve on the 3-point formula (1.3b) by relating f' to lattice points further removed from $x=0$. For example, using Eqs. (1.2), it is easy to show that the "5-point" formula

$$f' \approx \frac{1}{12h}[f_{-2}-8f_{-1}+8f_1-f_2]+O(h^4) \qquad (1.5)$$

cancels all derivatives in the Taylor series through fourth order. Computing the derivative in this way assumes that f is well-approximated by a fourth-degree polynomial over the 5-point interval $[-2h, 2h]$. Although requiring more computation, this approximation is considerably more accurate, as can be seen from Table 1.1. In fact, an accuracy comparable to Eq. (1.3b) is obtained with a step some 10 times larger. This can be an important consideration when many values of f must be stored in the computer, as the greater accuracy allows a sparser tabulation and so saves storage space. However, because (1.5) requires more mathematical operations than does (1.3b) and there is considerable cancellation among the various terms (they have both positive and negative coefficients), precision problems show up at a larger value of h.

Formulas for higher derivatives can be constructed by taking appropriate combinations of Eqs. (1.2). For example, it is easy to see that

$$f_1-2f_0+f_{-1}=h^2f''+O(h^4), \qquad (1.6)$$

so that an approximation to the second derivative accurate to order h^2 is

$$f'' \approx \frac{f_1-2f_0+f_{-1}}{h^2}. \qquad (1.7)$$

Difference formulas for the various derivatives of f that are accurate to a higher order in h can be derived straightforwardly. Table 1.2 is a summary of the 4- and 5-point expressions.

Exercise 1.1 Using any function for which you can evaluate the derivatives analytically, investigate the accuracy of the formulas in Table 1.2 for various values of h.   □ □ □

1.2 Numerical quadrature

In quadrature, we are interested in calculating the definite integral of f between two limits, $a<b$. We can easily arrange for these values to be points of the lattice separated by an even number of lattice spacings; i.e.,

$$N=\frac{(b-a)}{h}$$

Table 1.2 4- and 5-point difference formulas for derivatives

	4-point	5-point
hf'	$\pm\frac{1}{6}(-2f_{\mp1}-3f_0+6f_{\pm1}-f_{\pm2})$	$\frac{1}{12}(f_{-2}-8f_{-1}+8f_1-f_2)$
h^2f''	$f_{-1}-2f_0+f_1$	$\frac{1}{12}(-f_{-2}+16f_{-1}-30f_0+16f_1-f_2)$
h^3f'''	$\pm(-f_{\mp1}+3f_0-3f_{\pm1}+f_{\pm2})$	$\frac{1}{2}(-f_{-2}+2f_{-1}-2f_1+f_2)$
$h^4f^{(iv)}$		$f_{-2}-4f_{-1}+6f_0-4f_1+f_2$

is an even integer. It is then sufficient for us to derive a formula for the integral from $-h$ to $+h$, since this formula can be composed many times:

$$\int_a^b f(x)\,dx = \int_a^{a+2h} f(x)\,dx + \int_{a+2h}^{a+4h} f(x)\,dx + \cdots + \int_{b-2h}^b f(x)\,dx. \quad (1.8)$$

The basic idea behind all of the quadrature formulas we will discuss (technically of the closed Newton-Cotes type) is to approximate f between $-h$ and $+h$ by a function that can be integrated exactly. For example, the simplest approximation can be had by considering the intervals $[-h,0]$ and $[0,h]$ separately, and assuming that f is linear in each of these intervals (see Figure 1.1). The error made by this interpolation is of order h^2f'', so that the approximate integral is

$$\int_{-h}^h f(x)\,dx = \frac{h}{2}(f_{-1}+2f_0+f_1)+O(h^3), \quad (1.9)$$

which is the well-known trapezoidal rule.

A better approximation can be had by realizing that the Taylor series (1.1) can provide an improved interpolation of f. Using the difference formulas (1.3b) and (1.7) for f' and f'', respectively, for $|x|<h$ we can put

$$f(x)=f_0+\frac{f_1-f_{-1}}{2h}x+\frac{f_1-2f_0+f_{-1}}{2h^2}x^2+O(x^3), \quad (1.10)$$

which can be integrated readily to give

$$\int_{-h}^h f(x)\,dx = \frac{h}{3}(f_1+4f_0+f_{-1})+O(h^5). \quad (1.11)$$

This is Simpson's rule, which can be seen to be accurate to two orders higher than the trapezoidal rule (1.9). Note that the error is actually better than would be expected naively from (1.10) since the x^3 term gives no contribution to the integral. Composing this formula according to Eq. (1.8) gives

$$\int_a^b f(x)\,dx = \frac{h}{3}[f(a)+4f(a+h)+2f(a+2h)+4f(a+3h)+$$

$$\cdots + 4f(b-h)+f(b)]. \tag{1.12}$$

As an example, the following BASIC program calculates

$$\int_0^1 e^x\,dx = e-1 = 1.718282$$

using Simpson's rule for the value of $N=1/h$ input. (Source code for programs like this that are embedded in the text are *not* contained on the *Computational Physics* diskette, but can be easily entered into the reader's computer from the keyboard.)

```
5   DEF FNF(X)=EXP(X)                      'function to integrate
10  EXACT=EXP(1)-1
15  INPUT "enter N (even,>=2)",N%
20  IF N%<2 THEN STOP
25  '
30  H=1/N%
35  SUM=FNF(0)                             'contribution from X=0
40  FAC=2                                  'factor for Simpson's rule
45  '
50  FOR I%=1 TO N%-1                       'loop over lattice points
55      IF FAC=2 THEN FAC=4 ELSE FAC=2     'factors alternate
60      X=I%*H                             'X at this point
65      SUM=SUM+FNF(X)*FAC                 'contribution to integral
70  NEXT I%
75  '
80  SUM=SUM+FNF(1)                         'contribution from X=1
85  INTERGAL=H*SUM/3
90  DIFF=EXACT-INTEGRAL
95  PRINT USING "N=#### ERROR=#.######";N%,DIFF
100 GOTO 15                                'get another value of N%
```

Results are shown in Table 1.3 for various values of N, together with the values obtained using the trapezoidal rule. The improvement from the higher-order formula is evident. Note that the results are stable in the sense that a well-defined limit is obtained as N becomes very large and the mesh spacing h becomes small; round-off errors are unimportant because all values of f enter into the quadrature formula with the same sign, in contrast to what happens in numerical differentiation.

Table 1.3 Errors in evaluating $\int_0^1 e^x dx = 1.718282$

N	h	Trapezoidal Eq. (1.9)	Simpson's Eq. (1.12)	Bode's Eq. (1.13b)
4	0.2500000	-0.008940	-0.000037	-0.000001
8	0.1250000	-0.002237	0.000002	0.000000
16	0.0625000	-0.000559	0.000000	0.000000
32	0.0312500	-0.000140	0.000000	0.000000
64	0.0156250	-0.000035	0.000000	0.000000
128	0.0078125	-0.000008	0.000000	0.000000

An important issue in quadrature is how small an h is necessary to compute the integral to a given accuracy. Although it is possible to derive rigorous error bounds for the formulas we have discussed, the simplest thing to do in practice is to run the computation again with a smaller h and observe the changes in the results.

Higher-order quadrature formulas can be derived by retaining more terms in the Taylor expansion (1.10) used to interpolate f between the mesh points and, of course, using commensurately better finite-difference approximations for the derivatives. The generalizations of Simpson's rule using cubic and quartic polynomials to interpolate (Simpson's $\frac{3}{8}$ and Bode's rule, respectively) are:

$$\int_{x_0}^{x_3} f(x)dx = \frac{3h}{8}[f_0 + 3f_1 + 3f_2 + f_3] + O(h^5); \qquad (1.13a)$$

$$\int_{x_0}^{x_4} f(x)dx = \frac{2h}{45}[7f_0 + 32f_1 + 12f_2 + 32f_3 + 7f_4] + O(h^7). \qquad (1.13b)$$

The results of applying Bode's rule are also given in Table 1.3, where the improvement is evident, although at the expense of a more involved computation. (Note that for this method to be applicable, N must be a multiple of 4.) Although one might think that formulas based on interpolation using polynomials of a very high degree would be even more suitable, this is not the case; such polynomials tend to oscillate violently and lead to an inaccurate interpolation. Moreover, the coefficients of the values of f at the various lattice points can have both positive and negative signs in higher-order formulas, making round-off error a potential problem. It is therefore usually safer to improve accuracy by using a low-order method and making h smaller rather than by resorting to a higher-order formula. Quadrature formulas accurate to a very high order can be derived if we give up the requirement of equally-spaced abscissae; these are discussed in Chapter 4.

Exercise 1.2 Using any function whose definite integral you can compute analytically, investigate the accuracy of the various quadrature methods discussed above for different values of h.   ◻ ◻ ◻

Some care and common sense must be exercised in the application of the numerical quadrature formulas discussed above. For example, an integral in which the upper limit is very large is best handled by a change in variable. Thus, the Simpson's rule evaluation of

$$\int_{1}^{b} dx \, x^{-2} \, g(x)$$

with $g(x)$ constant at large x, would result in a (finite) sum converging very slowly as b becomes large for fixed h (and taking a very long time to compute!). However, changing variables to $t = x^{-1}$ gives

$$\int_{b^{-1}}^{1} g(t^{-1}) \, dt ,$$

which can then be evaluated by any of the formulas we have discussed.

Integrable singularities, which cause the naive formulas to give nonsense, can also be handled in a simple way. For example,

$$\int_{0}^{1} dx \, (1-x^2)^{-\frac{1}{2}} \, g(x)$$

has an integrable singularity at $x = 1$ (if g is regular there) and is a finite number. However, since $f(x=1) = \infty$, the quadrature formulas discussed above give an infinite result. An accurate result can be obtained by changing variables to $t = (1-x)^{\frac{1}{2}}$ to obtain

$$2 \int_{0}^{1} dt \, (2-t^2)^{-\frac{1}{2}} \, g(1-t^2),$$

which is then approximated with no trouble.

Integrable singularities can also be handled by deriving quadrature formulas especially adapted to them. Suppose we are interested in

$$\int_{0}^{1} f(x) \, dx = \int_{0}^{h} f(x) \, dx + \int_{h}^{1} f(x) \, dx ,$$

where $f(x)$ behaves as $Cx^{-1/2}$ near $x=0$, with C a constant. The integral from h to 1 is regular and can be handled easily, while the integral from 0 to h can be approximated as $2Ch^{1/2}=2hf(h)$.

□ □ □   **Exercise 1.3** Write a program to calculate

$$\int_0^1 t^{-2/3}(1-t)^{-1/3}\,dt = 2\pi/3^{1/2}$$

using one of the quadrature formulas discussed above and investigate its accuracy for various values of h. (Hint: Split the range of integration into two parts and make a different change of variable in each integral to handle the singularities.)

1.3 Finding roots

The final elementary operation that is commonly required is to find a root of a function $f(x)$ that we can compute for arbitrary x. One surefire method, when the approximate location of a root (say at $x=x_0$) is known, is to guess a trial value of x guaranteed to be less than the root, and then to increase this trial value by small positive steps, backing up and halving the step size every time f changes sign. The values of x generated by this procedure evidently converge to x_0, so that the search can be terminated whenever the step size falls below the required tolerance. Thus, the following BASIC program finds the positive root of the function $f(x)=x^2-5$, $x_0=5^{1/2}=2.236068$, to a tolerance of 10^{-6} using $x=1$ as an initial guess and an initial step size of 0.5:

```
 5   DEF FNF(X)=X*X-5          'function whose root is sought
10   TOLX=1.E-06               'tolerance for the search
15   X=1: FOLD=FNF(X): DX=.5   'initial guess, function, and step
20   ITER%=0                   'initialize iteration count
25   '
30   WHILE ABS(DX)>TOLX
35      ITER%=ITER%+1          'increment iteration count
40      X=X+DX                 'step X
45      PRINT ITER%,X,SQR(5)-X 'output current values
50      IF FOLD*FNF(X)>0 THEN GOTO 60  'if no sign change, take another step
55      X=X-DX: DX=DX/2        'back up and halve the step
60   WEND
65
70   STOP
```

Results for the sequence of x values are shown in Table 1.4, evidently converging to the correct answer, although only after some 33 iterations. One must be careful when using this method, since if the initial step size is too large, it is possible to step over the root desired when f has several roots.

Table 1.4 Error in finding the positive root of $f(x)=x^2-5$

Iteration	Search	Newton Eq. (1.14)	Secant Eq. (1.15)
0	1.236076	1.236076	1.236076
1	0.736068	-0.763932	-1.430599
2	0.236068	-0.097265	0.378925
3	-0.263932	-0.002027	0.098137
4	-0.013932	-0.000001	-0.009308
5	0.111068	0.000000	0.000008
6	-0.013932	0.000000	0.000000
. . .	. . .	. . .	. . .
33	0.000001	0.000000	0.000000

Exercise 1.4 Run the code above for various tolerances, initial guesses, and initial step sizes. Note that sometimes you might find convergence to the negative root. What happens if you start with an initial guess of -3 with a step size of 6?

A more efficient algorithm, Newton-Raphson, is available if we can evaluate the derivative of f for arbitrary x. This method generates a sequence of values, x^i, converging to x_0 under the assumption that f is locally linear near x_0 (see Figure 1.2). That is,

$$x^{i+1}=x^i - \frac{f(x^i)}{f'(x^i)}. \tag{1.14}$$

The application of this method to finding $5^{1/2}$ is also shown in Table 1.4, where the rapid convergence (5 iterations) is evident. This is the algorithm usually used in the computer evaluation of square roots; a linearization of (1.14) about x_0 shows that the number of significant digits doubles with each iteration, as is evident in Table 1.4.

The secant method provides a happy compromise between the efficiency of Newton-Raphson and the bother of having to evaluate the derivative. If the derivative in Eq. (1.14) is approximated by the difference formula related to (1.4b),

$$f'(x^i) \approx \frac{f(x^i)-f(x^{i-1})}{x^i-x^{i-1}},$$

we obtain the following 3-term recursion formula giving x^{i+1} in terms of x^i and x^{i-1} (see Figure 1.2):

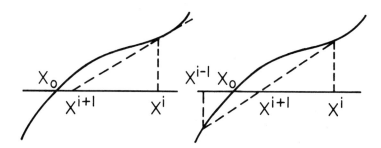

Figure 1.2 Geometrical bases of the Newton-Raphson (left) and secant (right) methods

$$x^{i+1}=x^i-f\left(x^i\right)\frac{\left(x^i-x^{i-1}\right)}{f\left(x^i\right)-f\left(x^{i-1}\right)}\ .\qquad(1.15)$$

Any two approximate values of x_0 can be used for x^0 and x^1 to start the algorithm, which is terminated when the change in x from one iteration to the next is less than the required tolerance. The results of the secant method for our model problem, starting with values $x^0=0.5$ and $x^1=1.0$, are also shown in Table 1.4. Provided that the initial guesses are close to the true root, convergence to the exact answer is almost as rapid as that of the Newton-Raphson algorithm.

□ □ □   **Exercise 1.5** Write programs to solve for the positive root of x^2-5 using the Newton-Raphson and secant methods. Investigate the behavior of the latter with changes in the initial guesses for the root.

When the function is badly behaved near its root (e.g., there is an inflection point near x_0) or when there are several roots, the "automatic" Newton-Raphson and secant methods can fail to converge at all or converge to the wrong answer if the initial guess for the root is poor. Hence, a safe and conservative procedure is to use the search algorithm to locate x_0 approximately and then to use one of the automatic methods.

□ □ □   **Exercise 1.6** The function $f\left(x\right)=\tanh x$ has a root at $x=0$. Write a program to show that the Newton-Raphson method does not converge for an initial guess of $x\gtrsim1$. Can you understand what's going wrong by considering a graph of $\tanh x$? From the explicit form of (1.14) for this problem, derive the critical value of the initial guess above which convergence will not occur. Try to solve the problem using the secant method. What happens for various initial guesses if you try to find the $x=0$ root of $\tan x$ using either method?

1.4 Semiclassical quantization of molecular vibrations

As an example combining several basic mathematical operations, we consider the problem of describing a diatomic molecule
such as O_2, which consists of two nuclei bound together by the
electrons that orbit about them. Since the nuclei are much
heavier than the electrons we can assume that the latter move
fast enough to readjust instantaneously to the changing position of
the nuclei (Born-Oppenheimer approximation). The problem is
therefore reduced to one in which the motion of the two nuclei is
governed by a potential, V, depending only upon r, the distance
between them. The physical principles responsible for generating
V will be discussed in detail in Project VIII, but on general grounds
one can say that the potential is attractive at large distances (van
der Waals interaction) and repulsive at short distances (Coulomb
interaction of the nuclei and Pauli repulsion of the electrons). A
commonly used form for V embodying these features is the
Lennard-Jones or 6-12 potential,

$$V(r)=4V_0\left[\left(\frac{a}{r}\right)^{12}-\left(\frac{a}{r}\right)^6\right],\qquad (1.16)$$

which has the shape shown in the upper portion of Figure 1.3, the
minimum occurring at $r_{min}=2^{1/6}a$ with a depth V_0. We will
assume this form in most of the discussion below. A thorough
treatment of the physics of diatomic molecules can be found in
[He50] while the Born-Oppenheimer approximation is discussed in
[Me68].

The great mass of the nuclei allows the problem to be
simplified even further by decoupling the slow rotation of the
nuclei from the more rapid changes in their separation. The
former is well described by the quantum mechanical rotation of a
rigid dumbbell, while the vibrational states of relative motion, with
energies E_n, are described by the bound state solutions, $\psi_n(r)$, of
a one-dimensional Schroedinger equation,

$$\left[-\frac{\hbar^2}{2m}\frac{d^2}{dr^2}+V(r)\right]\psi_n=E_n\psi_n.\qquad (1.17)$$

Here, m is the reduced mass of the two nuclei.

Our goal in this example is to find the energies E_n, given a particular potential. This can be done exactly by solving the
differential eigenvalue equation (1.17); numerical methods for
doing so will be discussed in Chapter 3. However, the great mass
of the nuclei implies that their motion is nearly classical, so that

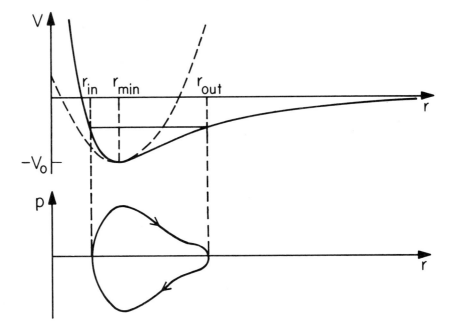

Figure 1.3 (Upper portion) The Lennard-Jones potential and the inner and outer turning points at a negative energy. The dashed line shows the parabolic approximation to the potential. (Lower portion) The corresponding trajectory in phase space.

approximate values of the vibrational energies E_n can be obtained by considering the classical motion of the nuclei in V and then applying "quantization rules" to determine the energies. These quantization rules, originally postulated by N. Bohr and Sommerfeld and Wilson, were the basis of the "old" quantum theory from which the modern formulation of quantum mechanics arose. However, they can also be obtained by considering the WKB approximation to the wave equation (1.17). (See [Me68] for details.)

Confined classical motion of the internuclear separation in the potential $V(r)$ can occur for energies $-V_0 < E < 0$. The distance between the nuclei oscillates periodically (but not necessarily harmonically) between inner and outer turning points, r_{in} and r_{out}, as shown in Figure 1.3. During these oscillations, energy is exchanged between the kinetic energy of relative motion and the potential energy such that the total energy,

$$E = \frac{p^2}{2m} + V(r),$$ (1.18)

is a constant (p is the relative momentum of the nuclei). We can therefore think of the oscillations at any given energy as defining a closed trajectory in phase space (coordinates r and p) along which Eq. (1.18) is satisfied, as shown in the lower portion of Figure 1.3. An explicit equation for this trajectory can be obtained by solving (1.18) for p:

$$p(r)=\pm[2m(E-V(r))]^{\frac{1}{2}}. \tag{1.19}$$

The classical motion described above occurs at *any* energy between $-V_0$ and 0. To quantize the motion, and hence obtain approximations to the eigenvalues E_n appearing in (1.17), we consider the dimensionless action at a given energy,

$$S(E)=\oint k(r)dr, \tag{1.20}$$

where $k(r)=\hbar^{-1}p(r)$ is the local de Broglie wave number and the integral is over one complete cycle of oscillation. This action is just the area (in units of $\hbar$) enclosed by the phase space trajectory. The quantization rules state that, at the allowed energies E_n, the action is a half-integral multiple of 2π. Thus, upon using (1.19) and recalling that the oscillation passes through each value of r twice (once with positive p and once with negative p), we have

$$S(E_n)=2\left[\frac{2m}{\hbar^2}\right]^{\frac{1}{2}}\int_{r_{in}}^{r_{out}}[E_n-V(r)]^{\frac{1}{2}}dr =(n+\tfrac{1}{2})2\pi, \tag{1.21}$$

where n is a nonnegative integer. At the limits of this integral, the turning points r_{in} and r_{out}, the integrand vanishes.

To specialize the quantization condition to the Lennard-Jones potential (1.16), we define the dimensionless quantities

$$\varepsilon=\frac{E}{V_0}, \quad x=\frac{r}{a}, \quad \gamma=\left[\frac{2ma^2V_0}{\hbar^2}\right]^{\frac{1}{2}},$$

so that (1.21) becomes

$$s(\varepsilon_n)\equiv\tfrac{1}{2}S(\varepsilon_n V_0)=\gamma\int_{x_{in}}^{x_{out}}[\varepsilon_n-v(x)]^{\frac{1}{2}}dx=(n+\tfrac{1}{2})\pi, \tag{1.22}$$

where

$$v(x)=4\left[\frac{1}{x^{12}}-\frac{1}{x^6}\right]$$

is the scaled potential.

The quantity γ is a dimensionless measure of the quantum nature of the problem. In the classical limit ($\hbar$ small or m large), γ becomes large. By knowing the moment of inertia of the molecule (from the energies of its rotational motion) and the dissociation energy (energy required to separate the molecule into its two constituent atoms), it is possible to determine from observation the parameters a and V_0 and hence the quantity γ. For the H_2 molecule, $\gamma=21.7$, while for the HD molecule, $\gamma=24.8$ (only m, but not V_0, changes when one of the protons is replaced by a deuteron), and for the much heavier O_2 molecule made of two ^{16}O nuclei, $\gamma = 150$. These rather large values indicate that a semiclassical approximation is a valid description of the vibrational motion.

The BASIC program for Example 1, whose source code is contained in Appendix B and in the file EXAM1.BAS on the *Computational Physics* diskette, finds, for the value of γ input, the values of the ε_n for which Eq. (1.22) is satisfied. After all of the energies have been found, the corresponding phase space trajectories are drawn. (Before attempting to run this code on your computer system, you should review the material on the programs in "How to use this book" and at the beginning of Appendix B.)

The following exercises are aimed at increasing your understanding of the physical principles and numerical methods demonstrated in this example.

□ □ □   **Exercise 1.7** One of the most important aspects of using a computer as a tool to do physics is knowing when to have confidence that the program is giving the correct answers. In this regard, an essential test is the detailed quantitative comparison of results with what is known in analytically soluble situations. Modify the code to use a parabolic potential (line 160, taking care to heed the warning on lines 170-180), for which the Bohr-Sommerfeld quantization gives the exact eigenvalues of the Schroedinger equation: a series of equally-spaced energies, with the lowest being one-half of the level spacing above the minimum of the potential. For several values of γ, compare the numerical results for this case with what you obtain by solving Eq. (1.22) analytically. Are the phase space trajectories what you expect?

□ □ □   **Exercise 1.8** Another important test of a working code is to compare its results with what is expected on the basis of physical intuition. Restore the code to use the Lennard-Jones potential and run it for $\gamma = 50$. Note that, as in the case of the purely parabolic potential discussed in the previous exercise, the first excited state is roughly three times as high above the bottom of the well as is the ground state and that the spacings between the few lowest states are roughly constant. This is because the Lennard-Jones potential is roughly parabolic about its minimum (see Figure 1.3).

By calculating the second derivative of V at the minimum, find the "spring constant" and show that the frequency of small-amplitude motion is expected to be

$$\frac{\hbar\omega}{V_0} = \frac{6\times2^{5/6}}{\gamma} \approx \frac{10.691}{\gamma}. \qquad (1.23)$$

Verify that this is consistent with the numerical results and explore this agreement for different values of γ. Can you understand why the higher energies are more densely spaced than the lower ones by comparing the Lennard-Jones potential with its parabolic approximation?

Exercise 1.9 Invariance of results under changes in the numerical algorithms or their parameters can give additional confidence in a calculation. Change the tolerances for the turning point and energy searches (line 120) or the number of Simpson's rule points (line 130) and observe the effects on the results. Note that because of the way in which the expected number of bound states is calculated (lines 1190-1200), this quantity can change if the energy tolerance is varied.

Exercise 1.10 Replace the searches for the inner and outer turning points by the Newton-Raphson method or the secant method. (When $N\%\neq0$, the turning points for $N\%-1$ are excellent starting values.) Replace the Simpson's rule quadrature for s by a higher-order formula (Eqs. (1.13a) or (1.13b)) and observe the improvement.

Exercise 1.11 Plot the ε_n of the Lennard-Jones potential as functions of γ for γ running from 20 to 200 and interpret the results. (As with many other short calculations, you may find it more efficient simply to run the code and plot the results by hand as you go along, rather than trying to automate the plotting operation.)

Exercise 1.12 For the H_2 molecule, observations show that the depth of the potential is $V_0=4.747$ eV and the location of the potential minimum is $r_{min}=0.74166$ Å. These two quantities, together with Eq. (1.23), imply a vibrational frequency of

$$\hbar\omega=0.492V_0=2.339 \text{ eV},$$

more than four times larger than the experimentally observed energy difference between the ground and first vibrational state, 0.515 eV. The Lennard-Jones shape is therefore not a very good description of the potential of the H_2 molecule. Another defect is

Table 1.5 Experimental vibrational energies of the H_2 molecule

n	E_n (eV)	n	E_n (eV)
0	-4.477	8	-1.151
1	-3.962	9	-0.867
2	-3.475	10	-0.615
3	-3.017	11	-0.400
4	-2.587	12	-0.225
5	-2.185	13	-0.094
6	-1.811	14	-0.017
7	-1.466		

that it predicts 6 bound states, while 15 are known to exist. (See Table 1.5, whose entries are derived from the data quoted in [Wa67].) A better analytic form of the potential, with more parameters, is required to reproduce simultaneously the depth and location of the minimum, the frequency of small amplitude vibrations about it, and the total number of bound states. One such form is the Morse potential,

$$V(r) = V_0[(1 - e^{-\beta(r - r_{min})})^2 - 1], \tag{1.24}$$

which also can be solved analytically. The Morse potential has a minimum at the expected location and the parameter β can be adjusted to fit the curvature of the minimum to the observed excitation energy of the first vibrational state. Find the value of β appropriate for the H_2 molecule, modify the program above to use the Morse potential, and calculate the spectrum of vibrational states. Show that a much more reasonable number of levels is now obtained. Compare the energies with experiment and with those of the Lennard-Jones potential and interpret the latter differences.

Project I: Scattering by a central potential

In this project, we will investigate the classical scattering of a particle of mass m by a central potential, in particular the Lennard-Jones potential considered in Section 1.4 above. In a scattering event, the particle, with initial kinetic energy E and impact parameter b, approaches the potential from a large distance. It is deflected during its passage near the force center and eventually emerges with the same energy, but moving at an angle Θ with respect to its original direction. Since the potential

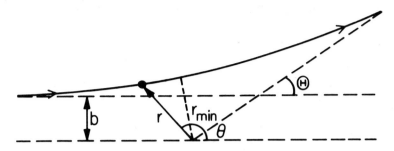

Figure I.1 Quantities involved in the scattering of a particle by a central potential

depends upon only the distance of the particle from the force center, the angular momentum is conserved and the trajectory lies in a plane. The polar coordinates of the particle, (r, θ), are a convenient way to describe its motion, as shown in Figure I.1. (For details, see any textbook on classical mechanics, such as [Go80].)

Of basic interest is the deflection function, $\Theta(b)$, giving the final scattering angle, Θ, as a function of the impact parameter; this function also depends upon the incident energy. The differential cross section for scattering at an angle Θ, $d\sigma/d\Omega$, is an experimental observable that is related to the deflection function by

$$\frac{d\sigma}{d\Omega} = \frac{b}{\sin\Theta}\left|\frac{db}{d\Theta}\right|. \tag{I.1}$$

Thus, if $d\Theta/db = (db/d\Theta)^{-1}$ can be computed, then the cross section is known.

Expressions for the deflection function can be found analytically for only a very few potentials, so that numerical methods usually must be employed. One way to solve the problem would be to integrate the equations of motion in time (i.e., Newton's law relating the acceleration to the force) to find the trajectories corresponding to various impact parameters and then to tabulate the final directions of the motion (scattering angles). This would involve integrating four coupled first-order differential equations for two coordinates and their velocities in the scattering plane, as discussed in Section 2.5 below. However, since angular momentum is conserved, the evolution of Θ is related directly to the radial motion, and the problem can be reduced to a one-dimensional one, which can be solved by quadrature. This latter approach, which is simpler and more accurate, is the one we will pursue here.

To derive an appropriate expression for Θ, we begin with the conservation of angular momentum, which implies that

$$L = mvb = mr^2 \frac{d\theta}{dt} , \qquad (I.2)$$

is a constant of the motion. Here, $d\theta/dt$ is the angular velocity and v is the asymptotic velocity, related to the bombarding energy by $E = \frac{1}{2}mv^2$. The radial motion occurs in an effective potential that is the sum of V and the centrifugal potential, so that energy conservation implies

$$\frac{1}{2}m \left[\frac{dr}{dt} \right]^2 + \frac{L^2}{2mr^2} + V = E . \qquad (I.3)$$

If we use r as the independent variable in (I.2), rather than the time, we can write

$$\frac{d\theta}{dr} = \frac{d\theta}{dt} \left[\frac{dr}{dt} \right]^{-1} = \frac{bv}{r^2} \left[\frac{dr}{dt} \right]^{-1} , \qquad (I.4)$$

and solving (I.3) for dr/dt then yields

$$\frac{d\theta}{dr} = \pm \frac{b}{r^2} \left[1 - \frac{b^2}{r^2} - \frac{V}{E} \right]^{-\frac{1}{2}} . \qquad (I.5)$$

Recalling that $\theta = \pi$ when $r = \infty$ on the incoming branch of the trajectory and that θ is always decreasing, this equation can be integrated immediately to give the scattering angle,

$$\Theta = \pi - 2 \int_{r_{min}}^{\infty} \frac{b\,dr}{r^2} \left[1 - \frac{b^2}{r^2} - \frac{V}{E} \right]^{-\frac{1}{2}} , \qquad (I.6)$$

where r_{min} is the distance of closest approach (the turning point, determined by the outermost zero of the argument of the square root) and the factor of 2 in front of the integral accounts for the incoming and outgoing branches of the trajectory, which give equal contributions to the scattering angle.

One final transformation is useful before beginning a numerical calculation. Suppose that there exists a distance r_{max} beyond which we can safely neglect V. In this case, the integrand in (I.6) vanishes as r^{-2} for large r, so that numerical quadrature could be very inefficient. In fact, since the potential has no effect for $r > r_{max}$, we would just be "wasting time" describing straight-line motion. To handle this situation efficiently, note that since $\Theta = 0$ when $V = 0$, Eq. (I.6) implies that

$$\pi = 2 \int_{b}^{\infty} \frac{b\,dr}{r^2} \left[1 - \frac{b^2}{r^2} \right]^{-\frac{1}{2}} , \qquad (I.7)$$

which, when substituted into (I.6), results in

$$\Theta = 2b \left[\int_{b}^{r_{max}} \frac{dr}{r^2} \left[1 - \frac{b^2}{r^2} \right]^{-\frac{1}{2}} - \int_{r_{min}}^{r_{max}} \frac{dr}{r^2} \left[1 - \frac{b^2}{r^2} - \frac{V}{E} \right]^{-\frac{1}{2}} \right]. \qquad (1.8)$$

The integrals here extend only to r_{max} since the integrands become equal when $r > r_{max}$.

Our goal will be to study scattering by the Lennard-Jones potential (1.16), which we can safely set to zero beyond $r_{max} = 3a$ if we are not interested in energies smaller than about

$$V(r = 3a) \approx 5 \times 10^{-3} V_0.$$

The study is best done in the following sequence of steps:

Step 1 Before beginning *any* numerical computation, it is impor- □ □ □
tant to have some idea of what the results should look like. Sketch what you think the deflection function is at relatively low energies, $E \lesssim V_0$, where the peripheral collisions at large $b \lesssim r_{max}$ will take place in a predominantly attractive potential and the more central collisions will "bounce" against the repulsive core. What happens at much higher energies, $E \gg V_0$, where the attractive pocket in V can be neglected? Note that for values of b where the deflection function has a maximum or a minimum, Eq. (I.1) shows that the cross section will be infinite, as occurs in the rainbow formed when light scatters from water drops.

Step 2 To have analytically soluble cases against which to test your □ □ □
program, calculate the deflection function for a square potential, where $V(r) = U_0$ for $r < r_{max}$ and vanishes for $r > r_{max}$. What happens when U_0 is negative? What happens when U_0 is positive and $E < U_0$? when $E > U_0$?

Step 3 Write a program that calculates, for a specified energy E, □ □ □
the deflection function by a numerical quadrature to evaluate both integrals in Eq. (I.8) at a number of equally spaced b values between 0 and r_{max}. (Note that the singularities in the integrands require some special treatment.) Check that the program is working properly and is accurate by calculating deflection functions for the square-well potential discussed in Step 2. Compare the accuracy with that of an alternative procedure in which the first integral in (I.8) is evaluated analytically, rather than numerically.

Step 4 Use your program to calculate the deflection function for □ □ □
scattering from the Lennard-Jones potential at selected values of E ranging from $0.1 V_0$ to $100 V_0$. Reconcile your answers in Step 1 with the results you obtain. Calculate the differential cross

section as a function of Θ at these energies.

□ □ □   **Step 5** If your program is working correctly, you should observe, for energies $E \lesssim V_0$, a singularity in the deflection function where Θ appears to approach $-\infty$ at some critical value of b, b_{crit}, that depends on E. This singularity, which disappears when E becomes larger than about V_0, is characteristic of "orbiting". In this phenomenon, the integrand in Eq. (I.6) has a linear, rather than a square root, singularity at the turning point, so that the scattering angle becomes logarithmically infinite. That is, the effective potential,

$$V + E \left(\frac{b}{r} \right)^2 ,$$

has a parabolic maximum and, when $b = b_{crit}$, the peak of this parabola is equal to the incident energy. The trajectory thus spends a very long time at the radius where this parabola peaks and the particle spirals many times around the force center. By tracing b_{crit} as a function of energy and by plotting a few of the effective potentials involved, convince yourself that this is indeed what's happening. Determine the maximum energy for which the Lennard-Jones potential exhibits orbiting, either by a solution of an appropriate set of equations involving V and its derivatives or by a systematic numerical investigation of the deflection function. If you pursue the latter approach, you might have to reconsider the treatment of the singularities in the numerical quadratures.

Chapter 2

Ordinary Differential Equations

Many of the laws of physics are most conveniently formulated in terms of differential equations. It is therefore not surprising that the numerical solution of differential equations is one of the most common tasks in modeling physical systems. The most general form of an ordinary differential equation is a set of M coupled first-order equations

$$\frac{d\mathbf{y}}{dx}=\mathbf{f}(x,\mathbf{y}), \qquad (2.1)$$

where x is the independent variable and $\mathbf{y}$ is a set of M dependent variables ($\mathbf{f}$ is thus an M-component vector). Differential equations of higher order can be written in this first-order form by introducing auxiliary functions. For example, the one-dimensional motion of a particle of mass m under a force field $F(z)$ is described by the second-order equation

$$m\frac{d^2z}{dt^2}=F(z) . \qquad (2.2)$$

If we define the momentum

$$p(t)=m\frac{dz}{dt},$$

then (2.2) becomes the two coupled first-order (Hamilton's) equations

$$\frac{dz}{dt}=\frac{p}{m}; \quad \frac{dp}{dt}=F(z) , \qquad (2.3)$$

which are in the form of (2.1). It is therefore sufficient to consider in detail only methods for first-order equations. Since the matrix structure of coupled differential equations is of the most natural form, our discussion of the case where there is only one independent variable can be generalized readily. Thus, we need be concerned only with solving

$$\frac{dy}{dx}=f(x,y) \qquad (2.4)$$

for a single dependent variable $y(x)$.

In this chapter, we will discuss several methods for solving ordinary differential equations, with emphasis on the initial value problem. That is, find $y(x)$ given the value of y at some initial point, say $y(x=0)=y_0$. This kind of problem occurs, for example, when we are given the initial position and momentum of a particle and we wish to find its subsequent motion using Eqs. (2.3). In Chapter 3, we will discuss the equally important boundary value and eigenvalue problems.

2.1 Simple methods

To repeat the basic problem, we are interested in the solution of the differential equation (2.4) with the initial condition $y(x=0)=y_0$. More specifically, we are usually interested in the value of y at a particular value of x, say $x=1$. The general strategy is to divide the interval $[0,1]$ into a large number, N, of equally spaced subintervals of length $h=1/N$ and then to develop a recursion formula relating y_n to $\{y_{n-1}, y_{n-2}, \cdots \}$, where y_n is our approximation to $y(x_n=nh)$. Such a recursion relation will then allow a step-by-step integration of the differential equation from $x=0$ to $x=1$.

One of the simplest algorithms is Euler's method, in which we consider Eq. (2.4) at the point x_n and replace the derivative on the left-hand side by its forward difference approximation (1.4a). Thus,

$$\frac{y_{n+1}-y_n}{h} + O(h) = f(x_n, y_n), \tag{2.5}$$

so that the recursion relation expressing y_{n+1} in terms of y_n is

$$y_{n+1}=y_n+hf(x_n, y_n)+O(h^2) . \tag{2.6}$$

This formula has a local error (that made in taking the single step from y_n to y_{n+1}) that is $O(h^2)$ since the error in (1.4a) is $O(h)$. The "global" error made in finding $y(1)$ by taking N such steps in integrating from $x=0$ to to $x=1$ is then $NO(h^2)\approx O(h)$. This error decreases only linearly with decreasing step size so that half as large an h (and thus twice as many steps) is required to halve the inaccuracy in the final answer. The numerical work for each of these steps is essentially a single evaluation of f.

As an example, consider the differential equation and boundary condition

$$\frac{dy}{dx}=-xy ; \; y(0)=1, \tag{2.7}$$

Table 2.1 Error in integrating $dy/dx = -xy$ with $y(0)=1$

| | Euler's method Eq. (2.6) | | Taylor series Eq. (2.10) | | Implicit method Eq. (2.18) | |
h	$y(1)$	$y(3)$	$y(1)$	$y(3)$	$y(1)$	$y(3)$
0.500	-.143469	.011109	.032312	-.006660	-.015691	.001785
0.200	-.046330	.006519	.005126	-.000712	-.002525	.000255
0.100	-.021625	.003318	.001273	-.000149	-.000631	.000063
0.050	-.010453	.001665	.000317	-.000034	-.000157	.000016
0.020	-.004098	.000666	.000051	-.000005	-.000025	.000003
0.010	-.002035	.000333	.000013	-.000001	-.000006	.000001
0.005	-.001014	.000167	.000003	.000000	-.000001	.000000
0.002	-.000405	.000067	.000001	.000000	.000000	.000000
0.001	-.000203	.000033	.000000	.000000	.000000	.000000

whose solution is

$$y = e^{-\frac{1}{2}x^2}.$$

The following BASIC program integrates forward from $x=0$ to $x=3$ using Eq. (2.6) with the step size input, printing the result and its error as it goes along.

```
10  DEF FNF(X,Y)=-X*Y          'function giving dy/dx
20  INPUT "enter h (<=0 to stop)";H
30  IF H<=0 THEN STOP
40  N%=3/H                     'number of steps to reach X=3
50  Y=1                        'initial value of Y
60  FOR I%=0 TO N%-1           'loop over steps
70      X=I%*H                 'X we step from
80      Y=Y+H*FNF(X,Y)         'step Y by Eq. (2.6)
90      DIFF=EXP(-0.5*(X+H)^2)-Y   'compare with exact at new X
100     PRINT I%,X,Y,DIFF      'output the current values
110 NEXT I%
120 GOTO 20                    'get new value of H
```

Errors in the results obtained for

$$y(1)=e^{-\frac{1}{2}}=0.606531, \; y(3)=e^{-9/2}=0.011109$$

with various step sizes are shown in the first two columns of Table 2.1. As expected from (2.6), the errors decrease linearly with smaller h. However, the fractional error (error divided by y) increases with x as more steps are taken in the integration and y becomes smaller.

□ □ □   **Exercise 2.1** A simple and often stringent test of an accurate numerical integration is to use the final value of y obtained as the initial condition to integrate backward from the final value of x to the starting point. The extent to which the resulting value of y differs from the original initial condition is then a measure of the inaccuracy. Apply this test to the example above.

Although Euler's method seems to work quite well, it is generally unsatisfactory because of its low-order accuracy. This prevents us from reducing the numerical work by using a larger value of h and so taking a smaller number of steps. This deficiency becomes apparent in the example above as we attempt to integrate to larger values of x, where some thought shows that we obtain the absurd result that $y=0$ (exactly) for $x>h^{-1}$. One simple solution is to change the step size as we go along, making h smaller as x becomes larger, but this soon becomes quite inefficient.

Integration methods with a higher-order accuracy are usually preferable to Euler's method. They offer a much more rapid increase of accuracy with decreasing step size and hence greater accuracy for a fixed amount of numerical effort. One class of simple higher order methods can be derived from a Taylor series expansion for y_{n+1} about y_n:

$$y_{n+1}=y(x_n+h)=y_n+hy_n'+\tfrac{1}{2}h^2y_n''+O(h^3). \qquad (2.8)$$

From (2.4), we have

$$y_n'=f(x_n,y_n), \qquad (2.9a)$$

and

$$y_n''=\frac{df}{dx}(x_n,y_n)=\frac{\partial f}{\partial x}+\frac{\partial f}{\partial y}\frac{dy}{dx}=\frac{\partial f}{\partial x}+\frac{\partial f}{\partial y}f, \qquad (2.9b)$$

which, when substituted into (2.8), results in

$$y_{n+1}=y_n+hf+\tfrac{1}{2}h^2\left[\frac{\partial f}{\partial x}+f\frac{\partial f}{\partial y}\right]+O(h^3), \qquad (2.10)$$

where f and its derivatives are to be evaluated at (x_n,y_n). This recursion relation has a local error $O(h^3)$ and hence a global error $O(h^2)$, one order more accurate than Euler's method (2.6). It is most useful when f is known analytically and is simple enough to differentiate. If we apply Eq. (2.10) to the example (2.7), we obtain the results shown in the middle two columns of Table 2.1; the improvement over Euler's method is clear. Algorithms with an even greater accuracy can be obtained by retaining more terms in the Taylor expansion (2.8), but the algebra soon becomes prohibitive in all but the simplest cases.

2.2 Multistep and implicit methods

Another way of achieving higher accuracy is to use recursion relations that relate y_{n+1} not just to y_n, but also to points further "in the past", say y_{n-1}, y_{n-2}, $\cdots$. To derive such formulas, we can integrate one step of the differential equation (2.4) *exactly* to obtain

$$y_{n+1} = y_n + \int_{x_n}^{x_{n+1}} f(x,y)\,dx. \tag{2.11}$$

The problem, of course, is that we don't know f over the interval of integration. However, we can use the values of y at x_n and x_{n-1} to provide a linear extrapolation of f over the required interval:

$$f \approx \frac{(x-x_{n-1})}{h} f_n - \frac{(x-x_n)}{h} f_{n-1} + O(h^2), \tag{2.12}$$

where $f_i \equiv f(x_i, y_i)$. Inserting this into (2.11) and doing the x integral then results in the Adams-Bashforth two-step method,

$$y_{n+1} = y_n + h\left(\tfrac{3}{2} f_n - \tfrac{1}{2} f_{n-1}\right) + O(h^3). \tag{2.13}$$

Related higher-order methods can be derived by extrapolating with higher-degree polynomials. For example, if f is extrapolated by a cubic polynomial fitted to f_n, f_{n-1}, f_{n-2}, and f_{n-3}, the Adams-Bashforth four-step method results:

$$y_{n+1} = y_n + \frac{h}{24}\left(55 f_n - 59 f_{n-1} + 37 f_{n-2} - 9 f_{n-3}\right) + O(h^4). \tag{2.14}$$

Note that because the recursion relations (2.13) and (2.14) involve several previous steps, the value of y_0 alone is not sufficient information to get them started, and so the values of y at the first few lattice points must be obtained from some other procedure, such as the Taylor series (2.8) or the Runge-Kutta methods discussed below.

Exercise 2.2 Apply the Adams-Bashforth two- and four-step algorithms to the example defined by Eq. (2.7) using Euler's method (2.6) to generate the values of y needed to start the recursion relation. Investigate the accuracy of $y(x)$ for various values of h by comparing with the analytical results and by applying the reversibility test described in Exercise 2.1.   □ □ □

The methods we have discussed so far are all "explicit" in that the y_{n+1} is given directly in terms of the already known value of

y_n. "Implicit" methods, in which an equation must be solved to determine y_{n+1}, offer yet another means of achieving higher accuracy. Suppose we consider Eq. (2.4) at a point $x_{n+\frac{1}{2}} \equiv (n+\frac{1}{2})h$ midway between two lattice points:

$$\left. \frac{dy}{dx} \right|_{x_{n+\frac{1}{2}}} = f(x_{n+\frac{1}{2}}, y_{n+\frac{1}{2}}). \qquad (2.15)$$

If we then use the symmetric difference approximation for the derivative (the analog of (1.3b) with $h \to \frac{1}{2}h$) and replace $f_{n+\frac{1}{2}}$ by the average of its values at the two adjacent lattice points (the error in this replacement is $O(h^2)$), we can write

$$\frac{y_{n+1} - y_n}{h} + O(h^2) = \frac{1}{2}[f_n + f_{n+1}] + O(h^2), \qquad (2.16)$$

which corresponds to the recursion relation

$$y_{n+1} = y_n + \frac{1}{2}h[f(x_n, y_n) + f(x_{n+1}, y_{n+1})] + O(h^3). \qquad (2.17)$$

This is all well and good, but the appearance of y_{n+1} on both sides of this equation (an implicit equation) means that, in general, we must solve a non-trivial equation (for example, by the Newton-Raphson method discussed in Section 1.3) at each integration step; this can be very time consuming. A particular simplification occurs if f is linear in y, say $f(x, y) = g(x)y$, in which case (2.17) can be solved to give

$$y_{n+1} = \left[\frac{1 + \frac{1}{2}g(x_n)h}{1 - \frac{1}{2}g(x_{n+1})h} \right] y_n. \qquad (2.18)$$

When applied to the problem (2.7), where $g(x) = -x$, this method gives gives the results shown in the last two columns of Table 2.1; the quadratic behavior of the error with h is clear.

□ □ □   **Exercise 2.3** Apply the Taylor series method (2.10) and the implicit method (2.18) to the example of Eq. (2.7) and obtain the results shown in Table 2.1. Investigate the accuracy of integration to larger values of x.

The Adams-Moulton methods are both multistep and implicit. For example, the Adams-Moulton two-step method can be derived from Eq. (2.11) by using a quadratic polynomial passing through f_{n-1}, f_n, and f_{n+1},

$$f \approx \frac{(x-x_n)(x-x_{n-1})}{h^2} f_{n+1} - \frac{(x-x_{n+1})(x-x_{n-1})}{h^2} f_n$$
$$+ \frac{(x-x_{n+1})(x-x_n)}{h^2} f_{n-1} + O(h^3),$$

to interpolate f over the region from x_n to x_{n+1}. The implicit recursion relation that results is

$$y_{n+1}=y_n+\frac{h}{12}(5f_{n+1}+8f_n-f_{n-1})+O(h^4).\qquad(2.19)$$

The corresponding three-step formula, obtained with a cubic polynomial interpolation, is

$$y_{n+1}=y_n+\frac{h}{24}(9f_{n+1}+19f_n-5f_{n-1}+f_{n-2})+O(h^5).\qquad(2.20)$$

Implicit methods are rarely used by solving the implicit equation to take a step. Rather, they serve as bases for "predictor-corrector" algorithms, in which a "prediction" for y_{n+1} based only on an explicit method is then "corrected" to give a better value by using this prediction in an implicit method. Such algorithms have the advantage of allowing a continuous monitoring of the accuracy of the integration, for example by making sure that the correction is small. A commonly used predictor-corrector algorithm with local error $O(h^5)$ is obtained by using the explicit Adams-Bashforth four-step method (2.14) to make the prediction, and then calculating the correction with the Adams-Moulton three-step method (2.20), using the predicted value of y_{n+1} to evaluate f_{n+1} on the right-hand side.

2.3 Runge-Kutta methods

As you might gather from the preceding section, there is quite a bit of freedom in writing down algorithms for integrating differential equations and, in fact, a large number of them exist, each having it own peculiarities and advantages. One very convenient and widely used class of methods are the Runge-Kutta algorithms, which come in varying orders of accuracy. We derive here a second-order version to give the spirit of the approach and then simply state the equations for the third- and commonly used fourth-order methods.

To derive a second-order Runge-Kutta algorithm (there are actually a whole family of them characterized by a continuous parameter), we approximate f in the integral of (2.11) by its Taylor series expansion about the mid-point of the integration interval. Thus,

$$y_{n+1}=y_n+hf(x_{n+\frac{1}{2}},y_{n+\frac{1}{2}})+O(h^3),\qquad(2.21)$$

where the error arises from the quadratic term in the Taylor series, as the linear term integrates to zero. Although it seems as if we need to know the value of $y_{n+\frac{1}{2}}$ appearing in f in the right-hand side of this equation for it to be of any use, this is not quite true. Since the error term is already $O(h^3)$, an approximation to y_{n+1} whose error is $O(h^2)$ is good enough. This is just what is provided by the simple Euler's method, Eq. (2.6). Thus, if we define k to be an intermediate approximation to twice the difference between $y_{n+\frac{1}{2}}$ and y_n, the following two-step procedure gives y_{n+1} in terms of y_n:

$$k = hf(x_n, y_n); \tag{2.22a}$$

$$y_{n+1} = y_n + hf(x_n + \tfrac{1}{2}h, y_n + \tfrac{1}{2}k) + O(h^3). \tag{2.22b}$$

This is a second-order Runge-Kutta algorithm. It embodies the general idea of substituting approximations for the values of y into the right-hand side of implicit expressions involving f. It is as accurate as the Taylor series or implicit methods (2.10) or (2.17), respectively, but places no special constraints on f, such as easy differentiability or linearity in y. It also uses the value of y at only one previous point, in contrast to the multipoint methods discussed above. However, (2.22) does require the evaluation of f twice for each step along the lattice.

Runge-Kutta schemes of higher-order can be derived in a relatively straightforward way. Any of the quadrature formulas discussed in Chapter 1 can be used to approximate the integral (2.11) by a finite sum of f values. For example, Simpson's rule yields

$$y_{n+1} = y_n + \frac{h}{6}[f(x_n, y_n) + 4f(x_{n+\frac{1}{2}}, y_{n+\frac{1}{2}}) + f(x_{n+1}, y_{n+1})]$$

$$+ O(h^5). \tag{2.23}$$

Schemes for generating successive approximations to the y's appearing in the right-hand side of a commensurate accuracy then complete the algorithms. A third-order algorithm with a local error $O(h^4)$ is

$$k_1 = hf(x_n, y_n);$$
$$k_2 = hf(x_n + \tfrac{1}{2}h, y_n + \tfrac{1}{2}k_1);$$
$$k_3 = hf(x_n + h, y_n - k_1 + 2k_2);$$
$$y_{n+1} = y_n + \tfrac{1}{6}(k_1 + 4k_2 + k_3) + O(h^4). \tag{2.24}$$

It is based on (2.23) and requires three evaluations of f per step. A fourth-order algorithm, which requires f to be evaluated four times for each integration step and has a local accuracy of $O(h^5)$, has been found by experience to give the best balance between accuracy and computational effort. It can be written as follows, with the k_i as intermediate variables:

$k_1 = hf(x_n, y_n);$

$k_2 = hf(x_n + \frac{1}{2}h, y_n + \frac{1}{2}k_1);$

$k_3 = hf(x_n + \frac{1}{2}h, y_n + \frac{1}{2}k_2);$

$k_4 = hf(x_n + h, y_n + k_3);$

$$y_{n+1} = y_n + \frac{1}{6}(k_1 + 2k_2 + 2k_3 + k_4) + O(h^5). \tag{2.25}$$

Exercise 2.4 Try out the second-, third-, and fourth-order Runge-Kutta methods discussed above on the problem defined by Eq. (2.7). Compare the computational effort for a given accuracy with that of other methods.

□ □ □

Exercise 2.5 The two coupled first-order equations

□ □ □

$$\frac{dy}{dt} = p \; ; \; \frac{dp}{dt} = -4\pi^2 y \tag{2.26}$$

define simple harmonic motion with period 1. By generalizing one of the single-variable formulas given above to this two-variable case, integrate these equations with any particular initial conditions you choose and investigate the accuracy with which the system returns to its initial state at integral values of t.

2.4 Stability

A major consideration in integrating differential equations is the numerical stability of the algorithm used; i.e., the extent to which round-off or other errors in the numerical computation can be amplified, in many cases enough for this "noise" to dominate the results. To illustrate the problem, let us attempt to improve the accuracy of Euler's method and approximate the derivative in (2.4) directly by the symmetric difference approximation (1.3b). We thereby obtain the three-term recursion relation

$$y_{n+1} = y_{n-1} + 2hf(x_n, y_n) + O(h^3), \tag{2.27}$$

which superficially looks about as useful as either of the third-order formulas (2.10) or (2.18). However, consider what happens when this method is applied to the problem

$$\frac{dy}{dx} = -y; \; y(x=0) = 1, \tag{2.28}$$

whose solution is $y = e^{-x}$. To start the recursion relation (2.27), we need the value of y_1 as well as $y_0 = 1$. This can be obtained by using (2.10) to get

Table 2.2 Integration of $dy/dx = -y$ with $y(0) = 1$ using Eq. (2.27)

x	Exact	Error	x	Exact	Error	x	Exact	Error
0.2	.818731	-.000269	3.3	.036883	-.000369	5.5	.004087	-.001533
0.3	.740818	-.000382	3.4	.033373	-.000005	5.6	.003698	.001618
0.4	.670320	-.000440	3.5	.030197	-.000380	5.7	.003346	-.001858
0.5	.606531	-.000517	3.6	.027324	.000061	5.8	.003028	.001989
0.6	.548812	-.000538	3.7	.024724	-.000400	5.9	.002739	-.002257
			3.8	.022371	.000133	6.0	.002479	.002439

$$y_1 = 1 - h + \tfrac{1}{2}h^2 + O(h^3).$$

(This is just the Taylor series for e^{-h}.) The following BASIC program then uses the method (2.27) to find y for values of x up to 6 using the value of h input:

```
10 INPUT "enter value of h (<=0 to stop)";H
20 YM=1: YZ=1-H+0.5*H^2            'starting values for y(0) and y(h)
30 FOR J%=2 TO 6/H                 'loop over steps
35    X=H*J%                       'X at this step
40    YP=YM-2*H*YZ                  'step Y by Eq. (2.15)
45    YM=YZ: YZ=YP                  'roll the Y values being saved
47    EXACT=EXP(-X)                 'analytic value at this point
50    PRINT X,EXACT,EXACT-YZ        'output analytic value and error
60 NEXT J%
70 GOTO 10                         'get another H
```

Note how the three-term recursion is implemented by keeping track of only three local variables, YP(lus), YZ(ero), and YM(inus).

A portion of the output of this code run for $h = 0.1$ is shown in Table 2.2. For small values of x, the numerical solution is only slightly larger than the exact value, the error being consistent with the $O(h^3)$ estimate. Then, near $x = 3.5$, an oscillation begins to develop in the numerical solution, which becomes alternately higher and lower than the exact values lattice point by lattice point. This oscillation grows larger as the equation is integrated further (see values near $x = 6$), eventually overwhelming the exponentially decreasing behavior expected.

The phenomenon observed above is a symptom of an instability in the algorithm (2.27). It can be understood as follows. For the problem (2.28), the recursion relation (2.27) reads

$$y_{n+1} = y_{n-1} - 2hy_n. \qquad (2.29)$$

We can solve this equation by assuming an exponential solution of the form $y_n = Ar^n$ where A and r are constants. Substituting into (2.29) then results in an equation for r,

$$r^2 + 2hr - 1 = 0,$$

the constant A being unimportant since the recursion relation is linear. The solutions of this equation are

$$r_+ = (1+h^2)^{1/2} - h \approx 1 - h \; ; \quad r_- = -(1+h^2)^{1/2} - h \approx -(1+h),$$

where we have indicated approximations valid for $h \ll 1$. The positive root is slightly less than one and corresponds to the exponentially decreasing solution we are after. However, the negative root is slightly less than -1, and so corresponds to a spurious solution

$$y_n \sim (-)^n (1+h)^n,$$

whose magnitude increases with n and which oscillates from lattice point to lattice point.

The general solution to the linear difference equation (2.27) is a linear combination of these two exponential solutions. Even though we might carefully arrange the initial values y_0 and y_1 so that only the decreasing solution is present for small x, numerical round-off during the recursion relation (Eq. (2.29) shows that two positive quantities are subtracted to obtain a smaller one) will introduce a small admixture of the "bad" solution that will eventually grow to dominate the results. This instability is clearly associated with the three-term nature of the recursion relation (2.29). A good rule of thumb is that instabilities and round-off problems should be watched for whenever integrating a solution that decreases strongly as the iteration proceeds; such a situation should therefore be avoided, if possible. We will see the same sort of instability phenomenon again in our discussion of second-order differential equations in Chapter 3.

Exercise 2.6 Investigate the stability of several other integration   □ □ □ methods discussed in this chapter by applying them to the problem (2.28). Can you give analytical arguments to explain the results you obtain?

2.5 Order and chaos in two-dimensional motion

A fundamental advantage of using computers in physics is the ability to treat systems that cannot be solved analytically. In the usual situation, the numerical results generated agree qualitatively with the intuition we have developed by studying soluble models and it is the quantitative values that are of real interest. However, in a few cases computer results defy our intuition (and thereby reshape it) and numerical work is then essential for a proper understanding. Surprisingly, such cases include the dynamics of simple classical systems, where the generic behavior differs *qualitatively* from that of the models covered in a traditional Mechanics course. In this example, we will study some of this surprising behavior by integrating numerically the trajectories of a particle moving in two dimensions. General discussions of these systems can be found in [He80], [Ri80], and [Ab78].

We consider a particle of unit mass moving in a potential, V, in two dimensions and assume that V is such that the particle remains confined for all times if its energy is low enough. If the momenta conjugate to the two coordinates (x, y) are (p_x, p_y), then the Hamiltonian takes the form

$$H = \tfrac{1}{2}(p_x^2 + p_y^2) + V(x,y). \tag{2.30}$$

Given any particular initial values of the coordinates and momenta, the particle's trajectory is specified by their time evolution, which is governed by four coupled first-order differential equations (Hamilton's equations):

$$\frac{dx}{dt} = \frac{\partial H}{\partial p_x} = p_x \;, \quad \frac{dy}{dt} = \frac{\partial H}{\partial p_y} = p_y \;;$$

$$\frac{dp_x}{dt} = -\frac{\partial H}{\partial x} = -\frac{\partial V}{\partial x} \;, \quad \frac{dp_y}{dt} = -\frac{\partial H}{\partial y} = -\frac{\partial V}{\partial y}\;. \tag{2.31}$$

For any V, these equations conserve the energy, E, so that the constraint

$$H(x,y,p_x,p_y) = E$$

restricts the trajectory to lie in a three-dimensional manifold embedded in the four-dimensional phase space. Apart from this, there are very few other general statements that can be made about the evolution of the system.

One important class of two-dimensional Hamiltonians for which additional statements about the trajectories *can* be made are those that are *integrable*. For these potentials, there is a second function of the coordinates and momenta, apart from the energy, that is a constant of the motion; the trajectory is thus constrained to a two-dimensional manifold of the phase space. Two familiar kinds of integrable systems are separable and central potentials. In the separable case,

$$V(x,y) = V_x(x) + V_y(y), \tag{2.32}$$

where the $V_{x,y}$ are two independent functions, so that the Hamiltonian separates into two parts, each involving only one coordinate and its conjugate momentum,

$$H = H_x + H_y \; ; \; H_{x,y} = \tfrac{1}{2} p^2_{x,y} + V_{x,y} .$$

The motions in x and y therefore decouple from each other and each of the Hamiltonians $H_{x,y}$ is separately a constant of the motion (Equivalently, $H_x - H_y$ is the second quantity conserved in addition to $E = H_x + H_y$.) In the case of a central potential,

$$V(x,y) = V(r); \; r = (x^2 + y^2)^{\frac{1}{2}}, \tag{2.33}$$

so that the angular momentum, $p_\theta = x p_y - y p_x$, is the second constant of the motion and the Hamiltonian can be written as

$$H = \tfrac{1}{2} p_r^2 + V(r) + \frac{p_\theta^2}{2r^2} ,$$

where p_r is the momentum conjugate to r. The additional constraint on the trajectory present in integrable systems allows the equations of motion to be "solved" by reducing the problem to one of evaluating certain integrals, much as we did for one-dimensional motion in Chapter 1. All of the familiar analytically soluble problems of classical mechanics are those that are integrable.

Although the dynamics of integrable systems are simple, it is often not at all easy to make this simplicity apparent. There is no general analytical method for deciding if there is a second constant of the motion in an arbitrary potential or for finding it if there is one. Numerical calculations are not obviously any better, as these supply only the trajectory for given initial conditions and this trajectory can be quite complicated in even familiar cases, as can be seen by recalling the Lissajous patterns of the (x,y) trajectories that arise when the motion in both coordinates is harmonic,

$$V_x = \tfrac{1}{2} \omega_x^2 x^2 , \; V_y = \tfrac{1}{2} \omega_y^2 y^2 . \tag{2.34}$$

An analysis in phase space suggests one way to detect integrability from the trajectory alone. Consider, for example, the case of a separable potential. Because the motions of each of the two coordinates are independent, plots of a trajectory in the (x,p_x) and (y,p_y) planes might look as shown in Figure 2.1. Here, we have assumed that each potential $V_{x,y}$ has a single minimum value of 0 at particular values of x and y, respectively. The particle moves on a closed contour in each of these two-dimensional projections of the four-dimensional phase space, each contour looking like that for ordinary one-dimensional motion shown in Figure 1.3. The areas of these contours depend upon how much energy is associated with each coordinate (i.e., E_x and E_y) and, as we consider trajectories with the same energy but with different initial conditions, the area of one contour will shrink as the other grows.

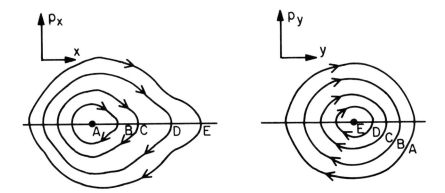

Figure 2.1 Trajectories of a particle in a two-dimensional separable potential as they appear in the (x,p_x) and (y,p_y) planes. Several trajectories corresponding to the same energy but different initial conditions are shown. Trajectories A and E are the limiting ones having vanishing E_x and E_y, respectively.

In each plane there is a limiting contour that is approached when all of the energy is in that particular coordinate. These contours are the intersections of the two-dimensional phase-space manifolds containing each of the trajectories and the (y,p_y) or (x,p_x) plot is therefore termed a "surface of section". The existence of these closed contours signals the integrability of the system.

Although we are able to construct Figure 2.1 only because we understand the integrable motion involved, a similar plot can be obtained from the trajectory alone. Suppose that every time we observe one of the coordinates, say x, to pass through zero, we plot the location of the particle on the (y,p_y) plane. In other words, crossing through $x=0$ triggers a "stroboscope" with which we observe the (y,p_y) variables. If the periods of the x and y motions are incommensurate (i.e., their ratio is an irrational number), then, as the trajectory proceeds, these observations will gradually trace out the full (y,p_y) contour; if the periods are commensurate (i.e., a rational ratio), then a series of discrete points around the contour will result. In this way, we can study the topology of the phase space associated with any given Hamiltonian just from the trajectories alone.

The general topology of the phase space for an integrable Hamiltonian can be illustrated by considering motion in a central potential. For fixed values of the energy and angular momentum, the radial motion is bounded between two turning points, r_{in} and

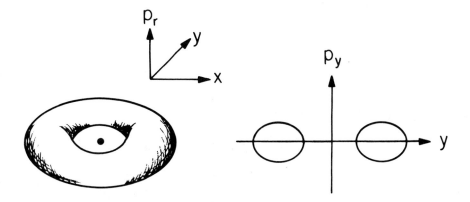

Figure 2.2 (Left) Toroidal manifold containing the trajectory of a particle in a central potential. (Right) Surface of section through this manifold at $x=0$.

r_{out}, which are the solutions of the equation

$$E-V(r)-\frac{p_\theta^2}{2r^2}=0 \ .$$

These two radii define an annulus in the (x,y) plane to which the trajectory is confined, as shown in the left-hand side of Figure 2.2. Furthermore, for a fixed value of r, energy conservation permits the radial momentum to take on only one of two values,

$$p_r=\pm\left[2E-2V(r)-\frac{p_\theta^2}{r^2}\right]^{\frac{1}{2}} \ .$$

These momenta define the two-dimensional manifold in the (x,y,p_r) space that contains the trajectory; it clearly has the topology of a torus, as shown in the left-hand side of Figure 2.2. If we were to construct a (y,p_y) surface of section by considering the $x=0$ plane, we would obtain two closed contours, as shown in the right-hand side of Figure 2.2. (Note that $y=r$ when $x=0$.) If the energy is fixed but the angular momentum is changed by varying the initial conditions, the dimensions of this torus change, as does the area of the contour in the surface of section.

    The toroidal topology of the phase space of a central potential can be shown to be common to all integrable systems. The manifold on which the trajectory lies for given values of the constants of the motion is called an "invariant torus", and there are many of them for a given energy. The general surface of section of such

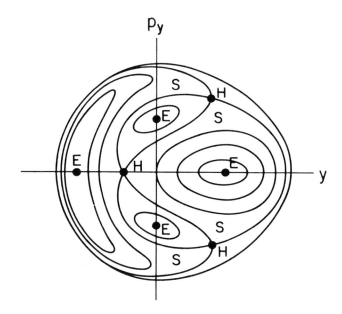

Figure 2.3 Possible features of the surface of section of
a general integrable system. E and H label elliptic and
hyperbolic fixed points, respectively, while the curve
labeled S is a separatrix.

tori looks like that shown in Figure 2.3. There are certain fixed
points associated with trajectories that repeat themselves exactly
after some period. The elliptic fixed points correspond to trajectories that are stable under small perturbations. Around each of
them is a family of tori, bounded by a separatrix. The hyperbolic
fixed points occur at the intersections of separatrices and are
stable under perturbations along one of the axes of the hyperbola,
but unstable along the other.

An interesting question is *"What happens to the tori of an
integrable system under a perturbation that destroys the integrability of the Hamitonian?"*. For small perturbations, "most" of the
tori about an elliptic fixed point become slightly distorted but
retain their topology (the KAM theorem due to Kolmogorov,
Arnold, and Moser, [Ar68]). However, adjacent regions of phase
space become "chaotic", giving surfaces of section that are a
seemingly random splatter of points. Within these chaotic regions
are nested yet other elliptic fixed points and other chaotic regions
in a fantastic heirarchy. (See Figure 2.4.)

Large deviations from integrability must be investigated
numerically. One convenient case for doing so is the potential

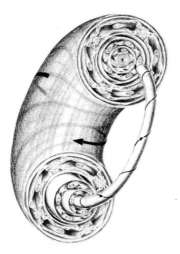

Figure 2.4 Nested tori for a slightly perturbed integrable system. Note the heirarchy of elliptic orbits interspersed with chaotic regions. A magnification of this heirarchy would show the same pattern repeated on a smaller scale and so on, *ad infinitum.* (Reproduced from [Ab78].)

$$V(x,y)=\tfrac{1}{2}(x^2+y^2)+x^2y-\tfrac{1}{3}y^3,\qquad(2.35)$$

which was originally introduced by Hénon and Heiles in the study of stellar orbits through a galaxy [He64]. This potential can be thought of as a perturbed harmonic oscillator potential (a small constant multiplying the cubic terms can be absorbed through a rescaling of the coordinates and energy, so that the magnitude of the energy becomes a measure of the deviation from integrability) and has the three-fold symmetry shown in Figure 2.5. The potential is zero at the origin and becomes unbounded for large values of the coordinates. However, for energies less than 1/6, the trajectories remain confined within the equilateral triangle shown.

The BASIC program for Example 2, whose source code is contained in Appendix B and in the file EXAM2.BAS on the *Computational Physics* diskette, constructs surfaces of section for the Hénon-Heiles potential. The method used is to integrate the equations of motion (2.31) using the fourth-order Runge-Kutta algorithm (2.25). Initial conditions are specified by putting $x=0$ and by giving the energy, y, and p_y; p_x is then fixed by energy conservation. The input of y and p_y is either analog using the cursor (if graphics is available) or digital from the keyboard. As the integration proceeds, the (x,y) trajectory and the (y,p_y) surface of section are displayed. Points on the latter are calculated by watching

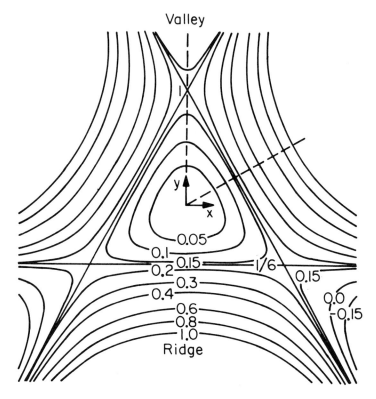

Figure 2.5 Equipotential contours of the Hénon -Heiles potential, Eq. (2.35).

for a time step during which x changes sign. When this happens, the precise location of the point on the surface of section plot is determined by switching to x as the independent variable, so that the equations of motion (2.31) become

$$\frac{dx}{dx} = 1 \, , \quad \frac{dy}{dx} = \frac{1}{p_x} p_y \, ;$$

$$\frac{dp_x}{dx} = -\frac{1}{p_x} \frac{\partial V}{\partial x} \, , \quad \frac{dp_y}{dx} = -\frac{1}{p_x} \frac{\partial V}{\partial y} \, ,$$

and then integrating one step backward in x from its value after the time step to 0 [He82]. If the value of the energy is not changed when new initial conditions are specified, all previous surface of section points can be plotted, so that plots like those in Figures 2.3 and 2.4 can be built up after some time.

The following exercises are aimed at improving your under- standing of the physical principles and numerical methods demonstrated in this example.

Exercise 2.7 One necessary (but not sufficient) check of the accuracy of the integration in Hamiltonian systems is the conservation of energy. Change the time step used (line 70) and observe how this affects the accuracy. Replace the integration method by one of the other algorithms discussed in this chapter and observe the change in accuracy and efficiency. Note that we require an algorithm that is both accurate and efficient because of the very long integration times that must be considered.

Exercise 2.8 Change the potential to a central one (lines 230-250) and observe the character of the (x,y) trajectories and surfaces of section for various initial conditions. Compare the results with Figure 2.2. and verify that the qualitative features don't change if you use a different central potential. Note that you will have to adjust the energy and scale of the potential so that the (x,y) trajectory does not go beyond the equilateral triangle of the Hénon -Heiles potential (Figure 2.5) or else the graphics subroutine will fail.

Exercise 2.9 If the sign of the y^3 term in the Hénon-Heiles potential (2.35) is reversed, the Hamiltonian becomes integrable. Verify analytically that this is so by making a canonical transformation to the variables $x \pm y$ and showing that the potential is separable in these variables. Make the corresponding change in sign in the code and observe the character of the surfaces of section that result for various initial conditions at a given energy. (You should keep the energy below 1/12 to avoid having the trajectory become unbounded.) Verify that there are no qualitative differences if you use a different separable potential, say the harmonic one of Eq. (2.34).

Exercise 2.10 Use the code to construct surfaces of section for the Hénon-Heiles potential (2.35) at energies ranging from 0.025 to 0.15 in steps of 0.025. For each energy, consider various initial conditions and integrate each trajectory long enough in time (some will require going to $t \approx 1000$) to map out the surface-of-section adequately. For each energy, see if you can find the elliptic fixed points, the tori (and tori of tori) around them, and the chaotic regions of phase space and observe how the relative proportions of each change with increasing energy. With some patience and practice, you should be able to generate a plot that resembles the schematic representation of Figure 2.4 around each elliptic trajectory.

Project II: The structure of white dwarf stars

White dwarf stars are cold objects composed largely of heavy nuclei and their associated electrons. These stars are one possible end result of the conventional nuclear processes that build the elements by binding nucleons into nuclei. They are often composed of the most stable nucleus, ^{56}Fe, with 26 protons and 30 neutrons, but, if the nucleosynthesis terminates prematurely, ^{12}C nuclei might predominate. The structure of a white dwarf star is determined by the interplay between gravity, which acts to compress the star, and the degeneracy pressure of the electrons, which tends to resist compression. In this project, we will investigate the structure of a white dwarf by integrating the equations defining this equilibrium. We will determine, among other things, the relation between the star's mass and its radius, quantities that can be determined from observation. Discussions of the physics of white dwarfs can be found in [Ch57], [Ch84], and [Sh83].

II.1 The equations of equilibrium

We assume that the star is spherically symmetric (i.e., the state of matter at any given point in the star depends only upon the distance of that point from the star's center), that it is not rotating, and that the effects of magnetic fields are not important. If the star is in mechanical (hydrostatic) equilibrium, the gravitational force on each bit of matter is balanced by the force due to spatial variation of the pressure, P. The gravitational force acting on a unit volume of matter at a radius r is

$$F_{grav} = -\frac{Gm}{r^2}\rho,$$ (II.1)

where G is the gravitational constant, $\rho(r)$ is the mass density, and $m(r)$ is the mass of the star interior to the radius r:

$$m(r) = 4\pi \int_0^r \rho(r')r'^2 dr'.$$ (II.2)

The force per unit volume of matter due to the changing pressure is $-dP/dr$. When the star is in equilibrium, the net force (gravitational plus pressure) on each bit of matter vanishes, so that, using (II.1), we have

$$\frac{dP}{dr} = -\frac{Gm(r)}{r^2}\rho(r).$$ (II.3)

A differential relation between the mass and the density can be obtained by differentiating (II.2):

$$\frac{dm}{dr} = 4\pi r^2 \rho(r).$$ (II.4)

The description is completed by specifying the "equation of state", an intrinsic property of the matter giving the pressure, $P(\rho)$, required to maintain it at a given density. Upon using the identity

$$\frac{dP}{dr} = \left[\frac{d\rho}{dr}\right]\left[\frac{dP}{d\rho}\right],$$

Eq. (II.3) can be written as

$$\frac{d\rho}{dr} = -\left[\frac{dP}{d\rho}\right]^{-1}\frac{Gm}{r^2}\rho. \qquad (II.5)$$

Equations (II.4) and (II.5) are two coupled first-order differential equations that determine the structure of the star for a given equation of state. The values of the dependent variables at $r=0$ are $\rho=\rho_c$, the central density, and $m=0$. Integration outward in r then gives the density profile, the radius of the star, R, being determined by the point at which ρ vanishes. (On very general grounds, we expect the density to decrease with increasing distance from the center.) The total mass of the star is then $M=m(R)$. Since both R and M depend upon ρ_c, variation of this parameter allows stars of different mass to be studied.

II.2 The equation of state

We must now determine the equation of state appropriate for a white dwarf. As mentioned above, we will assume that the matter consists of large nuclei and their electrons. The nuclei, being heavy, contribute nearly all of the mass but make almost no contribution to the pressure, since they hardly move about at all. The electrons, however, contribute virtually all of the pressure but essentially none of the mass. We will be interested in densities far greater than that of ordinary matter, where the electrons are no longer bound to individual nuclei, but rather move freely through the material. A good model is then a free Fermi gas of electrons at zero temperature, treated with relativistic kinematics.

For matter at a given mass density, the number density of electrons is

$$n = Y_e \frac{\rho}{M_p}, \qquad (II.6)$$

where M_p is the proton mass (we neglect the small difference between the neutron and proton masses) and Y_e is the number of electrons per nucleon. If the nuclei are all ^{56}Fe, then

$$Y_e = \frac{26}{56} = 0.464,$$

while $Y_e = \frac{1}{2}$ if the nuclei are ^{12}C; electrical neutrality of the matter requires one electron for every proton.

The free Fermi gas is studied by considering a large volume V containing N electrons that occupy the lowest energy plane-wave states with momentum $p < p_f$. Remembering the two-fold spin degeneracy of each plane wave, we have

$$N = 2V \int_0^{p_f} \frac{d^3p}{(2\pi)^3},\qquad (\text{II}.7)$$

which leads to

$$p_f = (3\pi^2 n)^{1/3},\qquad (\text{II}.8)$$

where $n = N/V$. (We put $\hbar = c = 1$ unless indicated explicitly.) The total energy density of these electrons is

$$\frac{E}{V} = 2 \int_0^{p_f} \frac{d^3p}{(2\pi)^3}(p^2 + m_e^2)^{1/2},\qquad (\text{II}.9)$$

which can be integrated to give

$$\frac{E}{V} = n_0 m_e x^3 \varepsilon(x);\qquad (\text{II}.10\text{a})$$

$$\varepsilon(x) = \frac{3}{8x^3}\left[x(1+2x^2)(1+x^2)^{1/2} - \log[x + (1+x^2)^{1/2}]\right]\qquad (\text{II}.10\text{b})$$

where

$$x \equiv \frac{p_f}{m_e} = \left(\frac{n}{n_0}\right)^{1/3};\quad n_0 = \frac{m_e^3}{3\pi^2}.\qquad (\text{II}.10\text{c})$$

The variable x characterizes the electron density in terms of

$$n_0 = 5.89 \times 10^{29}\ \text{cm}^{-3},$$

the density at which the Fermi momentum is equal to the electron mass, m_e.

In the usual thermodynamic manner, the pressure is related to how the energy changes with volume at fixed N:

$$P = -\frac{\partial E}{\partial V} = -\frac{\partial E}{\partial x}\frac{\partial x}{\partial V}.\qquad (\text{II}.11)$$

Using (II.10c) to find

$$\frac{\partial x}{\partial V} = -\frac{x}{3V}$$

and differentiating (II.10a,b) results in

$$P = \tfrac{1}{3}n_0 m_e x^4 \varepsilon',\qquad (\text{II}.12)$$

where $\varepsilon' = d\varepsilon/dx$.

It is now straightforward to calculate $dP/d\rho$. Since P is most naturally expressed in terms of x, we must relate x to ρ. This is most easily done using (II.10c) and (II.6):

$$x = \left[\frac{n}{n_0}\right]^{1/3} = \left[\frac{\rho}{\rho_0}\right]^{1/3} ; \qquad \text{(II.13a)}$$

$$\rho_0 = \frac{M_p n_0}{Y_e} = 9.79 \times 10^5 Y_e^{-1} \text{ gm cm}^{-3}. \qquad \text{(II.13b)}$$

Thus, ρ_0 is the mass density of matter in which the electron density is n_0. Differentiating the pressure and using (II.12, II.13a) then yields, after some algebra,

$$\frac{dP}{d\rho} = \frac{dP}{dx}\frac{dx}{d\rho} = Y_e \frac{m_e}{M_p}\gamma(x);$$

$$\gamma(x) = \frac{1}{9x^2}\frac{d}{dx}(x^4 \varepsilon') = \frac{x^2}{3(1+x^2)^{1/2}} . \qquad \text{(II.14)}$$

II.3 Scaling the equations

It is often useful to reduce equations describing a physical system to dimensionless form, both for physical insight and for numerical convenience (i.e., to avoid dealing with very large or very small numbers in the computer). To do this for the equations of white dwarf structure, we introduce dimensionless radius, density, and mass variables:

$$r = R_0 \bar{r}, \ \rho = \rho_0 \bar{\rho}, \ m = M_0 \bar{m} , \qquad \text{(II.15)}$$

with the radius and mass scales, R_0 and M_0, to be determined for convenience. Substituting into Eqs. (II.4, II.5) and using (II.14) yields, after some rearrangement,

$$\frac{d\bar{m}}{d\bar{r}} = \left[\frac{4\pi R_0^3 \rho_0}{M_0}\right]\bar{r}^2 \bar{\rho}; \qquad \text{(II.16a)}$$

$$\frac{d\bar{\rho}}{d\bar{r}} = -\left[\frac{GM_0}{R_0 Y_e (m_e/M_p)}\right]\frac{\bar{m}\bar{\rho}}{\gamma \bar{r}^2} . \qquad \text{(II.16b)}$$

If we now choose M_0 and R_0 so that the coefficients in parentheses in these two equations are unity, we find

$$R_0 = \left[\frac{Y_e (m_e/M_p)}{4\pi G \rho_0}\right]^{1/2} = 7.72 \times 10^8 Y_e \text{ cm}, \qquad \text{(II.17a)}$$

$$M_0 = 4\pi R_0^3 \rho_0 = 5.67 \times 10^{33} Y_e^2 \text{ gm}, \qquad \text{(II.17b)}$$

and the dimensionless differential equations are

$$\frac{d\bar{p}}{d\bar{r}} = -\frac{\bar{m}\bar{p}}{\gamma\bar{r}^2}; \tag{II.18a}$$

$$\frac{d\bar{m}}{d\bar{r}} = \bar{r}^2\bar{p}. \tag{II.18b}$$

These equations are completed by recalling that γ is given by Eq. (II.14) with $x = \bar{p}^{1/3}$.

The scaling quantities (II.13b) and (II.17) tell us how to relate the structures of stars with different Y_e and hence reduce the numerical work to be done. Specifically, if we have the solution for $Y_e = 1$ at some value of $\bar{p}_c = \rho_c/\rho_0$, then the solution at any other Y_e is obtained by scaling the density as Y_e^{-1}, the radius as Y_e, and the mass as Y_e^2.

To give some feeling for the scales (II.17), recall that the solar radius and mass are

$$R_\odot = 6.95 \times 10^{10}\,\text{cm} \;;\; M_\odot = 1.98 \times 10^{33}\,\text{gm},$$

while the density at the center of the sun is ≈ 150 gm cm^{-3}. We therefore expect white dwarf stars to have masses comparable to a solar mass, but their radii will be considerably smaller and their densities will be much higher.

II.4 Solving the equations

We are now in a position to solve the equations and determine the properties of a white dwarf. This can be done in the following sequence of steps.

□ □ □   **Step 1** Verify the steps in the derivations given above. Determine the leading behavior of ε and γ in the extreme non-relativistic ($x \ll 1$) and relativistic ($x \gg 1$) limits and verify that these are what you expect from simple arguments. Analyze Eqs. (II.18a,b) when $\bar{m}$ and $\bar{r}$ are finite and $\bar{p}$ is small to show that $\bar{p}$ vanishes at a finite radius; the star therefore has a well-defined surface. Determine the functional form with which $\bar{p}$ vanishes near the surface.

□ □ □   **Step 2** Write a program to integrate Eqs. (II.18) outward from $\bar{r}=0$ for $Y_e=1$. Calculate the density profiles, total masses, and radii of stars with selected values of the dimensionless central density $\bar{p}_c$ ranging from 10^{-1} to 10^6. By changing the radial step and the integration algorithm, verify that your solutions are accurate. Show that the total kinetic and rest energy of the electrons in the star is

$$U = \int_0^R \left[\frac{E}{V}\right] 4\pi r^2 dr,$$

where E/V is the energy density given by (II.10) and that the total gravitational energy of the star can be written as

$$W = -\int_0^R \frac{Gm(r)}{r} \rho(r) 4\pi r^2 dr \,.$$

Use the scaling discussed above to cast these integrals in dimensionless form and calculate these energies for the solutions you've generated. Try to understand all of the trends you observe through simple physical reasoning.

Step 3 A commonly used description of stellar structure involves a   □ □ □
polytropic equation of state,

$$P = P_0 \left[\frac{\rho}{\rho_0} \right]^\Gamma \,,$$

where P_0, ρ_0, and Γ are constants, the latter being the adiabatic index. By suitably scaling Eqs. (II.4,5), show that a star with a polytropic equation of state obeys Eqs. (II.18) with

$$\gamma = \bar{\rho}^{\Gamma-1} = x^{3\Gamma-3} \,.$$

The degenerate electron gas therefore corresponds to $\Gamma = 4/3$ in the extreme relativistic (high-density) limit and $\Gamma = 5/3$ in the extreme non-relativistic (low-density) limit.

Polytropic equations of state provide a simple way to study the effects of varying the equation of state (by changing Γ) and also give, for two special values of Γ, analytical solutions against which numerical solutions can be checked. For this latter purpose, it is most convenient to recast Eqs. (II.18) as a single second-order differential equation for $\bar{\rho}$ by solving (II.18a) for $\bar{m}$ and then differentiating with respect to $\bar{r}$ and using (II.18b). Show that the differential equation that results (the Lane-Emden equation, [Ch57]) is

$$\frac{1}{\bar{r}^2} \frac{d}{d\bar{r}} \left[\bar{r}^2 \bar{\rho}^{\Gamma-2} \frac{d\bar{\rho}}{d\bar{r}} \right] = -\bar{\rho},$$

and that for $\Gamma = 2$ the solution is

$$\bar{\rho} = \bar{\rho}_c \frac{\sin \bar{r}}{\bar{r}} \,,$$

while for $\Gamma = 6/5$ the solution is

$$\bar{\rho} = \bar{\rho}_c \left[1 + \frac{\bar{r}^2}{3a^2} \right]^{-5/2} \quad ; \quad a = 5^{1/2} \bar{\rho}_c^{-2/5} \,.$$

Use these analytical solutions to check the code you wrote in Step

2. Then study the structure of stars with different adiabatic indexes Γ and interpret the results.

□ □ □ **Step 4** If things are working correctly, you should find that, as the central density of a white dwarf increases, its mass approaches a limiting value, (the Chandrasekhar mass, M_{Ch}) and the star becomes very small. To understand this and to get an estimate of M_{Ch}, we can follow the steps in an argument originally given by Landau in 1932. The total energy of the star is composed of the gravitational energy ($W<0$) and the internal energy of the matter ($U>0$). Assume that a star of a given total mass has a *constant* density profile. (This is not a very good assumption for quantitative purposes, but it is valid enough to be useful in understanding the situation.) The radius is therefore related simply to the total mass and the constant density. Calculate U and W for this density profile, assuming that the density is high enough so that the electrons can be treated in the relativistic limit. Show that both energies scale as $1/R$ and that W dominates U when the total mass exceeds a certain critical value. It then becomes energetically favorable for the star to collapse (shrink its radius to zero). Estimate the Chandrasekhar mass in this way and compare it to the result of your numerical calculation. Also verify the correctness of this argument by showing that U and W for your solutions found in Step 2 become equal and opposite as the star approaches the limiting mass.

□ □ □ **Step 5** Scale the mass-radius relation you found in Step 1 to the cases corresponding to ^{56}Fe and ^{12}C nuclei. Three white dwarf stars, Sirius B, 40 Eri B, and Stein 2051, have masses and radii (in units of the solar values) determined from observation to be (1.053±0.028, 0.0074±0.0006), (0.48±0.02, 0.0124±0.0005), and (0.50±0.05 or 0.72±0.08, 0.0115±0.0012), respectively [Sh83]. Verify that these values are consistent with the model we have developed. What can you say about the compositions of these stars?

Chapter 3

Boundary Value and Eigenvalue Problems

Many of the important differential equations of physics can be cast in the form of a linear, second-order equation:

$$\frac{d^2y}{dx^2} + k^2(x)y = S(x), \tag{3.1}$$

where S is an inhomogeneous ("driving") term and k^2 is a real function. When k^2 is positive, the solutions of the homogeneous equation (i.e., $S=0$) are oscillatory with local wavenumber k, while when k^2 is negative, the solutions grow or decay exponentially at a local rate $(-k^2)^{1/2}$. For example, consider trying to find the electrostatic potential, Φ, generated by a localized charge distribution, $\rho(\mathbf{r})$. Poisson's equation is

$$\nabla^2\Phi = -4\pi\rho, \tag{3.2}$$

which, for a spherically symmetric ρ and Φ, simplifies to

$$\frac{1}{r^2}\frac{d}{dr}\left[r^2\frac{d\Phi}{dr}\right] = -4\pi\rho. \tag{3.3}$$

The standard substitution

$$\Phi(r) = r^{-1}\varphi(r)$$

then results in

$$\frac{d^2\varphi}{dr^2} = -4\pi r\rho, \tag{3.4}$$

which is of the form (3.1) with $k^2=0$ and $S=-4\pi r\rho$. In a similar manner, the quantum mechanical wavefunction for a particle of mass m and energy E moving in a central potential $V(r)$ can be written as

$$\Psi(\mathbf{r}) = r^{-1}R(r)Y_{LM}(\hat{\mathbf{r}}),$$

where Y_{LM} is a spherical harmonic and the radial wavefunction R satisfies

$$\frac{d^2R}{dr^2} + k^2(r)R = 0; \quad k^2(r) = \frac{2m}{\hbar^2}\left[E - \frac{L(L+1)\hbar^2}{2mr^2} - V(r)\right]. \tag{3.5}$$

This is also of the form (3.1), with $S=0$.

The equations discussed above appear unremarkable and readily treated by the methods discussed in Chapter 2, except for two points. First, the boundary conditions imposed by the physics often appear as constraints on the dependent variable at two *separate* points of the independent variable, so that solution as an initial value problem is not obviously possible. Moreover, the Schroedinger equation (3.5) is an eigenvalue equation in which we must *find* the energies that lead to physically acceptable solutions satisfying the appropriate boundary conditions. This chapter is concerned with methods for treating such problems. We begin by deriving an integration algorithm particularly well suited to equations of the form (3.1), and then discuss boundary value and eigenvalue problems in turn.

3.1 The Numerov algorithm

There is a particularly simple and efficient method for integrating second-order differential equations having the form of (3.1). To derive this method, commonly called the Numerov or Cowling's method, we begin by approximating the second derivative in (3.1) by the three-point difference formula (1.7),

$$\frac{y_{n+1}-2y_n+y_{n-1}}{h^2}=y_n''+\frac{h^2}{12}y_n''''+O(h^4), \tag{3.6}$$

where we have written out explicitly the $O(h^2)$ "error" term, which is derived easily from the Taylor expansion (1.1, 1.2a). From the differential equation itself, we have

$$y_n''''=\frac{d^2}{dx^2}(-k^2y+S)|_{x=x_n}$$

$$=-\frac{(k^2y)_{n+1}-2(k^2y)_n+(k^2y)_{n-1}}{h^2}$$

$$+\frac{S_{n+1}-2S_n+S_{n-1}}{h^2}+O(h^2). \tag{3.7}$$

When this is substituted into (3.6), we can write, after some rearrangement,

$$(1+\frac{h^2}{12}k_{n+1}^2)y_{n+1}-2(1-\frac{5h^2}{12}k_n^2)y_n+(1+\frac{h^2}{12}k_{n-1}^2)y_{n-1}=$$

$$\frac{h^2}{12}(S_{n+1}+10S_n+S_{n-1})+O(h^6). \tag{3.8}$$

Solving this linear equation for either y_{n+1} or y_{n-1} then provides a recursion relation for integrating either forward or backward in x, with a local error $O(h^6)$. Note that this is one order more accurate than the fourth-order Runge-Kutta method (2.25), which might be used to integrate the problem as two coupled first-order equations. The Numerov scheme is also more efficient, as each step requires the computation of k^2 and S at only the lattice points.

Exercise 3.1 Apply the Numerov algorithm to the problem

$$\frac{d^2y}{dx^2}=-4\pi^2 y; \;\; y(0)=1, \;\; y'(0)=0 .$$

Integrate from $x=0$ to $x=1$ with various step sizes and compare the efficiency and accuracy with some of the methods discussed in Chapter 2. Note that you will have to use some special procedure (e.g., a Taylor series) to generate the value of $y_1 \equiv y(h)$ needed to start the three-term recursion relation.

3.2 Direct integration of boundary value problems

As a concrete illustration of boundary value problems, consider trying to solve Poisson's equation (3.4) when the charge distribution is

$$\rho(r)=\frac{1}{8\pi}e^{-r},\tag{3.9}$$

which has a total charge

$$Q=\int \rho(r)d^3r = \int_0^\infty \rho(r)4\pi r^2 dr = 1 .$$

The exact solution to this problem is

$$\varphi(r)=1-\tfrac{1}{2}(r+2)e^{-r},\tag{3.10}$$

from which $\Phi=r^{-1}\varphi$ follows immediately. This solution has the expected behavior at large r, $\varphi \to 1$, which corresponds to $\Phi \to r^{-1}$, the Coulomb potential from a unit charge.

Suppose that we try to solve this example as an ordinary initial value problem. Since ρ has no singular behavior at the origin (e.g., there is no point charge), Φ is regular there, which implies that $\varphi=r\Phi$ vanishes at $r=0$; this is indeed the case for the explicit solution (3.10). We could then integrate (3.4) outward from the origin using the appropriate rearrangement of (3.8) (recall $k^2=0$ here):

$$\varphi_{n+1}=2\varphi_n -\varphi_{n-1}+\frac{h^2}{12}(S_{n+1}+10S_n +S_{n-1}),\tag{3.11}$$

with

$$S = -4\pi r \rho = -\tfrac{1}{2} r e^{-r} .$$

However, to do so we must know the value of φ_1 (or, equivalently, $d\varphi/dr = \Phi$ at $r = 0$) in addition to $\varphi_0 = 0$. This is not a very happy situation, since φ_1 is part of the very function we're trying to find, and so is not known *a priori*. We will discuss below what to do in the general case, but in the present example, since we have an analytical solution, we can find $\varphi_1 = \varphi(r = h)$ from (3.10). The following BASIC program does the outward integration to $r = 20$, storing the solution in an array and printing the exact result and error as it goes along.

```
5  DIM PHI(200)                                'array for the solution
10 DEF FNS(R)=-.5*R*EXP(-R)                     'function for S
15 H=.1                                         'radial step size
20 N%=20/H                                      'number of points to r=20
25 CON=H^2/12                                   'constant in Numerov method
30 SM=0: SZ=FNS(H)                              'S at first two points
35 PHI(0)=0                                      'boundary condition at r=0
40 PHI(1)=1-.5*(H+2)*EXP(-H)                     'exact value at 1'st point
45 '
50 FOR J%=1 TO N%-1                              'loop for outward integration
55    R=(J%+1)*H:  SP=FNS(R)                     'radius and S at next point
60    '                                          'Numerov formula,Eq.(3.11)
65    PHI(J%+1)=2*PHI(J%)-PHI(J%-1)+CON*(SP+10*SZ+SM)
70    SM=SZ: SZ=SP                               'roll the values of S
75    EXACT=1-.5*(R+2)*EXP(-R)                   'analytical value, Eq. (3.10)
80    DIFF=EXACT-PHI(J%+1)                       'error at this point
85    PRINT R,EXACT,DIFF                         'output current values
90 NEXT J%
```

Note how computation is minimized by keeping track of the values of S at the current and adjacent lattice points (SZ, SM, and SP) and by computing the constant $h^2/12$ appearing in the Numerov formula (3.11) outside of the integration loop.

Results generated by this program are given in the first three columns of Table 3.1, which show that the numerical solution is rather accurate for small r. All is not well, however. The error per step gets larger at large r (the error after the 20 steps from $r = 0$ to $r = 2$ is 3×10^{-6}, while during the 20 steps from $r = 18$ to $r = 20$, it grows by 1.2×10^{-5}, 4 times as much), and the solution will become quite inaccurate if continued to even larger radii. This behavior is quite surprising because the errors in the Numerov integration should be getting smaller at large r as φ becomes a constant.

Further symptoms of a problem can be found by considering a more general case. We then have no analytical formula to give us Φ near the origin, which we need to get the three-term recursion relation started. One way to proceed is find $\Phi(0)$ by direct numerical quadrature of the Coulomb potential,

Table 3.1 Errors in solving the Poisson problem defined by Eqs. (3.4,3.9)

r	Exact $\varphi(r)$	Analytical $\varphi(h)$	5% Error in $\varphi(h)$	Linear correction
2	0.729330	-0.000003	0.049919	-0.000016
4	0.945053	-0.000006	0.099838	-0.000011
6	0.990085	-0.000005	0.149762	-0.000007
8	0.998323	0.000001	0.199690	-0.000003
10	0.999728	0.000010	0.249622	-0.000001
12	0.999957	0.000022	0.299556	0.000002
14	0.999993	0.000036	0.349493	0.000003
16	0.999999	0.000052	0.399431	0.000000
18	1.000000	0.000065	0.449366	0.000000
20	1.000000	0.000077	0.499301	0.000000

$$\Phi(0)=\int \frac{\rho(r)}{r}\,d^3r=4\pi\int_0^\infty r\rho\,dr,$$

perhaps using Simpson's rule. There will, however, always be some error associated with the value obtained. We can simulate such an error in the code above (suppose it is 5%) by inserting the line

```
43 PHI(1)=0.95*PHI(1) .
```

The code then gives the errors listed in the fourth column of Table 3.1. Evidently disaster has struck, for a 5% change in the initial conditions has induced a 50% error in the solution at large r.

It is simple to understand what has happened. Solutions to the homogeneous version of (3.4),

$$\frac{d^2\varphi}{dr^2}=0,$$

can be added to any particular solution of (3.4) to give yet another solution. There are two linearly independent homogeneous solutions,

$$\varphi\sim r; \quad \Phi\sim\text{constant},$$

and

$$\varphi\sim\text{constant}; \quad \Phi\sim r^{-1}.$$

The general solution to (3.4) in the asymptotic region (where ρ vanishes and the equation is homogeneous) can be written as a linear combination of these two functions, but the latter, subdominant solution is the physical one, since we know the potential at large r is given by $\Phi \to 1/r$. Imprecision in the specification of Φ at the origin or any numerical round-off error in the integration process can introduce a small admixture of the $\varphi \sim r$ solution, which eventually dominates at large r.

The cure for this difficulty is straightforward: subtract a multiple of the "bad", unphysical solution to the homogeneous equation from the numerical result to guarantee the physical behavior in the asymptotic region. It is easy to see that the "bad" results shown in the fourth column of Table 3.1 vary linearly with r for large r. The following lines of code then fit the last 10 points of the numerical solution to the form

$$\varphi = mr + b$$

and subtract mr from the numerical results to guarantee the appropriate large-r behavior.

```
95  M=(PHI(N%)-PHI(N%-10))/(10*H)
100 B=PHI(N%)-M*N%*H
105 FOR J%=1 TO N%
110    R=J%*H
115    PHI(J%)=PHI(J%)-M*R
120 NEXT J%
```

The errors in φ so obtained are shown in the final column of Table 3.1; the solution is even more accurate at large r than the uncorrected one found when the exact value of PHI(1) is used to start the integration.

In this simple example, the instabilities are not too severe; satisfactory results for moderate values of r are obtained with outward integration when the exact (or reasonably accurate approximate) value of PHI(1) is used. Alternatively, it is also feasible to integrate inward, starting at large r with $\varphi = Q$, independent of r. This results in a solution that often satisfies accurately the boundary condition at $r = 0$ and avoids having to perform a quadrature to determine the (approximate) starting value of PHI(1).

□ □ □ **Exercise 3.2** Solve the problem defined by Eqs. (3.4, 3.9) by Numerov integration inward from large r using the known asymptotic behavior of φ for the starting values. How well does your solution satisfy the boundary condition $\varphi(r=0)=0$?

3.3 Green's function solution of boundary value problems

When the two solutions to the homogeneous equation have very different behaviors, some extra precautions must be taken. For example, in describing the potential from a charge distribution of a multipole order $l > 0$, the monopole equation (3.4) is modified to

$$\left[\frac{d^2}{dr^2} - \frac{l(l+1)}{r^2}\right]\varphi = -4\pi r\rho, \qquad (3.12)$$

which has the two homogeneous solutions

$$\varphi \sim r^{l+1} \; ; \; \varphi \sim r^{-l}.$$

For large r, the first of these solutions is much larger than the second, so that ensuring the correct asymptotic behavior by subtracting a multiple of this dominant homogeneous solution from a particular solution we have found by outward integration is subject to large round-off errors. Inward integration is also unsatisfactory, in that the unphysical r^{-l} solution is likely to dominate at small r.

One possible way to generate an accurate solution is by combining the two methods. Inward integration can be used to obtain the potential for r greater than some intermediate radius, r_m, and outward integration can be used for the potential when $r < r_m$. As long as r_m is chosen so that neither homogeneous solution is dominant, the outer and inner potentials obtained respectively from these two integrations will match at r_m and, together, describe the entire solution. Of course, if the inner and outer potentials don't quite match, a multiple of the homogeneous solution can be added to the former to correct for any deficiencies in our knowledge of $\varphi'(r=0)$.

Sometimes the two homogeneous solutions have such different behaviors that it is impossible to find a value of r_m that permits satisfactory integration of the inner and outer potentials. Such cases can be solved by the Green's function of the homogeneous equation. To illustrate, let us consider Eq. (3.1) with the boundary condition $\varphi(x=0) = \varphi(x=\infty) = 0$. Since the problem is linear, we can write the solution as

$$\varphi(x) = \int_0^\infty G(x,x')S(x')dx', \qquad (3.13)$$

where G is a Green's function satisfying

$$\left[\frac{d^2}{dx^2} + k^2(x)\right]G(x,x') = \delta(x-x'). \qquad (3.14)$$

It is clear that G satisfies the homogeneous equation for $x \neq x'$. However, the derivative of G is discontinuous at $x = x'$, as can be seen by integrating (3.14) from $x = x' - \varepsilon$ to $x = x' + \varepsilon$, where ε is an infinitesimal:

$$\left.\frac{dG}{dx}\right|_{x=x'+\varepsilon} - \left.\frac{dG}{dx}\right|_{x=x'-\varepsilon} = 1. \qquad (3.15)$$

The problem, of course, is to find G. This can be done by considering two solutions to the homogeneous problem, $\varphi_<$ and $\varphi_>$, satisfying the boundary conditions at $x=0$ and $x=\infty$, respectively, and normalized so that their Wronskian,

$$W = \frac{d\varphi_>}{dx}\varphi_< - \frac{d\varphi_<}{dx}\varphi_>, \qquad (3.16)$$

is unity. (It is easy to use the homogeneous equation to show that W is independent of x). Then the Green's function is given by

$$G(x,x') = \varphi_<(x_<)\varphi_>(x_>), \qquad (3.17)$$

where $x_<$ and $x_>$ are the smaller and larger of x and x', respectively. It is evident that this expression for G satisfies the homogeneous equation and the discontinuity condition (3.15). From (3.15), we then have the explicit solution

$$\varphi(x) = \varphi_>(x)\int_0^x \varphi_<(x')S(x')dx' + \varphi_<(x)\int_x^\infty \varphi_>(x')S(x')dx'. \quad (3.18)$$

This expression can be evaluated by a numerical quadrature and is not subject to any of the stability problems we have seen associated with a direct integration of the inhomogeneous equation.

In the case of arbitrary k^2, the homogeneous solutions $\varphi_<$ and $\varphi_>$ can be found numerically by outward and inward integrations, respectively, of initial value problems and then normalized to satisfy (3.16). However, for simple forms of $k^2(x)$, they are known analytically. For example, for the problem defined by Eq. (3.12), it is easy to show that

$$\varphi_<(r) = r^{l+1}; \; \varphi_>(r) = -\frac{1}{2l+1}r^{-l}$$

are one possible set of homogeneous solutions satisfying the appropriate boundary conditions and Eq. (3.16).

□ □ □ **Exercise 3.3** Solve the problem defined by Eqs. (3.9, 3.12) for $l=0$ using the Green's function method. Compare your results with those obtained by direct integration and with the analytical solution.

□ □ □ **Exercise 3.4** Spherically symmetric solutions to the equation

$$(\nabla^2 - a^2)\Phi = -4\pi\rho$$

lead to the ordinary differential equation

$$\left[\frac{d^2}{dr^2} - a^2\right]\varphi = -4\pi r\rho$$

with the boundary conditions $\varphi(r=0)=\varphi(r\to\infty)=0$. Here, a is a constant. Write a program to solve this problem using the Green's function when ρ is given by Eq. (3.9). Compare your numerical results for various values of a with the analytical solution,

$$\varphi=\left[\frac{1}{1-a^2}\right]^2\left\{e^{-ar}-e^{-r}\left[1+\tfrac{1}{2}(1-a^2)r\right]\right\}.$$

What happens if you try to solve this problem by integrating only inward or only outward? What happens if you try a solution by integrating inward and outward to an intermediate matching radius? How do your results change when you vary the matching radius?

3.4 Eigenvalues of the wave equation

Eigenvalue problems involving differential equations often arise in finding the normal-mode solutions of wave equations. As a simple example with which to illustrate a method of solution, we consider the normal modes of a stretched string of uniform mass density. After a suitable scaling of the physical quantities, the equation and boundary conditions defining these modes can be written as

$$\frac{d^2\varphi}{dx^2}=-k^2\varphi; \quad \varphi(x=0)=\varphi(x=1)=0 . \qquad (3.19)$$

Here, $0<x<1$ is the scaled coordinate along the string, φ is the transverse displacement of the string, and k is the constant wavenumber, linearly related to the frequency of vibration. This equation is an eigenvalue equation in the sense that solutions satisfying the boundary conditions exist only for particular values of k, $\{k_n\}$, which we must find. Furthermore, it is linear and homogeneous, so that the normalization of the eigenfunction corresponding to any of the k_n, φ_n, is not fixed, but can be chosen for convenience.

The (un-normalized) eigenfunctions and eigenvalues of this problem are well-known analytically:

$$k_n=n\pi; \quad \varphi_n\sim\sin n\pi x, \qquad (3.20)$$

where n is a positive integer. These provide a useful check of the numerical methods for solving this problem.

One suitable general strategy for numerical solution of an eigenvalue problem is an iterative one. We guess a trial eigenvalue and generate a solution by integrating the differential equation as a initial value problem. If the resulting solution does not satisfy the boundary conditions, we change the trial eigenvalue and integrate again, repeating the process until a trial eigenvalue is found for which the boundary conditions are satisfied to within a predetermined tolerance.

For the problem at hand, this strategy (known picturesquely as the "shooting" method) can be implemented as follows. For each trial value of k, we integrate forward from $x=0$ with the initial conditions

$$\varphi(x=0)=0, \quad \varphi'(x=0)=\delta.$$

The number δ is arbitrary and can be chosen for convenience, since the problem we are solving is a homogeneous one and the normalization of the solutions is not specified. Upon integrating to $x=1$, we will find, in general, a non-vanishing value of φ, since the trial eigenvalue will not be one of the true eigenvalues. We must then readjust k and integrate again, repeating the process until we find $\varphi(x=1)=0$ to within a specified tolerance; we will have then found an eigenvalue and the corresponding eigenfunction.

The problem of finding a value of k for which $\varphi(1)$ vanishes is a root-finding problem of the type discussed in Chapter 1. Note that the Newton-Raphson method is inappropriate since we cannot differentiate explicitly the numerically determined value of $\varphi(1)$ with respect to k and the secant method could be dangerous, as there are many eigenvalues and it might be difficult to control the one to which the iterations will ultimately converge. Therefore, it is safest to use a simple search to locate an approximate eigenvalue and then, if desired, switch to the more efficient secant method.

The following BASIC program finds the lowest eigenvalue of stretched string problem (3.19) by the shooting method described above, printing the trial eigenvalue as it goes along. The search (lines 140-190) is terminated when the eigenvalue is determined within a precision of 10^{-5}. The initial trial eigenvalue and the search step size are set in line 120.

```
100 N%=100: H=1/N%                         'define lattice parameters
110 TOLK=9.999999E-06                       'tolerance for K
120 K=1: DK=1                               'initial values to start search
130 GOSUB 1000: PHIOLD=PHIP                 'find PHIP at first guess for K
140 WHILE ABS(DK)>TOLK                      'simple search to zero PHIP
150     K=K+DK                              'take a step in K
160     GOSUB 1000                          'calculate PHIP at new value of K
170     IF PHIP*PHIOLD>0 GOTO 190           'if PHIP changes sign,
180     K=K-DK: DK=DK/2                      ' back-up and halve step
190 WEND
200 PRINT USING "eigenvalue=###.#####";K    'print the eigenvalue found
210 END
220 '
1000 'subroutine to calculate phi(x=1)=PHIP for the value of k input
1010 PRINT K                                'print current trial eigenvalue
```

```
1020 PHIM=0: PHIZ=.01              'initial conditions
1030 CON=(K*H)^2/12                'constant in the Numerov method
1040 FOR I%=1 TO N%-1              'forward integration to x=1
1050    PHIP=2*(1-5*CON)*PHIZ-(1+CON)*PHIM  'Numerov formula, Eq. (3.8)
1060    PHIP=PHIP/(1+CON)
1070    PHIM=PHIZ: PHIZ=PHIP       'roll the values of phi
1080 NEXT I%
1090 RETURN
1100 '
```

When run, this program generates results that converge to a value close to the exact answer, π, the error being caused by the finite integration step and the value of TOLK.

Exercise 3.5 Use the program above to find some of the higher eigenvalues. Note that the numerical inaccuracies become greater as the eigenvalue increases and the integration of the more rapidly oscillating eigenfunction becomes inaccurate. Change the search algorithm to the more efficient secant method. How close does your initial guess have to be in order to converge to a given eigenvalue? Change the code to correspond to the boundary conditions

$$\varphi'(x=0)=0, \quad \varphi(x=1)=0,$$

and verify that the numerical eigenvalues agree with the analytical values expected.

Exercise 3.6 The wave equation in cylindrical geometry often leads to the eigenvalue problem

$$\left[\frac{d^2}{dr^2} + \frac{1}{r}\frac{d}{dr} \right] \Phi(r) = -k^2\Phi; \quad \Phi(r=0)=1, \; \Phi(r=1)=0.$$

The analytical eigenfunctions are the regular cylindrical Bessel function of order zero, the eigenvalues being the zeros of this function:

$$k_1=2.404826, \; k_2=5.520078, \; k_3=8.653728, \; k_4=11.791534, \; \cdots$$

Show that the substitution $\Phi=r^{-\frac{1}{2}}\varphi$ changes this equation into one for which the Numerov algorithm is suitable and modify the code above to solve this problem. Compare the numerical eigenvalues with the exact values.

3.5 Stationary solutions of the one-dimensional Schroedinger equation

A rich example of the shooting method for eigenvalue problems is the task of finding the stationary quantum states of a particle of mass m moving in a one-dimensional potential $V(x)$. We'll assume that $V(x)$ has roughly the form shown in Figure 3.1: the potential becomes infinite at $x=x_{min}$ and $x=x_{max}$ (i.e., there are "walls" at these positions) and has a well somewhere in between. The time-independent Schroedinger equation and boundary conditions defining the stationary states are [Me68]

$$\frac{d^2\psi}{dx^2}+k^2(x)\psi(x)=0; \ \psi(x_{min})=\psi(x_{max})=0, \qquad (3.21)$$

which is of the form (3.1) with

$$k^2(x)=\frac{2m}{\hbar^2}[E-V(x)].$$

We must find the energies E (eigenvalues) for which there is a non-zero solution to this problem. At one of these eigenvalues, we expect the eigenfunction to oscillate in the classically allowed regions where $E>V(x)$ and to behave exponentially in the classically forbidden regions where $E<V(x)$. Thus, there will be "bound" solutions with $E<0$, which are localized within the well and decay exponentially toward the walls and "continuum" solutions with $E>0$, which have roughly constant magnitude throughout the entire region between the walls.

The eigenvalue problem defined by Eq. (3.21) can be solved by the shooting method. Suppose that we are seeking a bound state and so take a negative trial eigenvalue. Upon integrating toward larger x from x_{min}, we can generate a solution, $\psi_<$, which increases exponentially through the classically forbidden region and then oscillates beyond the left turning point in the classically allowed region (see the lower portion of Figure 3.1). If we were to continue integrating past the right turning point, the integration would become numerically unstable since, even at an exact eigenvalue where $\psi_<(x_{max})=0$, there can be an admixture of the undesirable exponentially growing solution. As a general rule, integration *into* a classically forbidden region is likely to be inaccurate. Therefore, at each energy it is wiser to generate a second solution, $\psi_>$, by integrating from x_{max} toward smaller x. To determine whether the energy is an eigenvalue, $\psi_<$ and $\psi_>$ can be compared at a matching point, x_m, chosen so that neither integration will be inaccurate. (A convenient choice for x_m is the left turning point.) Since both $\psi_<$ and $\psi_>$ satisfy a homogeneous equation, their normalizations can always be chosen so that the two

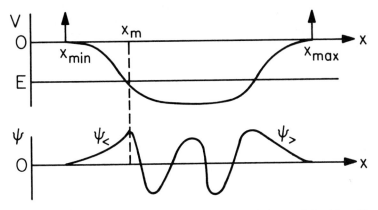

Figure 3.1 (Upper) Schematic potential used in the discussion of the one-dimensional Schroedinger equation. (Lower) Solutions $\psi_<$ and $\psi_>$ of the Schroedinger equation at an arbitrary energy $E<0$. The left turning point is used as the matching point. When E is an eigenvalue, the derivative is continuous at the matching point.

functions are equal at x_m. An eigenvalue is then signaled by equality of the derivatives at x_m; i.e., the solutions match smoothly, as is invoked in analytical solutions of such problems. Thus,

$$\frac{d\psi_<}{dx}\bigg|_{x_m} - \frac{d\psi_>}{dx}\bigg|_{x_m} = 0. \tag{3.22}$$

If we approximate the derivatives by their simplest finite difference approximations (1.14) using the points x_m and x_m-h, an equivalent condition is

$$f \equiv \frac{1}{\psi}[\psi_<(x_m-h)-\psi_>(x_m-h)]=0, \tag{3.23}$$

since the normalizations have been chosen to guarantee $\psi_<(x_m)=\psi_>(x_m)$. The quantity ψ in Eq. (3.23) is a convenient scale for the difference, which can be chosen to make f typically of order unity. It might be the value of $\psi_<$ at x_m or the maximum value of $\psi_<$ or $\psi_>$. Note that if there are no turning points (e.g., if $E>0$ for a potential like that shown in Figure 3.1), then x_m can be chosen anywhere, while if there are more than two turning points, three or more homogeneous solutions, each accurate in different regions, must be patched together.

The program for Example 3, whose BASIC source code is given in Appendix B and in the file EXAM3.BAS on the *Computational Physics* diskette, solves for the stationary states of a one-dimensional potential by the algorithm described above on a 160 point lattice. The potential is assumed to be of the form $V(x)=V_0 v(x)$, where the dimensionless function $v(x)$ has a minimum value of -1 and a maximum value of +1. (This can always be guaranteed by a suitable linear scaling of the energies.) If the coordinate is scaled by a physical length a, the Schroedinger equation (3.21) can be written as

$$\left[-\frac{1}{\gamma^2}\frac{d^2}{dx^2}+v(x)-\varepsilon\right]\psi(x)=0,$$

where

$$\gamma=\left[\frac{2ma^2V_0}{\hbar^2}\right]^{\frac{1}{2}}$$

is a dimensionless measure of the classical nature of the system and $\varepsilon=E/V_0$ is the dimensionless energy. All eigenvalues therefore satisfy $\varepsilon>-1$. The functional form of the potential defined by the program can be any one of three analytical types (square well, parabolic well, and Lennard-Jones potential); any of these can also be modified in an analog manner using a cursor controlled from the keyboard. For the value of γ input, a number of states are sought using an initial trial energy and energy increment. For each state, a simple search on the energy is made to try to zero the function f defined by (3.23) and, when an approximate eigenvalue is located, the secant method is employed until $|f|$ becomes less than 5×10^{-5}. For each trial eigenvalue, the Schroedinger equation is integrated forward and backward, and the two solutions are matched at the left-most turning point where the behavior of the wavefunction changes from oscillatory to exponential (near x_{max} if there is no such turning point). As the search proceeds, each trial wavefunction is displayed (graphics only), as is the trial eigenvalue, the current step in the energy, the current value of f, and the number of nodes in the trial wavefunction. When an eigenvalue is found, it is indicated on a graph of the potential by a line at the appropriate level between the left-most and right-most turning points. As the solution is likely to be inaccurate where there are three or more turning points, a warning is printed in these cases.

The following exercises can help to improve your understanding of the physical principles and numerical methods illustrated in this example.

□ □ □   **Exercise 3.7** Verify that the code gives the expected answers for the eigenvalues of the square-well and parabolic-well potentials ($\gamma=50$ might be a convenient value to use). Observe how the discontinuity in the derivative at the matching point is smoothed out as the energy converges to each eigenvalue. Also note the increase in the number of nodes with increasing eigenvalue and, for the parabolic well, the behavior of the solution near the turning points. For states with large quantum numbers, the amplitude of the oscillations becomes larger near the turning points, consistent with the behavior of the WKB wavefunction, which is proportional to the inverse square-root of the classical velocity, $(E-V)^{-1/4}$. Find some solutions in these potentials also for values of γ that are small (say 10) and large (say 200), corresponding to the extreme quantum and classical limits, respectively.

Exercise 3.8 For the analytically soluble square and parabolic wells, investigate the effects of changing the integration method from the Numerov algorithm to the "naive" one obtained by approximating the second derivative by just the three-point formula (1.7); that is, neglecting the $O(h^2)$ term in (3.6).

Exercise 3.9 Change the program so that the eigenvalues are found by integrating only forward and demanding that the wavefunction vanish at x_{max}. Observe the problems that arise in trying to integrate an exponentially dying solution into a classically forbidden region. (It is wise to keep γ relatively small here so that the instabilities don't become *too* large.)

Exercise 3.10 When the potential is reflection symmetric about $x=0$, the eigenfunctions will have a definite parity (symmetric or anti-symmetric about $x=0$) and that parity will alternate as the quantum number (energy) increases. Verify that this is the case for the numerical solutions generated by the code. Can you think of a way in which the parity can be exploited to halve the numerical effort involved in finding the eigenvalues of a symmetric potential? If so, modify the code to try it out.

Exercise 3.11 If we consider a situation where $v(x)=0$ for $x<x_{min}$ and $x>x_{max}$ (i.e., we remove the walls), then the zero boundary conditions at the ends of the lattice are inappropriate for weakly bound states ($\varepsilon \lesssim 0$) since the wavefunction decays very slowly as $|x|$ becomes large. More appropriate boundary conditions at x_{min} and x_{max} are therefore

$$\frac{1}{\psi}\frac{d\psi}{dx}=\pm\gamma(-\varepsilon)^{1/2}.$$

Change the code to implement these boundary conditions and observe the effect on the wavefunctions and energies of the states near zero energy. Note that if we were to normalize the wavefunction in the conventional way, the contributions from these exponential tails would have to be included. Can you derive, in the style of the Numerov algorithm, a numerical realization of these boundary conditions accurate to a high order in DX?

□ □ □   **Exercise 3.12** Check numerically that, for a given potential, two eigenfunctions, ψ_E and $\psi_E{}'$, corresponding to different eigenvalues E and E', are orthogonal,

$$\int \psi_E(x)\psi_E{}'(x)dx = 0,$$

as is required by the general principles of quantum mechanics.

□ □ □   **Exercise 3.13** For small, intermediate, and large values of γ, compare the exact eigenvalues of the Lennard-Jones potential with the semiclassical energies generated by the code in for Example 1.

□ □ □   **Exercise 3.14** Investigate the eigenfunctions and eigenvalues for potentials you might have encountered in learning elementary quantum mechanics: the δ—function potential, a finite square-well, double-well potentials, periodic potentials, etc. Interpret the wavefunctions and eigenvalues you find. With a little imagination, the analog input of the potential can be used to generate a variety of interesting situations. (Note that the code will sometimes have trouble finding two eigenvalues that are nearly degenerate.)

Project III: Atomic structure in the Hartree-Fock approximation

The self-consistent field approximation (Hartree-Fock) is known to be an accurate description of many of the properties of multi-electron atoms and ions. In this approximation, each electron is described by a separate single-particle wavefunction (as distinct from the many-electron wavefunction) that solves a Schroedinger-like equation. The potential appearing in this equation is that generated by the average motion of all of the other electrons, and so depends on their single-particle wavefunctions. The result is a set of non-linear eigenvalue equations, which can be solved by the methods introduced in this chapter. In this project, we will solve the self-consistent field equations to determine the ground-state structure of small atomic systems (e.g., the atoms and ions of the elements in the periodic table from Hydrogen through Neon). The total energies calculated can be compared directly with experimental values. The brief derivation we give here can be supplemented with the material found in [Be68], [Me68], and [We80].

III.1 Basis of the Hartree-Fock approximation

The Hamiltonian for N electrons moving about a heavy nucleus of charge Z located at the origin can be written as

$$H = \sum_{i=1}^{N} \frac{p_i^2}{2m} - \sum_{i=1}^{N} \frac{Ze^2}{r_i} + \tfrac{1}{2} \sum_{i \neq j=1}^{N} \frac{e^2}{r_{ij}}. \tag{III.1}$$

Here, the $\{r_i\}$ are the locations of the electrons, m and $-e$ are the electron mass and charge, and $r_{ij} \equiv r_i - r_j$ is the separation between electrons i and j. The three sums in (III.1) embody the electron kinetic energy, the electron-nucleus attraction, and the inter-electron repulsion. As is appropriate to the level of accuracy of the self-consistent field approximation, we have neglected much smaller terms, such as those associated with the spin-orbit interaction, hyperfine interactions, recoil motion of the nucleus, and relativity.

A proper quantum mechanical description requires that we specify the spin state of each electron, in addition to its location. This can be done by giving its spin projection on some fixed quantization axis, $\sigma_i = \pm \tfrac{1}{2}$. For convenience, we will use the notation $x_i \equiv (r_i, \sigma_i)$ to denote all of the coordinates (space and spin) of electron i.

The self-consistent field methods are based on the Rayleigh-Ritz variational principle, which states that the ground state eigenfunction of the Hamiltonian, $\Psi(x_1, x_2, \cdots, x_N)$, is that wavefunction that minimizes the expectation value of H,

$$E = \langle \Psi | H | \Psi \rangle, \tag{III.2}$$

subject to the constraints that Ψ obey the Pauli principle (i.e., it be anti-symmetric under the interchange of any two of the x's) and that it be normalized to unity:

$$\int |\Psi|^2 d^N x = 1. \tag{III.3}$$

(The notation $d^N x$ means integration over all of the spatial coordinates and summation over all of the spin coordinates of the N electrons.) Furthermore, this minimum value of E *is* the ground state energy. A calculation of (III.2) for *any* normalized and anti-symmetric trial function Ψ therefore furnishes an upper bound to the ground state energy.

The Hartree-Fock approximation is based on restricting the trial wavefunction to be a Slater determinant:

$$\Psi(x_1, x_2, \cdots, x_N) = (N!)^{-\tfrac{1}{2}} \det \psi_\alpha(x_j). \tag{III.4}$$

Here, the $\psi_\alpha(x)$ are a set of N orthonormal single-particle wavefunctions; they are functions of the coordinates of only a single electron. The determinant is that of the $N \times N$ matrix formed as α and x_i each take on their N possible values, while the factor $(N!)^{-\frac{1}{2}}$ ensures that Ψ is normalized according to (III.3). The physical interpretation of this wavefunction is that each of the electrons moves independently in an orbital ψ_α under the average influence of all the other electrons. This turns out to be a good approximation to the true wavefunction of an atom because the smooth Coulomb interaction between the electrons averages out many of the details of their motion.

Using the properties of determinants, it is easy to see that Ψ has the required anti-symmetry under interchange of any two electrons (a determinant changes sign whenever any two of its columns are interchanged) and that Ψ is properly normalized according to Eq. (III.3) if the single-particle wavefunctions are orthonormal:

$$\int \psi_\alpha^*(x)\psi_{\alpha'}(x)\,dx = \delta_{\alpha\alpha'}. \qquad (III.5)$$

Since the Hamiltonian (III.1) does not involve the electron spin variables, the spins decouple from the space degrees of freedom, so that it is useful to write each single-particle wavefunction as a product of space and spin functions:

$$\psi_\alpha(x) = \chi_\alpha(\mathbf{r})|\sigma_\alpha\rangle, \qquad (III.6)$$

where $\sigma_\alpha = \pm\frac{1}{2}$ is the spin projection of the orbital α. The orthonormality constraint (III.5) then takes the form

$$\delta_{\sigma_\alpha \sigma_{\alpha'}} \int \chi_\alpha^*(\mathbf{r})\chi_{\alpha'}(\mathbf{r})\,d^3r = \delta_{\alpha\alpha'}, \qquad (III.7)$$

so that orbitals can be orthogonal by either their spin or space dependence.

The computation of the energy (III.2) using the wavefunction defined by (III.4-6) is straightforward but tedious. After some algebra, we have

$$E = \sum_{\alpha=1}^{N} \langle \alpha | \frac{p^2}{2m} | \alpha \rangle + \int [-\frac{Ze^2}{r} + \frac{1}{2}\Phi(\mathbf{r})]\rho(\mathbf{r})\,d^3r$$

$$-\frac{1}{2}\sum_{\alpha,\alpha'=1}^{N} \delta_{\sigma_\alpha \sigma_{\alpha'}} \langle \alpha\alpha' | \frac{e^2}{r_{ij}} | \alpha'\alpha \rangle. \qquad (III.8)$$

In this expression, the one-body matrix elements of the kinetic energy are

$$\langle \alpha | \frac{p^2}{2m} | \alpha \rangle = -\frac{\hbar^2}{2m} \int \chi_\alpha^*(\mathbf{r})\nabla^2 \chi_\alpha(\mathbf{r})\,d^3r, \qquad (III.9)$$

the electron density is the sum of the single-particle densities,

$$\rho(\mathbf{r}) = \sum_{\alpha=1}^{N} |\chi_\alpha(\mathbf{r})|^2, \qquad (III.10)$$

the electrostatic potential generated by the electrons is

$$\Phi(\mathbf{r})=e^2 \int \frac{1}{|\mathbf{r}-\mathbf{r}'|}\rho(\mathbf{r}')d^3r', \tag{III.11a}$$

so that

$$\nabla^2\Phi=-4\pi e^2\rho(\mathbf{r}), \tag{III.11b}$$

and the exchange matrix elements of the inter-electron repulsion are

$$\langle\alpha\alpha'|\frac{e^2}{r_{ij}}|\alpha'\alpha\rangle$$

$$=e^2\int\chi_\alpha^*(\mathbf{r})\chi_{\alpha'}^*(\mathbf{r}')\frac{1}{|\mathbf{r}-\mathbf{r}'|}\chi_{\alpha'}(\mathbf{r})\chi_\alpha(\mathbf{r}')d^3rd^3r'. \tag{III.12}$$

The interpretation of the various terms in (III.8) is straightforward. The kinetic energy is the sum of the kinetic energies of the single particle orbitals, while the electron-nucleus attraction and direct inter-electron repulsion are just what would be expected from a total charge of $-Ne$ distributed in space with density $\rho(\mathbf{r})$. The final term in (III.8) is the exchange energy, which arises from the anti-symmetry of the trial wavefunction (III.4). It is a sum over all pairs of orbitals with the same spin projection; pairs of orbitals with different spin projections are "distinguishable" and therefore do not contribute to this term.

The strategy of the self-consistent field approach should now be clear. The variational wavefunction (III.4) depends on a set of "parameters": the values of the single-particle wavefunctions at each point in space. Variation of these parameters so as to minimize the energy (III.8) while respecting the constraints (III.7) results in a set of Euler-Lagrange equations (the Hartree-Fock equations) that define the "best" determinental wavefunction and give an optimal bound on the total energy. Because these equations are somewhat complicated in detail, we consider first the two-electron problem, and then turn to situations with three or more electrons.

III.2 The two-electron problem

For two electrons that don't interact with each other, the ground state of their motion around a nucleus is the $1s^2$ configuration; i.e., both electrons are in the same real, spherically symmetric spatial state, but have opposite spin projections. It is therefore natural to take a trial wave function for the interacting system that realizes this same configuration; the corresponding two single-particle wavefunctions are

$$\psi(x) = \frac{1}{(4\pi)^{\frac{1}{2}}r}R(r)|\pm\frac{1}{2}\rangle. \qquad (\text{III.13})$$

so that the many-body wavefunction (III.4) is

$$\Psi = \frac{1}{\sqrt{2}}\frac{1}{4\pi r_1 r_2}R(r_1)R(r_2)[|+\frac{1}{2}\rangle|-\frac{1}{2}\rangle - |-\frac{1}{2}\rangle|+\frac{1}{2}\rangle], \qquad (\text{III.14})$$

This trial wavefunction is anti-symmetric under the interchange of the electron spins but is symmetric under the interchange of their space coordinates. It respects the Pauli principle, since it is antisymmetric under the interchange of all variables describing the two electrons. The normalization condition (III.5) becomes

$$\int_0^\infty R^2(r)dr = 1, \qquad (\text{III.15})$$

while the energy (III.8) becomes

$$E = 2 \times \frac{\hbar^2}{2m}\int_0^\infty \left[\frac{dR}{dr}\right]^2 dr + \int_0^\infty \left[-\frac{Ze^2}{r} + \frac{1}{4}\Phi(r)\right]\rho(r)4\pi r^2 dr, \qquad (\text{III.16})$$

with (III.10) reducing to

$$\rho(r) = 2 \times \frac{1}{4\pi r^2}R^2(r); \quad \int_0^\infty \rho(r)4\pi r^2 dr = 2, \qquad (\text{III.17})$$

and (III.11b) becoming

$$\frac{1}{r^2}\frac{d}{dr}\left[r^2\frac{d\Phi}{dr}\right] = -4\pi e^2\rho. \qquad (\text{III.18})$$

Note that the exchange energy is attractive and has a magnitude of one-half of that of the direct inter-electron repulsion (resulting in a net factor of 1/4 in the final term of (III.16)) and that various factors of two have entered from the sum over the two spin projections.

A common variational treatment of the two-electron system ("poor man's Hartree-Fock") takes R to be a hydrogenic 1s orbital parametrized by an effective charge, Z^*:

$$R(r) = 2\left(\frac{Z^*}{a}\right)^{\frac{1}{2}}\frac{Z^*r}{a}e^{-Z^*r/a}, \qquad (\text{III.19})$$

where a is the Bohr radius. The energy (III.16) is then minimized as a function of Z^* to find an approximation to the wavefunction and energy. This procedure, which is detailed in many textbooks (see, for example, [Me68]), results in

$$Z^* = Z - \frac{5}{16}; \quad E = -\frac{e^2}{a}\left[Z^2 - \frac{5}{8}Z + \frac{25}{256}\right]. \qquad (\text{III.20})$$

In carrying out this minimization, it is amusing to note that the kinetic energy scales as Z^{*2}, while all of the potential energies scale as Z^*, so that, at the optimal Z^*, the kinetic energy is $-\frac{1}{2}$ of the potential. This is a specific case of a more general virial theorem pertaining to the Hartree-Fock approximation (see Step 1 below).

The full Hartree-Fock approximation for the two-electron problem is very much in this same variational spirit, but the most general class of normalized single-particle wavefunctions is considered. That is, we consider E in Eq. (III.16) to be a *functional* of R and require that it be stationary with respect to all possible norm-conserving variations of the single-particle wavefunction. If the normalization constraint (III.15) is enforced by the method of Lagrange multipliers, for an arbitrary variation of $\delta R(r)$ we require

$$\delta \left(E - 2\varepsilon \int_0^\infty R^2 dr \right) = 0, \tag{III.21}$$

where ε is a Lagrange multiplier to be determined after variation so that the solution is properly normalized. The standard techniques of variational calculus then lead to

$$\int_0^\infty \delta R(r) \left[-4 \frac{\hbar^2}{2m} \frac{d^2}{dr^2} - 4 \frac{Ze^2}{r} + 2\Phi(r) - 4\varepsilon \right] R(r) dr = 0, \tag{III.22}$$

which is satisfied if R solves the Schroedinger-like equation

$$\left[-\frac{\hbar^2}{2m} \frac{d^2}{dr^2} - \frac{Ze^2}{r} + \frac{1}{2}\Phi(r) - \varepsilon \right] R(r) = 0. \tag{III.23}$$

Choosing ε (the "single-particle energy") to be an eigenvalue of the single-particle hamiltonian appearing in (III.23) ensures that R is normalizable. Equations (III.18,23) are the two coupled non-linear differential equations in one dimension that form the Hartree-Fock approximation to the original six-dimensional Schroedinger equation. Note that only one-half of Φ appears in (III.23) since each electron interacts only with the other and not "with itself"; inclusion of the exchange term in the energy (III.16) is necessary to get this book-keeping right.

III.3 Many-electron systems

The assumption of spherical symmetry is an enormous simplification in the two-electron problem, as it allowed us to reduce the eigenvalue problem for the single-particle wavefunction and the Poisson equation for the potential from three-dimensional partial differential equations to ordinary differential equations. For the two-electron problem, it is plausible (and true) that a spherically symmetric solution has the lowest energy. However, for most many-electron systems, spherical symmetry of the density and potential are by no means guaranteed. In principle, non-spherical solutions should be considered, and such "deformed" wavefunctions are in fact the optimal ones for describing the structure of certain nuclei.

To understand what the problem is, let us assume that the potential Φ is spherically symmetric. The solutions to the single-particle Schroedinger equation in such a potential are organized into "shells", each characterized by an orbital angular momentum, l, and a radial quantum number, n. Within each shell, all $2(2l+1)$ orbitals associated with the various values of σ_α and the projection of the orbital angular momentum, m, are degenerate. The orbitals have the form

$$\chi_\alpha(\mathbf{r}) = \frac{1}{r} R_{nl}(r) Y_{lm}(\hat{\mathbf{r}}); \quad \int_0^\infty R_{nl}^2(r)\,dr = 1. \qquad \text{(III.24)}$$

However, we must decide which of these orbitals to use in constructing the Hartree-Fock determinant. Unless the number of electrons is such that all of the $2(2l+1)$ substates of a given shell are filled, the density as given by (III.10) will not be spherically symmetric. This, in turn, leads to a non-symmetric potential and a much more difficult single-particle eigenvalue equation; the general problem is therefore intrinsically three-dimensional.

A slight modification of the rigorous Hartree-Fock method (the filling or central-field approximation) is useful in generating a spherically symmetric approximation to such "open-shell" systems. The basic idea is to spread the valence electrons uniformly over the last occupied shell. For example, in discussing the neutral Carbon atom, there would be 2 electrons in the 1s shell, 2 electrons in the 2s shell, and 2 electrons spread out over the 6 orbitals of the 2p shell. (Note that we don't put 4 electrons in the 2p shell and none in the 2s shell since the single-particle energy of the latter is expected to be more negative.) Thus, we introduce the number of electrons in each shell, N_{nl}, which can take on integer values between 0 and $2(2l+1)$, and, using the wavefunctions (III.24), write the density (III.10) as

$$\rho(r) = \frac{1}{4\pi r^2} \sum_{nl} N_{nl} R_{nl}^2(r); \quad \int_0^\infty \rho(r) 4\pi r^2 dr = \sum_{nl} N_{nl} = N. \quad \text{(III.25)}$$

In writing this expression, we have used the identity

$$\sum_{m=-l}^{l} |Y_{lm}(\hat{\mathbf{r}})|^2 = \frac{2l+1}{4\pi}.$$

In the same spirit, the energy functional (III.8) can be generalized to open-shell situations as

$$E = \sum_{nl} N_{nl} \frac{\hbar^2}{2m} \int_0^\infty \left[\left(\frac{dR_{nl}}{dr} \right)^2 + \frac{l(l+1)}{r^2} R_{nl}^2 \right] dr$$

$$+ \int_0^\infty \left[-\frac{Ze^2}{r} + \tfrac{1}{2}\Phi(r) \right] \rho(r) 4\pi r^2 dr + E_{ex}, \qquad \text{(III.26a)}$$

with the exchange energy being

$$E_{ex} = -\frac{1}{4} \sum_{nln'l'} N_{nl} N_{n'l'} \sum_{\lambda = |l-l'|}^{l+l'} \begin{pmatrix} l & l' & \lambda \\ 0 & 0 & 0 \end{pmatrix}^2 I_{nl,n'l'}^\lambda. \qquad \text{(III.26b)}$$

In this expression, I is the integral

$$I_{nl,n'l'}^\lambda = e^2 \int_0^\infty dr \int_0^\infty dr' R_{nl}(r) R_{n'l'}(r') \frac{r_<^\lambda}{r_>^{\lambda+1}} R_{n'l'}(r) R_{nl}(r'), \text{(III.27)}$$

where $r_<$ and $r_>$ are the smaller and larger of r and r' and the $3{-}j$ symbol vanishes when $l + l' + \lambda$ is odd and otherwise has the value

$$\begin{pmatrix} l & l' & \lambda \\ 0 & 0 & 0 \end{pmatrix}^2 = \frac{(-l+l'+\lambda)!(l-l'+\lambda)!(l+l'-\lambda)!}{(l+l'+\lambda+1)!}$$

$$\times \left[\frac{p!}{(p-l)!(p-l')!(p-\lambda)!} \right]^2,$$

where $p = \tfrac{1}{2}(l+l'+\lambda)$. In deriving these expressions, we have used the multipole decomposition of the Coulomb interaction and the standard techniques of angular momentum algebra [Br68].

The Hartree-Fock equations defining the optimal radial wavefunctions now follow from the calculus of variations, as in the two-electron case. Lagrange multipliers ε_{nl} are introduced to keep each of the radial wavefunction normalized and, after some algebra, we have

$$\left[-\frac{\hbar^2}{2m} \frac{d^2}{dr^2} + \frac{l(l+1)\hbar^2}{2mr^2} - \frac{Ze^2}{r} + \Phi(r) - \varepsilon_{nl} \right] R_{nl}(r) = -F_{nl}(r), \quad \text{(III.28a)}$$

with

$$F_{nl}(r) = -\frac{e^2}{2} \sum_{n'l'} N_{n'l'} R_{n'l'}(r) \sum_{\lambda=|l-l'|}^{l+l'} \begin{pmatrix} l & l' & \lambda \\ 0 & 0 & 0 \end{pmatrix}^2 J_{nl,n'l'}^\lambda; \qquad \text{(III.28b)}$$

$$J_{nl,n'l'}^\lambda = \frac{1}{r^{\lambda+1}} \int_0^r R_{n'l'}(r') R_{nl}(r') r'^\lambda dr' + r^\lambda \int_r^\infty \frac{R_{n'l'}(r') R_{nl}(r')}{r'^{\lambda+1}} dr'. \text{(III.28c)}$$

The eigenvalue equation (III.28a) can be seen to be analogous to (III.23) for the two-electron problem, except that the exchange energy has introduced a non-locality (Fock potential) embodied in

F and has coupled together the eigenvalue equations for each of the radial wavefunctions; it is easy to show that these two equations are equivalent when there is a single orbital with $l=0$. It is useful to note that (III.26b, 28b) imply that the exchange energy can be also be written as

$$E_{ex} = \tfrac{1}{2} \sum_{nl} N_{nl} \int_0^\infty R_{nl}(r) F_{nl}(r) dr, \qquad (III.29)$$

and that, by multiplying (III.28a) by R_{nl} and integrating, we can express the single particle eigenvalue as

$$\varepsilon_{nl} = \frac{\hbar^2}{2m} \int_0^\infty \left[\left(\frac{dR_{nl}}{dr} \right)^2 + \frac{l(l+1)}{r^2} R_{nl}^2 \right] dr$$

$$+ \int_0^\infty \left[-\frac{Ze^2}{r} + \Phi(r) \right] R_{nl}^2(r) dr + \int_0^\infty R_{nl}(r) F_{nl}(r) dr . \qquad (III.30)$$

III.4 Solving the equations

For the numerical solution of the Hartree-Fock equations, we must first adopt a system of units. For comparison with experimental values, it is convenient to measure all lengths in Angstroms and all energies in electron volts. If we use the constants

$$\frac{\hbar^2}{m} = 7.6359 \text{ eV}-\text{Å}^2; \quad e^2 = 14.409 \text{ eV}-\text{Å}, \qquad (III.31)$$

then the Bohr radius and Rydberg have their correct values,

$$a = \frac{\hbar^2}{me^2} = 0.5299 \text{ Å}; \quad Ry = \frac{e^2}{2a} = 13.595 \text{ eV}. \qquad (III.32)$$

For a large atom with many electrons, the accurate solution of the Hartree-Fock equations is a considerable task. However, if we consider the ground states of systems with at most 10 electrons (requiring three shells: $1s$, $2s$, and $2p$), then the numerical work can be managed on a microcomputer. A lattice of several hundred points with a radial step size of ≤ 0.01 Å extending out to ≈ 3 Å should be sufficient for most cases.

The best approach to developing a program to solve the Hartree-Fock equations is to consider the two-electron problem first, for which there is only a single radial wavefunction solving a local eigenvalue equation, and then to consider the more complex case of several orbitals. The attack can be made through the following sequence of steps.

Step 1 Verify the algebra leading to the final equations presented   □ □ □
above for the two-electron system (Eqs. (III.16,18,23)) and for the
multi-electron system (Eqs. (III.18,26,28)) and make sure that you
understand the physical principles behind the derivations. Prove
the virial theorem that the kinetic energy is $-\frac{1}{2}$ of the potential
energy. This can be done by imagining that the single-particle
wavefunctions of a solution to the Hartree-Fock equations are subject to a norm-preserving scaling transformation,

$$\chi_\alpha(\mathbf{r}) \to \tau^{3/2}\chi(\tau\mathbf{r}),$$

where τ is a dimensionless scaling parameter. Show that the total
kinetic energy in (III.8) scales as τ^2, while all of the potential energies scale as τ. Since the energy at the Hartree-Fock solution is
stationary with respect to *any* variation of the wavefunctions, use

$$\left.\frac{\partial E}{\partial \tau}\right|_{\tau=1} = 0$$

to prove the theorem.

Step 2 Write a program to calculate the energy from (III.16) if R is   □ □ □
known at all of the lattice points. This will require writing a subroutine that calculates Φ by solving (III.18) (you might modify the
one given earlier in this chapter) and then evaluating suitable quadratures for the various terms in (III.16). Verify that your program is working by calculating the energies associated with the
hydrogenic orbital (III.19) and comparing it with the analytical
results (remember to normalize the wavefunction by the appropriate discretization of (III.15)).

Step 3 Write a subroutine that uses the shooting method to solve   □ □ □
the radial equation (III.23) for the lowest eigenvalue ε and
corresponding normalized wavefunction R if the potential Φ is
given at the lattice points. The zero boundary condition at the origin is easily implemented, but the boundary condition at large distances can be taken as $R(r=L)=0$, where L is the outer end of the
lattice. (Greater accuracy, particularly for weakly bound states,
can be had by imposing instead an exponential boundary condition
at the outer radius.) Note that the radial scale (i.e., R and the
radial step size) should change with the strength of the central
charge. Verify that your subroutine works by setting Φ to 0 and
comparing, for $Z=2$ and $Z=4$, the calculated wavefunction, eigenvalue, and energy of the 1s orbital with the analytical hydrogenic
values.

☐ ☐ ☐   **Step 4** Combine the subroutines developed in Steps 2 and 3 into a code that, given a value of Z, solves the two-electron Hartree-Fock equations by iteration. An iteration scheme is as follows, the organization into subroutines being obvious:

i)   "Guess" an initial wavefunction, say the hydrogenic one (III.19) with the appropriate value of Z^*.

ii) Solve (III.18) for the potential generated by the initial wavefunction and calculate the total energy of the system from Eq. (III.16).

iii) Find a new wavefunction and its eigenvalue by solving (III.23) and normalizing according to (III.15).

iv) Calculate the new potential and new total energy. Then go back to iii) and repeat iii) and iv) until the total energy has converged to within the required tolerance.

At each iteration, you should print out the eigenvalue, the total energy, and the three separate contributions to the energy appearing in (III.16); a plot of the wavefunction is also useful for monitoring the calculation. Note that the total energy should decrease as the iterations proceed and will converge relatively quickly to a minimum. The individual contributions to the energy will take longer to settle down, consistent with the fact that it is only the total energy that is stationary at the variational minimum, not the individual components; at convergence, the virial theorem discussed in Step 1 should be satisfied. Try beginning the iteration procedure with different single-particle wavefunctions and note that the converged solution is still the same. Vary the values of the lattice spacing and the boundary radius, L, and prove that your results are stable under these changes.

☐ ☐ ☐   **Step 5** Use your program to solve the Hartree-Fock equations for central charges $Z=1$-9. Compare the total energies obtained with the experimental values given in $N=2$ column of Table III.1. (These binding energies, which are the negative of the total energies, are obtained from the measured ionization potentials of atoms and ions given in [We71].) Compare your results also with the wavefunctions and associated variational energies given by Eqs. (III.19,20). Note that both approximations should give upper bounds to the exact energy. Give a physical explanation for the qualitative behavior of the discrepancies as a function of Z. Can you use second-order perturbation theory to show that the discrepancy between the Hartree-Fock and exact energies should become a constant for large Z? Show that for $Z=1$, the Hartree-

Table III.1: Binding energies (in eV) of small atomic systems

Z	Number of electrons, N						
	2	3	4	5	6	7	8
1	14.34						
2	78.88						
3	198.04	203.43					
4	371.51	389.71	399.03				
5	599.43	637.35	662.49	670.79			
6	881.83	946.30	994.17	1018.55	1029.81		
7	1218.76	1316.62	1394.07	1441.19	1471.09	1485.62	
8	1610.23	1743.31	1862.19	1939.58	1994.47	2029.58	2043.19
9	2054.80	2239.93	2397.05	2511.27	2598.41	2661.05	2696.03

Fock approximation predicts that the H⁻ ion is unbound in that its energy is greater than that of the H atom and so it is energetically favorable to shed the extra electron. As can be seen from Table III.1, this is not the case in the real world. In finding the $Z=1$ solution, you might discover that convergence is quite a delicate business; it is very easy for the density to change so much from iteration to iteration that the lowest eigenvalue of the single-particle Hamiltonian becomes positive. One way to alleviate this problem is to prevent the density from changing too much from one iteration to the next, for example by averaging the new density and the old following step *iii)* above.

Step 6 Modify your two-electron program to treat systems in which several orbitals are involved. It is easiest to first modify the calculation of the total energy for a given set of radial wavefunctions to include E_{ex}. This is most conveniently done by calculating and storing the F_{nl} of Eq. (III.28b) and using Eq. (III.29). Because of the Fock term, the eigenvalue equations (III.28a) cannot be treated by the shooting method we have discussed. However, one scheme is to treat the F_{nl} calculated from the previous set of wavefunctions as an inhomogeneous terms in solving for the new set of wavefunctions. For trial values of the ε_{nl} calculated from (III.30) using the previous set of wavefunctions, (III.28a) can be solved as uncoupled inhomogeneous boundary value problems using the Green's function method of Eq. (3.18); after normalization according to (III.24), the solutions serve as a new set of wavefunctions. The two-electron systems can be used to check the accuracy of your modifications; for these systems you should find that the exchange energy is $-\tfrac{1}{2}$ of the direct inter-electron interaction energy and that the solutions converge to the same results as those generated by the code in Step 4. Use this

Hartree-Fock code to calculate the wavefunctions and energies for some of the other systems listed in Table III.1 and compare your results with the experimental values; interpret what you find. A convenient set of initial wavefunctions are the hydrogenic orbitals, given by (III.19) for the 1s state and

$$R_{2s}(r) = 2\left(\frac{Z^*}{2a}\right)^{\frac{1}{2}}\left[1 - \frac{Z^*r}{2a}\right]\frac{Z^*r}{2a}e^{-Z^*r/2a},$$

$$R_{2p}(r) = \left(\frac{2Z^*}{3a}\right)^{\frac{1}{2}}\left(\frac{Z^*r}{2a}\right)^2 e^{-Z^*r/2a},$$

for the 2s and 2p states, respectively. The optimal common value of Z^* in these expression should be determined for any system by minimizing the total energy.

Chapter 4

Special Functions and Gaussian Quadrature

In this chapter, we discuss two loosely related topics: algorithms for computing the special functions of mathematical physics (Bessel functions, orthogonal polynomials, etc.) and efficient methods of quadrature based on orthogonal functions. In most scientific computing, large libraries supply almost all of the subroutines relevant to these tasks and so relieve the individual from the tedium of writing his own code. In fact, there is usually little need to know very much about how these subroutines work in detail. However, a rough idea of the methods used is useful; this is what we hope to impart in this chapter.

4.1 Special functions

The special functions of mathematical physics were largely developed long before large-scale numerical computation became feasible, when analytical methods were the rule. Nevertheless, they are still relevant today, for two reasons. One is the insight analytical solutions offer; they guide our intuition and provide a framework for the qualitative interpretation of more complicated problems. However, of particular importance to numerical work is the fact that special functions often allow part of a problem to be solved analytically and so dramatically reduce the amount of computation required for a full solution.

As an illustration, consider a one-dimensional harmonic oscillator moving under an external perturbation: its frequency, ω, is being changed with time. Suppose that the frequency has its unperturbed value, ω_0, for times before $t=0$ and for times after $t>T$, and that we are interested in the motion for times long after the perturbation ceases. Given the oscillator's initial coordinate and velocity, one straightforward method of solution is to integrate the equations of motion,

$$\frac{dx}{dt}=v(t); \quad \frac{dv}{dt}=-\omega^2(t)x(t),$$

as an initial-value problem using one of the methods discussed in

Chapter 2. However, this would be inefficient, as the motion after the perturbation stops $(t>T)$ is well-understood and is readily expressed in terms of the "special" sine and cosine functions involved,

$$x(t>T)=x(T)\cos\omega_0(t-T)+\omega_0^{-1}v(T)\sin\omega_0(t-T).$$

Since there are very efficient methods for computing the trigonometric functions, it is wiser to integrate numerically only the non-trivial part of the motion $(0<t<T)$ and then to use the velocity and coordinate at $t=T$ to compute directly the sinusoidal function given above. Although this example might seem trivial, the concept of using special functions to "do part of the work" is a general one.

A useful resource in dealing with special functions is the *Handbook of Mathematical Functions* [Ab64]. This book contains the definitions and properties of most of the functions one often needs. Methods for computing them are also given, as well as tables of their values for selected arguments. These last are particularly useful for checking the accuracy of the subroutines you are using.

Recursion is a particularly simple way of computing some special functions. Many functions are labeled by an order or index and satisfy recursion relations with respect to this label. If the function can be computed explicitly for the few lowest orders, then the higher orders can be found from these formulas. As an example, consider the computation of the Legendre polynomials, $P_l(x)$, for $|x|\le1$ and $l=0,1,2,\cdots$. These are important in the solution of wave equations in situations with a spherical symmetry. The recursion relation with respect to degree is

$$(l+1)P_{l+1}(x)+lP_{l-1}(x)-(2l+1)xP_l(x)=0. \qquad (4.1)$$

Using the explicit values $P_0(x)=1$ and $P_1(x)=x$, forward recursion in l yields P_l for any higher value of l required. The following BASIC program accomplishes this for any value for x and l input.

```
10  INPUT "ENTER x,l";X,L%
20  IF L%=0 THEN PL=1: GOTO 120        'explicit formula for l=0
30  IF L%=1 THEN PL=X: GOTO 120        'explicit formula for l=1
40  '
50  PM=1: PZ=X                         'values to start recursion
60  FOR LL%=1 TO L%-1                  'loop for forward recursion
70     PP=((2*LL%+1)*X*PZ-LL%*PM)/(LL%+1)   'Eq. (4.1)
80     PM=PZ: PZ=PP                    'roll the current values
90  NEXT LL%
100 PL=PZ
110 '
120 PRINT X,L%,PL                      'output the results
130 GOTO 10                            'get the next values of x and l
```

This code works with no problems, and the results agree with the values given in the tables to the arithmetic precision of the computer. We can also compute the derivatives of the Legendre polynomials with this algorithm using the relation

$$(1-x^2)P'_l = -lxP_l + lP_{l-1}. \tag{4.2}$$

Other sets of orthogonal polynomials, such as Hermite and Laguerre, can be treated similarly.

As a second example, consider the cylindrical Bessel functions, $J_n(x)$ and $Y_n(x)$, which arise as the regular and irregular solutions to wave equations in cylindrical geometries. These functions satisfy the recursion relation

$$C_{n-1}(x) + C_{n+1}(x) = \frac{2n}{x} C_n(x), \tag{4.3}$$

where C_n is either J_n or Y_n. To use these recursion relations in the forward direction, we need the values of C_0 and C_1. These are most easily obtained from the polynomial approximations given in [Ab64], formulas 9.4.1-3. For $|x| < 3$, we have

$$J_0(x) = 1 - 2.2499997\,y^2 + 1.2656208\,y^4$$

$$-0.3163866\,y^6 + 0.0444479\,y^8 - 0.039444\,y^{10}$$

$$+0.0002100\,y^{12} + \varepsilon; \quad |\varepsilon| \leq 5 \times 10^{-8}, \tag{4.4a}$$

where $y = x/3$ and

$$Y_0(x) = \frac{2}{\pi} \log(\tfrac{1}{2}x) J_0(x) + 0.36746691 + 0.605593666\,y^2$$

$$-0.74350384\,y^4 + 0.25300117\,y^6 - 0.04261214\,y^8$$

$$+0.00427916\,y^{10} - 0.00024846\,y^{12} + \varepsilon; \quad |\varepsilon| \leq 1.4 \times 10^{-8}; \tag{4.4b}$$

while for $x > 3$,

$$J_0(x) = x^{-\frac{1}{2}} f_0 \cos\theta \qquad Y_0(x) = x^{-\frac{1}{2}} f_0 \sin\theta \tag{4.4c}$$

where

$$f_0 = 0.79788456 - 0.00000077\,y^{-1} - 0.00552740\,y^{-2}$$

$$-0.00009512\,y^{-3} + 0.00137237\,y^{-4} - 0.00072805\,y^{-5}$$

$$+0.00014476\,y^{-6} + \varepsilon; \quad |\varepsilon| < 1.6 \times 10^{-8}, \tag{4.4d}$$

Table 4.1 Forward recursion for the irregular Bessel function $Y_n(2)$

n	$Y_n(2)$
0	+0.51037
1	-0.10703
2	-0.61741
3	-1.1278
4	-2.7659
5	-9.9360
6	-46.914
7	-271.55
8	-1853.9
9	-14560.

and

$$\theta = x - 0.78539816 - 0.04166397\,y^{-1} - 0.00003954\,y^{-2}$$
$$+ 0.00262573\,y^{-3} - 0.00054125\,y^{-4} - 0.00029333\,y^{-5}$$
$$+ 0.00013558\,y^{-6} + \varepsilon; \quad |\varepsilon| < 7 \times 10^{-8} \tag{4.4e}$$

Similar expressions for J_1 and Y_1 are given in Sections 9.4.5-6 of [Ab64]. Note that these formulas are *not* Taylor series, but rather polynomials whose coefficients have been adjusted to best represent the Bessel functions over the intervals given.

Let us now attempt to calculate the Y_n by forward recursion using (4.3) together with the polynomial approximations for the values of Y_0 and Y_1. Doing so leads to results that reproduce the values given in the tables of Chapter 9 of [Ab64] to within the arithmetic precision of the computer. For example, we find the results listed Table 4.1 for $x=2$.

It is natural to try to compute the regular solutions, J_n, with the same forward recursion procedure. Using Eq. (4.4a) and its analog for J_1, we find the errors listed in the third column of Table 4.2; the exact values are given in the second column. As can be seen, forward recursion gives good results for $n < 5$, but there are gross errors for the higher values of n.

It is relatively easy to understand what is going wrong. We can think about the recursion relation (4.3) for the Bessel functions as the finite difference analog of a second-order differential equation in n. In fact, if we subtract $2C_n$ from both sides of (4.3), we obtain

$$C_{n+1} - 2C_n + C_{n-1} = 2\left(\frac{n}{x} - 1\right)C_n, \tag{4.5}$$

Table 4.2 Computation of the regular Bessel function $J_n(2)$

n	Exact value	Error in forward recursion	Un-normalized backward recursion	Error in normalized backward recursion
0	0.223891E+00	0.000000E+00	0.150602E-10	0.000000E+00
1	0.576725E+00	0.000000E+00	0.387940E-10	0.000000E+00
2	0.352834E+00	0.000000E+00	0.237337E-10	0.000000E+00
3	0.128943E+00	0.000000E+00	0.867350E-11	0.000000E+00
4	0.339957E-01	-0.000002E-01	0.228676E-11	0.000000E-01
5	0.703963E-02	-0.000075E-02	0.473528E-12	0.000000E-02
6	0.120243E-02	-0.000355E-02	0.808826E-13	0.000000E-02
7	0.174944E-03	-0.020559E-03	0.117678E-13	0.000000E-03
8	0.221795E-04	-0.140363E-03	0.149193E-14	0.000000E-04
9	0.249234E-05	-0.110234E-02	0.167650E-15	0.000000E-05
10	0.251539E-06	-0.978959E-02	0.169200E-16	0.000000E-06
11	0.230428E-07		0.155000E-17	0.000000E-07
12	0.193270E-08		0.130000E-18	0.000007E-08
13	0.149494E-09		0.100000E-19	0.000830E-09
14	0.107295E-10		0.000000E-19	0.107295E-10

which, in the limit of continuous n, we can approximate by

$$\frac{d^2C}{dn^2} = -k^2(n)C; \quad k^2(n) = 2\left[1 - \frac{n}{x}\right]. \tag{4.6}$$

In deriving this equation, we have used the three-point finite difference formula (1.7) for the second derivative with $h=1$ and have identified the local wavenumber, $k^2(n)$. Equation (4.6) will have two linearly independent solutions, either both oscillatory in character (when k^2 is positive, or when $n<x$) or one exponentially growing and the other exponentially decreasing (when k^2 is negative, or $n>x$). As is clear from Table 4.1, Y_n is the solution that grows exponentially with increasing n, so that no loss of precision occurs in forward recursion. However, Table 4.2 shows that the exact values of J_n decrease rapidly with increasing n, and so precision is lost rapidly as forward recursion proceeds beyond $n=x$. This disease is the same as that encountered in Chapter 3 in integrating exponentially dying solutions of second-order differential equations; it's cure is also the same: avoid using the recursion relation in the direction of decreasing values of the function.

□ □ □   **Exercise 4.1** Use Eq. (4.1) to show that recursion of the Legendre polynomials is stable in either direction.

To compute the regular cylindrical Bessel functions accurately, we can exploit the linearity of the recursion relation and use Eq. (4.3) in the direction of decreasing n. Suppose we are interested in $J_n(2)$ for $n \leq 10$. Then, choosing $J_{14}=0$ and $J_{13}=1 \times 10^{-20}$, an arbitrarily small number, we can recur backwards to $n=0$. The resulting sequence of numbers will then reproduce the J_n, to within an arbitrary normalization, since, as long as we have chosen the initial value of n high enough, the required solution of the difference equation (4.3), which grows exponentially with decreasing n, will dominate for small n. The sequence can then be normalized through the identity

$$J_0(x)+2J_2(x)+2J_4(x)+ \cdots =1 . \qquad (4.7)$$

The following BASIC code evaluates the regular cylindrical Bessel functions using backward recursion.

```
 5 DIM J(50)
10 INPUT "Enter maximum value of n (<=50)";NMAX%
15 INPUT "Enter value of x";X
20 '                                     'backward recursion
25 J(NMAX%)=0: J(NMAX%-1)=1E-20          'initial conditions
30 FOR N%=NMAX%-1 TO 1 STEP -1
35    J(N%-1)=(2*N%/X)*J(N%)-J(N%+1)     'Eq. (4.3)
40 NEXT J%
45 '                                     'calculate sum in Eq. (4.7)
50 SUM=J(0)
55 FOR N%=2 TO NMAX% STEP 2
60    SUM=SUM+2*J(N%)
65 NEXT N%
70 FOR N%=0 TO NMAX%                     'normalize and output
75    J(N%)=J(N%)/SUM
80    PRINT N%,J(N%)
85 NEXT N%
90 GOTO 10
```

When run for NMAX%=14 and X=2, it gives the unnormalized values (i.e., after line 40) shown in the fourth column of Table 4.2 and the errors in the final values shown in the fifth column of that table. The results are surprisingly accurate, even for values of n close to 14. An alternative way of obtaining the constant with which to normalize the whole series is to calculate the value of $J_0(2)$ from the polynomial approximation (4.4a).

□ □ □   **Exercise 4.2** Run the code above for various values of NMAX% at fixed X. By comparing with tabulated values, verify that the results are accurate as long as NMAX% is large enough (somewhat greater than the larger of X and the maximum order desired). Change the normalization algorithm to use the approximations (4.4a,c) for J_0.

Exercise 4.3 The regular and irregular spherical Bessel functions,   □ □ □
j_l and n_l, satisfy the recursion relation

$$s_{l+1}+s_{l-1}=\frac{2l+1}{x}s_l,$$

where s_l is either j_l or n_l. The explicit formulas for the few lowest orders are

$$j_0=\frac{\sin x}{x}\,;\; j_1=\frac{\sin x}{x^2}-\frac{\cos x}{x}\,;\; j_2=\left[\frac{3}{x^3}-\frac{1}{x}\right]\sin x-\frac{3}{x^2}\cos x\,,$$

and

$$n_0=-\frac{\cos x}{x}\,;\; n_1=-\frac{\cos x}{x^2}-\frac{\sin x}{x}\,;\; n_2=\left[-\frac{3}{x^3}+\frac{1}{x}\right]\cos x-\frac{3}{x^2}\sin x\,.$$

At $x=0.5$, the exact values of the functions of order 2 are
$$n_2=-25.059923;\; j_2=1.6371107\times10^{-2}.$$

Show that n_2 can be calculated either by explicit evaluation or by forward recursion and convince yourself that the latter method will work for all l and x. Investigate the calculation of $j_2(0.5)$ by forward recursion, explicit evaluation, and by backward recursion and show that the first two methods can be quite inaccurate. Can you see why? Thus, even if explicit expressions for a function are available, the stability of backward recursion can make it the method of choice.

Our discussion has illustrated some pitfalls in computing the some commonly used special functions. Specific methods useful for other functions can be found in the appropriate chapters of [Ab64].

4.2 Gaussian quadrature
In Chapter 1, we discussed several methods for computing definite integrals that were most convenient when the integrand was known at a series of equally spaced lattice points. While such methods enjoy widespread use, especially when the integrand involves a numerically-generated solution to a differential equation, more efficient quadrature schemes exist if we can evaluate the integrand for arbitrary abscissae. One of the most useful of these is Gaussian quadrature.

Consider the problem of evaluating

$$I = \int_{-1}^{1} f(x)\,dx.$$

The formulas discussed in Chapter 1 were of the form

$$I \approx \sum_{n=1}^{N} w_i f(x_n), \qquad (4.8)$$

where

$$x_n = -1 + 2\frac{(n-1)}{(N-1)}$$

are the equally spaced lattice points. Here, we are referring to the "elementary" quadrature formulas (such as (1.9), (1.11), or (1.13a,b)), and not to compound formulas such as (1.12). For example, for Simpson's rule (1.11), $N=3$ and

$$x_1 = -1, \; x_2 = 0, \; x_3 = 1; \quad w_1 = w_3 = \frac{1}{3}, \quad w_2 = \frac{4}{3}.$$

From the derivation of Simpson's rule, it is clear that the formula is exact when f a polynomial of degree 3 or less, which is commensurate with the error estimate given in Eq. (1.11). More generally, if a quadrature formula based on a Taylor series uses N points, it will integrate exactly a polynomial of degree $N-1$ (degree N if N is odd). That is, the N weights w_n can be chosen to satisfy the N linear equations

$$\int_{-1}^{1} x^p\,dx = \sum_{n=1}^{N} w_n x_n^p; \quad p = 0, 1, \cdots, N-1 . \qquad (4.9)$$

(When N is odd, the quadrature formula is also exact for the odd monomial x^N.)

A greater precision for a given amount of numerical work can be achieved if we are willing to give up the requirement of equally-spaced quadrature points. That is, we will choose the x_n in some optimal sense, subject only to the constraint that they lie within the interval $[-1,1]$. We then have $2N$ parameters at our disposal in constructing the quadrature formula (the N x_n's and the N w_n's), and so we should be able to choose them so that Eq. (4.9) is satisfied for p ranging from 0 to $2N-1$. That is, the quadrature formula using only N carefully chosen points can be made exact for polynomials of degree $2N-1$ or less. This is clearly more efficient than using equally-spaced abscissae.

To see how to best choose the x_n, we consider the Legendre polynomials, which are orthogonal on the interval $[-1,1]$:

$$\int_{-1}^{1} P_i(x)P_j(x)\,dx = \frac{2}{2i+1}\delta_{ij} . \qquad (4.10)$$

It is easily shown that P_i is a polynomial of degree i with i roots in the interval $[-1,1]$. Any polynomial of degree $2N-1$ or less then can be written in the form

$$f(x) = Q(x)P_N(x) + R(x),$$

where Q and R are polynomials of degree $N-1$ or less. The exact value of the required integral (4.8) is then

$$I = \int_{-1}^{1}(QP_n + R)dx = \int_{-1}^{1}Rdx, \qquad (4.11)$$

where the second step follows from the orthogonality of P_N to all polynomials of degree $N-1$ or less. If we now take the x_n to be the N zeros of P_N, then application of (4.8) gives (exactly)

$$I = \sum_{n=1}^{N} w_n[Q(x_n)P_N(x_n) + R(x_n)] = \sum_{n=1}^{N} w_n R(x_n). \qquad (4.12)$$

It remains to choose the w_n so that R (a polynomial of degree $N-1$ or less) is integrated exactly. That is, the w_n satisfy the set of linear equations (4.9) when the x_n are the zeros of P_N. It can be shown that w_n is related to the derivative of P_N at the corresponding zero. Specifically,

$$w_n = \frac{2}{(1-x_n^2)[P'_N(x_n)]^2}$$

This completes the specification of what is known as the Gauss-Legendre quadrature formula. Note that it can be applied to any definite integral between finite limits by a simple linear change of variable. That is, for an integral between limits $x=a$ and $x=b$, a change of variable to

$$t = -1 + 2\frac{(x-a)}{(b-a)}$$

reduces the integral to the form required. Other non-linear changes of variable that make the integrand as smooth as possible will also improve the accuracy.

Other types of orthogonal polynomials provide useful Gaussian quadrature formulas when the integrand has a particular form. For example, the Laguerre polynomials, L_i, which are orthogonal on the interval $[0,\infty]$ with the weight function e^{-x}, lead to the Gauss-Laguerre quadrature formula

$$\int_{0}^{\infty} e^{-x}f(x)dx \approx \sum_{n=1}^{N} w_n f(x_n), \qquad (4.13)$$

where the x_n are the roots of L_N and the w_n are related to the values of L_{N+1} at these points. Similarly, the Hermite polynomials provide Gauss-Hermite quadrature formulas for integrals of the form

$$\int_{-\infty}^{\infty} e^{-x^2} f(x)\,dx.$$

These Gaussian quadrature formulas, and many others, are given in Section 25.4 of [Ab64], which also contains tables of the abscissae and weights.

In the practical application of Gaussian quadrature formulas, one does not need to write programs to calculate the abscissae and weights. Rather, there are usually library subroutines that can be used to establish arrays containing these numbers. For example, the subroutine beginning at line 12000 in Example 4 (see Appendix B) establishes the Gauss-Legendre abscissae and weights for many different values of N.

As a general rule, Gaussian quadrature is the method of choice when the integrand is smooth, or can be made smooth by extracting from it a function that is the weight for a standard set of orthogonal polynomials. We must, of course, also have the ability to evaluate the integrand at the required abscissae. If the integrand varies rapidly, we can compound the basic Gaussian quadrature formula by applying it over several sub-intervals of the range of integration. Of course, when the integrand can be evaluated only at equally-spaced abscissae (such as when it is generated by integrating a differential equation), then formulas of the type discussed in Chapter 1 must be used.

As an illustration of Gaussian quadrature, consider using a 3-point Gauss-Legendre quadrature to evaluate the integral

$$I = \int_0^3 (1+t)^{1/2} dt = 4.66667 . \tag{4.14}$$

Making the change of variable to

$$x = -1 + \frac{2}{3} t$$

results in

$$I = \frac{3}{2} \int_{-1}^1 (\frac{3}{2} x + \frac{5}{2})^{1/2} dx. \tag{4.15}$$

For $N=3$, the Gauss-Legendre abscissae and weights are

$x_1 = -x_3 = 0.774597,\ x_2 = 0;\quad w_1 = w_3 = 0.555556,\ w_2 = 0.888889.$

Straightforward evaluation of the quadrature formula (4.8) then results in $I=4.66683$, while a Simpson's rule evaluation of (4.14) with $h=1.5$ gives 4.66228. Gaussian quadrature is therefore more accurate than Simpson's rule by about a factor of 27, yet requires the same number of evaluations of the integrand (three).

Exercise 4.4 Consider the integral

$$\int_{-1}^{1}(1-x^2)^{\frac{1}{2}}dx = \frac{\pi}{2}.$$

Evaluate this integral using some of the quadrature formulas discussed in Chapter 1 and using Gauss-Legendre quadrature. Note that the behavior of the integrand near $x=\pm 1$ is cause for some caution. Compare the accuracy and efficiency of these various methods for different numbers of abscissae. Note that this integral can be evaluated exactly with a "one-point" Gauss-Chebyshev quadrature formula of the form

$$\int_{-1}^{1}(1-x^2)^{\frac{1}{2}}f(x)dx = \sum_{n=1}^{N} w_n f(x_n),$$

with

$$x_n = \cos\frac{n}{N+1}\pi; \quad w_n = \frac{\pi}{N+1}\sin^2\frac{n}{N+1}\pi.$$

(See Section 25.4.40 of [Ab64].)

4.3 Born and eikonal approximations to quantum scattering

In this example, we will investigate the Born and eikonal approximations suitable for describing quantum-mechanical scattering at high energies, and in particular calculate the scattering of fast electrons (energies greater than several 10's of eV) from neutral atoms. The following project deals with the exact partial-wave solution of this problem.

Extensive discussions of the quantum theory of scattering are given in many texts (see, for example, [Me68], [Ne66], or [Wu62]); we will only review the essentials here. For particles of mass m and energy

$$E = \frac{\hbar^2}{2m}k^2 > 0,$$

scattering from a central potential $V(r)$ is described by a wavefunction $\Psi(\mathbf{r})$ that satisfies the Schroedinger equation,

$$-\frac{\hbar^2}{2m}\nabla^2\Psi + V\Psi = E\Psi, \tag{4.16}$$

with the boundary condition at large distances

$$\Psi \underset{r \to \infty}{\to} e^{ikz} + f(\theta)\frac{e^{ikr}}{r}. \tag{4.17}$$

Here, the beam is incident along the z direction and θ is the scattering angle (angle between $\mathbf{r}$ and $\hat{\mathbf{z}}$). The complex scattering amplitude f embodies the observable scattering properties and is the basic function we seek to determine. The differential cross section is given by

$$\frac{d\sigma}{d\Omega} = |f(\theta)|^2, \tag{4.18}$$

and the total cross section is

$$\sigma = \int d\Omega \frac{d\sigma}{d\Omega} = 2\pi \int_0^\pi d\theta \sin\theta \, |f(\theta)|^2. \tag{4.19}$$

In general, f is a function of both E and θ.

At this point, many elementary treatments of scattering introduce a partial-wave decomposition of Ψ, express f in terms of the phase shifts, and then proceed to discuss the radial Schroedinger equation in each partial wave, from which the phase shift and hence the exact cross section, can be calculated. We will use this method, which is most appropriate when the energies are low and only a few partial waves are important, in Project IV below. However, in this example, we will consider two approximation schemes, the Born and eikonal, which are appropriate to high-energy situations when many partial waves contribute.

Both the Born and eikonal approximations are based on an exact integral expression for the scattering amplitude derived in many advanced treatments:

$$f(\theta) = -\frac{m}{2\pi\hbar^2} \int e^{-i\mathbf{k}_f \cdot \mathbf{r}} V(r)\Psi(r)d^3r. \tag{4.20}$$

Here, the wavenumber of the scattered particle is $\mathbf{k}_f$, so that $|\mathbf{k}_f| = k$ and $\hat{\mathbf{k}}_f \cdot \hat{\mathbf{z}} = \cos\theta$. It is also convenient to introduce the wavenumber of the incident particle, $\mathbf{k}_i = k\hat{\mathbf{z}}$.

The Born approximation (more precisely, the first Born approximation) consists of assuming that the scattering is weak, so that the full scattering wavefunction Ψ differs very little from the incident plane wave, $\exp(i\hat{\mathbf{k}}_i \cdot \mathbf{r})$. Making this replacement in (4.20) results in the Born scattering amplitude,

$$f_B(\theta) = -\frac{m}{2\pi\hbar^2} \int e^{-i\mathbf{q}\cdot\mathbf{r}} V(r)d^3r = -\frac{2m}{q\hbar^2} \int_0^\infty \sin qr \, V(r) \, rdr. \tag{4.21}$$

Here, we have introduced the momentum transfer, $\mathbf{q} = \mathbf{k}_f - \mathbf{k}_i$, so that

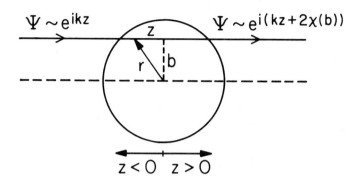

Figure 4.1 Geometry of the eikonal approximation

$$q = |\mathbf{k}_f - \mathbf{k}_i| = 2k \sin\tfrac{1}{2}\theta .$$

and have used the identity

$$\int e^{-i\mathbf{q}\cdot\mathbf{r}} d\hat{\mathbf{r}} = 4\pi j_0(qr) = 4\pi \frac{\sin qr}{qr}. \tag{4.22}$$

Note that the Born approximation to the scattering amplitude depends only upon q and not separately upon E and θ.

Better approximations to Ψ in Eq. (4.20) result in correspondingly better approximations to f. One possible improvement is the eikonal approximation, valid at high energies and small scattering angles. (See [Wa73] or [Ne66].) This approximation is semiclassical in nature; its essence is that each ray of the incident plane wave suffers a phase shift as it passes through the potential on a straight-line trajectory (see Figure 4.1). Since this phase shift depends upon the impact parameter of the ray, the wavefronts of the wavefunction are distorted after passing through the potential; it is this distortion that carries the scattering information.

To derive the eikonal approximation, we will put, without loss of generality,

$$\Psi(\mathbf{r}) = e^{i\mathbf{k}_i\cdot\mathbf{r}}\psi(\mathbf{r}), \tag{4.23}$$

where ψ is a slowly-varying function describing the distortion of the incident wave. Upon inserting this into the original Schroedinger equation (4.16), we obtain an equation for ψ:

$$-\frac{\hbar^2}{2m}(2i\,\mathbf{k}_i\cdot\nabla + \nabla^2)\psi + V\psi = 0. \tag{4.24}$$

If we now assume that ψ varies slowly enough so that the $\nabla^2\psi$ term can be ignored (i.e., k is very large), we have

$$ik\frac{\hbar^2}{m}\frac{\partial\psi(\mathbf{b},z)}{\partial z}=V(b,z)\psi(\mathbf{b},z). \qquad (4.25)$$

Here, we have introduced the coordinate $\mathbf{b}$ in the plane transverse to the incident beam, so that

$$V(\mathbf{b},z)=V(r); \quad r=(b^2+z^2)^{\frac{1}{2}} .$$

From symmetry considerations, we expect that ψ will be azimuthally symmetric and so independent of $\hat{\mathbf{b}}$. Equation (4.25) can be integrated immediately and, using the boundary condition that $\psi\rightarrow1$ as $z\rightarrow-\infty$ since there is no distortion of the wave before the particle reaches the potential, we have

$$\psi(b,z)=e^{2i\chi(b,z)}; \quad \chi(b,z)=-\frac{m}{2\hbar^2 k}\int_{-\infty}^{z} V(b,z')dz'. \qquad (4.26)$$

Having obtained the eikonal approximation to the scattering wavefunction, we can now obtain the eikonal scattering amplitude, f_e. Inserting Eq. (4.23) into (4.20), we have

$$f_e=-\frac{m}{2\pi\hbar^2}\int d^2b\int_{-\infty}^{\infty}dze^{-i\mathbf{q}\cdot\mathbf{r}}V(b,z)\psi(b,z). \qquad (4.27)$$

Using (4.25), we can relate $V\psi$ directly to $\partial\psi/\partial z$. Furthermore, if we restrict our consideration to relatively small scattering angles, so that $q_z\approx0$, then the z integral in (4.27) can be done immediately and, using (4.26) for ψ, we obtain

$$f_e=-\frac{ik}{2\pi}\int d^2be^{-i\mathbf{q}\cdot\mathbf{b}}(e^{2i\chi(b)}-1), \qquad (4.28)$$

with the "profile function"

$$\chi(b)=\chi(b,z=\infty)=-\frac{m}{2\hbar^2 k}\int_{-\infty}^{\infty} V(b,z)dz. \qquad (4.29)$$

Since χ is azimuthally symmetric, we can perform the azimuthal integration in (4.28) and obtain our final expression for the eikonal scattering amplitude,

$$f_e=-ik\int_0^{\infty} bdbJ_0(qb)(e^{2i\chi(b)}-1). \qquad (4.30)$$

In deriving this expression, we have used the identity (compare with Eq. (4.22))

$$J_0(qb)=\frac{1}{2\pi}\int_0^{2\pi}e^{-iqb\cos\varphi}d\varphi.$$

Note that in contrast to f_B, f_e depends upon both E (through k) and q.

An important property of the exact scattering amplitude is the optical theorem, which relates the total cross section to the imaginary part of the forward scattering amplitude. After a bit of algebra, one can show that f_e satisfies this relation in the limit that the incident momentum becomes large compared to the length scale over which the potential varies:

$$\sigma = \frac{4\pi}{k} \mathrm{Im} f (q=0) = 8\pi \int_0^\infty b\,db \sin^2\chi(b). \qquad (4.31)$$

The Born approximation cannot lead to a scattering amplitude that respects this relation, as Eq. (4.21) shows that f_B is purely real. It is also easy to show that, in the extreme high-energy limit, where $k \to \infty$ and χ becomes small, the Born and eikonal amplitudes become equal (see Exercise 4.5). The eikonal formula (4.30) also can be related to the usual partial wave expression for f (see Exercise 4.6).

One practical application of the approximations discussed above is in the calculation of the scattering of high-energy electrons from neutral atoms. In general, this is a complicated multi-channel scattering problem since there can be reactions leading to final states in which the atom is excited. However. as the reaction probabilities are small in comparison to elastic scattering, for many purposes the problem can be modeled by the scattering of an electron from a central potential. This potential represents the combined influence of the attraction of the central nuclear charge (Z) and the screening of this attraction by the Z atomic electrons. For a neutral target atom, the potential vanishes at large distances faster than r^{-1}. A very accurate approximation to this potential can be had by solving for the self-consistent Hartree-Fock potential of the neutral atom, as was done in Project III. However, a much simpler estimate can be obtained using an approximation to the Thomas-Fermi model of the atom given by Lenz and Jensen [Go49]:

$$V = -\frac{Ze^2}{r} e^{-x} (1 + x + b_2 x^2 + b_3 x^3 + b_4 x^4); \qquad (4.32a)$$

with

$$e^2 = 14.409; \quad b_2 = 0.3344; \quad b_3 = 0.0485; \quad b_4 = 2.647 \times 10^{-3}; \quad (4.32b)$$

and

$$x = 4.5397 Z^{1/6} r^{1/2}. \qquad (4.32c)$$

Here, the potential is measured in eV and the radius is measured in Å. Note that there is a possible problem with this potential, since it is singular as r^{-1} at the origin, and so leads to a divergent expression for χ at $b=0$. However, if the potential is regularized by taking it to be a constant within some small radius r_{min}, (say the radius of the atoms 1s shell), then the calculated cross section will be unaffected except at momentum transfers large enough so that $qr_{min} \gg 1$.

Our goal is to compute the Born and eikonal approximations to the differential and total cross sections for a given central potential at a specified incident energy, and in particular for the potential (4.32). To do this, we must compute the integrals (4.21), (4.29), and (4.30) defining the scattering amplitudes, as well as the integral (4.19) for the total cross section. The BASIC program for Example 4, whose source code is given in Appendix B, as well as in the file EXAM4.BAS on the *Computational Physics* diskette, does these calculations and graphs the results on a semi-log plot; the total cross section given by the optical theorem, Eq. (4.31), is also calculated.

The incident particle is assumed to have the mass of the electron, and, as is appropriate for atomic systems, all lengths are measured in Å and all energies in eV. The potential can be chosen to be a square well of radius 2 Å, a gaussian well of the form

$$V(r) = -V_0 e^{-2r^2},$$

or the Lenz-Jensen potential (4.32). All potentials are assumed to vanish beyond 2 Å. Furthermore, the r^{-1} singularity in the Lenz-Jensen potential is cutoff inside the radius of the 1s shell of the target atom.

Because the differential cross sections become very peaked in the forward direction at the high energies where the Born and eikonal approximations are valid, the integration over $\cos\theta$ in Eq. (4.19) is divided into two regions for a more accurate integration of the forward peak. One of these extends from $\theta=0$ to $\theta=\theta_{cut}$, where

$$q_{cut} R_{cut} = 2\pi = 2k R_{cut} \sin \tfrac{1}{2}\theta_{cut}$$

and R_{cut} is 1 Å for the Lenz-Jensen potential and 2 Å for either the square- or gaussian-well potentials, and the other extends from θ_{cut} to π. All integrals are done by Gauss-Legendre quadrature using the same number of points, and the Bessel function of order zero required by Eq. (4.30) is evaluated using the approximations (4.4).

The following exercises are aimed at improving your understanding of this program and the physics it describes.

Exercise 4.5 Verify the algebra in the derivations above. Show that ☐ ☐ ☐
in the limit of very high energies, where χ is small, so that $\sin\chi\approx\chi$,
the Born and eikonal results are identical. Also prove that the
eikonal amplitude satisfies the optical theorem (4.31) in the limit
where the incident momentum becomes large in comparison with
the length scale of the potential.

Exercise 4.6 Show that if the conventional expression for f in ☐ ☐ ☐
terms of a sum over partial waves (Eq. (IV.4) below) is approximated by an integral over l (or, equivalently, over $b=l/k$) and the
small-θ / large-l approximation

$$P_l(\cos\theta)\approx J_0(l\,\theta)$$

is used, Eq. (4.30) results, with the identification $\chi(b)=\delta_l$. Investigate, either numerically or analytically, the validity of this relation
between the Bessel function of order 0 and the Legendre polynomials.

Exercise 4.7 Test the Born approximation cross sections generated ☐ ☐ ☐
by the code by comparing the numerical values with the analytical
Born results for a square or gaussian well of depth 20 eV and for
varying incident energies from 1 eV to 10 keV. Verify for these
cases that $\chi(b)$ as computed by the code has the expected values.
Investigate the variation of the numerical results with changes in
the number of quadrature points. (Note that only particular values
of N are allowed by the subroutine generating the Gauss-Legendre
abscissae and weights.)

Exercise 4.8 Fix the depth of a square well at 20 eV and calculate ☐ ☐ ☐
for various incident energies to get a feeling for how the
differential and total cross sections vary. Compare the square well
cross sections with those of a gaussian well of comparable depth
and explain the differences. Show that the Born and eikonal
results approach each other at high energies for either potential
and that the optical relation for the eikonal amplitude becomes
better satisfied at higher energies.

Exercise 4.9 Using the Lenz-Jensen potential and a fixed charge ☐ ☐ ☐
and incident energy, say $Z=50$ and $E=1000$ eV, investigate the
sensitivity of the calculation to the number of quadrature points
used and to the small-r regularization of the potential. Calculate
the cross sections for various Z ranging from 20 to 100 and for E
ranging from 10 eV to 10 keV; interpret the trends you observe.

□ □ □   **Exercise 4.10** Use the program constructed in Project I to calculate the classical differential cross section for electrons scattering from the Lenz-Jensen potential for various Z and incident energies. Compare the results with the Born and eikonal cross sections. Can you establish an analytical connection between the classical description and the quantum approximations? (See [Ne66] for a detailed discussion.)

Project IV: Partial wave solution of quantum scattering

In this project, we will use the method of partial waves to solve the quantum scattering problem for a particle incident on a central potential, and in particular consider the low-energy scattering of electrons from neutral atoms. The strategy of the method is to employ a partial wave expansion of the scattering wavefunction to decompose the three-dimensional Schroedinger equation (4.16) into a set of uncoupled one-dimensional ordinary differential equations for the radial wavefunctions; each of these is then solved as a boundary-value problem to determine the phase shift, and hence the scattering amplitude.

IV.1 Partial wave decomposition of the wavefunction

The standard partial wave decomposition of the scattering wavefunction Ψ is

$$\Psi(\mathbf{r})= \sum_{l=0}^{\infty} (2l+1)i^l e^{i\delta_l} \frac{R_l(r)}{kr} P_l(\cos\theta), \tag{IV.1}$$

When this expansion is substituted into the Schroedinger equation (4.16), the radial wavefunctions R_l are found to satisfy the radial differential equations

$$\left[-\frac{\hbar^2}{2m}\frac{d^2}{dr^2} + V(r) + \frac{l(l+1)\hbar^2}{2mr^2} - E \right] R_l(r)=0. \tag{IV.2}$$

Although this is the same equation as that satisfied by a bound state wavefunction, the boundary conditions are different. In particular, R vanishes at the origin, but it has the large-r asymptotic behavior

$$R_l \to kr[\cos\delta_l j_l(kr) - \sin\delta_l n_l(kr)] \;, \tag{IV.3}$$

where j_l and n_l are the regular and irregular spherical Bessel functions of order l. (See Exercise 4.3.)

The scattering amplitude f is related to the phase shifts δ_l by

$$f(\theta)=\frac{1}{k} \sum_{l=0}^{\infty} (2l+1)e^{i\delta_l}\sin\delta_l P_l(\cos\theta), \tag{IV.4}$$

and the total cross section is easily found from Eq. (4.19) and the orthogonality of the Legendre polynomials, (4.10):

$$\sigma = \frac{4\pi}{k^2} \sum_{l=0}^{\infty} (2l+1)\sin^2\delta_l . \qquad (\text{IV}.5)$$

Although the sums in (IV.4,5) extend over all l, they are in practice limited to only a finite number of partial waves. This is because, for large l, the repulsive centrifugal potential in (IV.2) is effective in keeping the particle outside the range of the potential and so the phase shift is very small. If the potential is negligible beyond a radius r_{max} an estimate of the highest partial wave that is important, l_{max}, can be had by setting the turning point at this radius. Thus,

$$\frac{l_{max}(l_{max}+1)\hbar^2}{2mr_{max}^2} = E,$$

which leads to $l_{max} \sim kr_{max}$. This estimate is usually slightly low, since penetration of the centrifugal barrier leads to non-vanishing phase shifts in partial waves somewhat higher than this.

IV.2 Finding the phase shifts

To find the phase shift in a given partial wave, we must solve the radial equation (IV.2), using, for example, the Numerov method discussed in Chapter 3. Although the boundary conditions specified by (IV.3) and the vanishing of R at the origin are non-local, we can still integrate the equation as an initial value problem. This is because the equation is linear, so that the boundary condition at large r can be satisfied simply by appropriately normalizing the solution.

If we put $R_l(r=0)=0$ and take the value at the next lattice point, $R_l(r=h)$, to be any convenient small number (R will generally rise rapidly through the centrifugal barrier and so we must avoid overflows), we can integrate outward in r to a radius $r^{(1)} > r_{max}$. (See Figure IV.1.) Here, V vanishes and R must be a linear combination of the free solutions, $krj_l(kr)$ and $krn_l(kr)$:

$$R_l^{(1)} = Akr^{(1)}[\cos\delta_l j_l(kr^{(1)}) - \sin\delta_l n_l(kr^{(1)})]. \qquad (\text{IV}.6)$$

Although the constant A depends upon the value chosen for $R(r=h)$, it is largely irrelevant for our purposes; however, it must be kept small enough so that overflows are avoided.

Knowing only $R_l^{(1)}$ does not allow us to solve for the two unknowns, A and δ_l. However, if we continue integrating to a larger radius $r^{(2)} > r^{(1)}$, then we also have

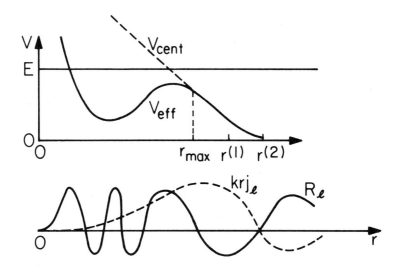

Figure IV.1 (Upper) Schematic centrifugal potential (dashed line) and effective total potential (solid line) in a partial wave $l>0$. The incident energy is also shown (horizontal line), as are the range of the potential (r_{max}) and the two matching radii, $r^{(1)}$ and $r^{(2)}$. (Lower) Schematic scattering wavefunction, R_l (solid line), and free wavefunction, krj_l (dashed line), for the same situation.

$$R_l^{(2)}=Akr^{(2)}[\cos\delta_l j_l(kr^{(2)})-\sin\delta_l n_l(kr^{(2)})]. \tag{IV.7}$$

Equations (IV.6,7) can then be solved for δ_l. After a bit of algebra, we have

$$\tan\delta_l=\frac{Gj_l^{(1)}-j_l^{(2)}}{Gn_l^{(1)}-n_l^{(2)}}\;;\;\;G=\frac{r^{(1)}R_l^{(2)}}{r^{(2)}R_l^{(1)}}\,, \tag{IV.8}$$

where $j_l^{(1)}=j_l(kr^{(1)})$, etc. Note that this equation determines δ_l only within a multiple of π, although this does not affect the physical observables (see Eqs. (IV.4,5)). The correct number of π's at a given energy can be determined by comparing the number of nodes in R and in the free solution, krj_l, which occur for $r<r_{max}$. With the conventional definitions we have used, the phase shift in each partial wave vanishes at high energies and approaches $N_l\pi$ at zero energy, where N_l is the number of bound states in the potential in the l'th partial wave.

IV.3 Solving the equations

Our goal in a numerical treatment of this problem is to investigate the scattering of electrons with energies from 0.5 eV to 10 keV from the potential (4.32); a radius $r_{max}=2$ Å is reasonable. For the charge Z and energy E specified, the program should calculate the phase shift and plot the radial wavefunction for each important partial wave, and then sum the contributions from each l to find the total cross section and the differential cross section as a function of θ from 0° to 180°, say in 5° steps. This program can be constructed and exploited in the following sequence of steps.

Step 1 Write a subroutine that computes the values of the Legendre polynomials for the degrees and angles required and stores them in an array. (See Eq. (4.1).) Also write a subroutine that computes j_l and n_l for a given value of x. (See Exercise 4.3). Check that these routines are working correctly by comparing their results with tabulated values.

Step 2 Write a subroutine that calculates the phase shift in a specified partial wave. Use the Numerov method to integrate to $r^{(1)}$ and then to $r^{(2)}$ and then determine the phase shift from Eq. (IV.8). Note that if $r^{(2)}$ is too close to $r^{(1)}$, problems with numerical precision may arise (both the numerator and denominator of (IV.8) vanish), while too great a distance between the matching radii leads to wasted computation. Check that your subroutine is working properly by verifying that the calculated phase shift vanishes when $V(r)$ is set to 0 and that it is independent of the choice of the matching radii, as long as they are greater than r_{max}. Also check that the phase shifts you find for a square-well potential agree with those that you can calculate analytically. Note that since quite a bit of numerical work is involved in the calculation of the Lenz-Jensen potential (4.32), it is most efficient if the values of V at the lattice points are stored in an array to be used in the integration of all important partial waves.

Step 3 Write a main program that, for a given charge and energy, calls the subroutines you have constructed in performing the sum over partial waves to calculate the total and differential cross sections. Note that since BASIC does not support complex arithmetic, you must compute the real and imaginary values of the scattering amplitude separately. Verify that your estimate of l_{max} is reasonable in that all non-vanishing phase shifts are computed but that no computational effort is wasted on partial waves for which the phase shift is negligible.

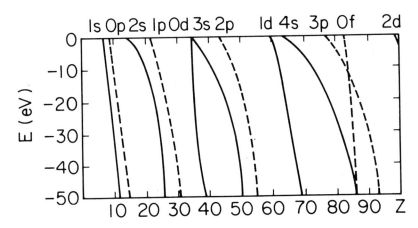

Figure IV.2 Weakly-bound levels of the Lenz-Jensen potential. Levels with a negative parity (p and f) are shown by dashed curves.

Step 4 Study the cross section as a function of energy for several Z, say 20, 50, and 100. Show that, at low energies, resonances occurring in particular partial waves make the angular distributions quite complex and the total cross sections very energy-dependent, but that all quantities become smooth and monotonic at high energies. Tabulations of experimental data with which to compare your results are given in [Ki71]. Also compare your high energy results with the Born and eikonal approximations generated by the program in Example 4.

Step 5 To understand the low-energy resonances, consider Figure IV.2, which gives the energies of weakly bound levels in the Lenz-Jensen potential as a function of Z. By studying the cross section at a fixed energy of 5 eV as a function of Z, show that the regions near $Z=46$ and $Z=59$ reflect the extensions of the $2p$ and $1d$ bound states into the continuum. By examining the $Z=59$ cross section, angular distribution, phase shifts, and radial wavefunctions as functions of energy near $E=5$ eV, show that the resonance is indeed in the $l=2$ partial wave. Note that since the Lenz-Jensen potential is based on the Thomas-Fermi approximation, it is not expected to be accurate in the outer region of the atom. The resonance energies and widths you find therefore differ quantitatively from those that are found in experiment.

Step 6 Use the Hartree-Fock code from Project III to generate the direct potential, Φ, for the neutral Ne atom and modify your program to calculate the scattering of electrons from this potential. Note that when the energy of the incident electron is large, it is easily "distinguishable" from the electrons in the atom, and so any effects of the Fock (exchange) potential will be small. Compare your results with those produced with the Lenz-Jensen potential. A further discussion of the electron-atom scattering problem can be found in [Bo74].

Chapter 5

Matrix Operations

Linearization is a common assumption or approximation in describing physical processes and so linear systems of equations are ubiquitous in computational physics. Indeed, the matrix manipulations associated with finding eigenvalues or with solving simultaneous linear equations are often the bulk of the work involved in solving many physical problems. In this chapter, we will discuss briefly two of the more non-trivial matrix operations: inversion and diagonalization. Our treatment here will be confined largely to "direct" methods appropriate for "dense" matrices (where most of the elements are non-zero) of dimension less than several hundred; iterative methods for treating the very large sparse matrices that arise in the discretization of ordinary and partial differential equations will be discussed in the following two chapters. As is the case with the special functions of the previous chapter, a variety of library subroutines employing several different methods for solving matrix problems are usually available on any large computer. Our discussions here are therefore limited to selected basic methods, to give a flavor of what has to be done. More detailed treatments can be found in many texts, for example [Ac70] and [Bu81].

5.1 Matrix inversion

Let us consider the problem of inverting a square $(N \times N)$ matrix $\mathbf{A}$. This might arise, for example, in solving the linear system

$$\mathbf{Ax=b},\tag{5.1}$$

for the unknown vector $\mathbf{x}$, where $\mathbf{b}$ is a known vector. The most natural way of doing this is to follow our training in elementary algebra and use Cramer's rule; i.e., to evaluate $\mathbf{A}^{-1}$ as being proportional to the transpose of the matrix of cofactors of $\mathbf{A}$. To do this, we would have to calculate N^2 determinants of matrices of dimension $(N-1) \times (N-1)$. If these were done in the most naive way, in which we would evaluate

$$\det\mathbf{A}=\sum_P (-)^P A_{1P1} A_{2P2} \cdots A_{NPN},\tag{5.2}$$

for the determinant of the $N \times N$ matrix **A**, where P is one of the $N!$ permutations of the N columns, Pi is the i'th element of P, and $(-)^P$ is the signature of P, then the numerical work would be horrendous. For example, of order $N!$ multiplications are required to evaluate (5.2). For $N=20$, this is some 2×10^{18} multiplications. On the fastest computer currently available, 10^8 multiplications per second might be possible, so that some 10^3 years are required just to invert this one 20×20 matrix! Clearly, this is not the way to evaluate determinants (or to invert matrices, as it turns out).

One of the simplest practical methods for evaluating $\mathbf{A}^{-1}$ is the Gauss-Jordan method. Here, the idea is to consider a class of elementary row operations on the matrix **A**. These involve multiplying a particular row of **A** by a constant, interchanging two rows, or adding a multiple of one row to another. Each of these three operations can be represented by left-multiplying **A** by a simple matrix, **T**. For example, when $N=3$, the matrices

$$
\begin{bmatrix} 1 & 0 & 0 \\ 0 & 1 & 0 \\ 0 & 0 & 2 \end{bmatrix}, \quad
\begin{bmatrix} 0 & 1 & 0 \\ 1 & 0 & 0 \\ 0 & 0 & 1 \end{bmatrix}, \quad \text{and} \quad
\begin{bmatrix} 1 & 0 & 0 \\ 0 & 1 & 0 \\ -\frac{1}{2} & 0 & 1 \end{bmatrix}
\tag{5.3}
$$

will multiply the third row by 2, interchange the first and second rows, and subtract one-half of the first row from the third row, respectively. The Gauss-Jordan strategy is to find a sequence of such operations,

$$\mathbf{T} = \cdots \mathbf{T}_3 \mathbf{T}_2 \mathbf{T}_1$$

which, when applied to **A**, reduces it to the unit matrix. That is,

$$\mathbf{TA} = (\cdots \mathbf{T}_3 \mathbf{T}_2 \mathbf{T}_1) \mathbf{A} = \mathbf{I}, \tag{5.4}$$

so that $\mathbf{T} = \mathbf{A}^{-1}$ is the required inverse. Equivalently, the same sequence of operations applied to the unit matrix yields the inverse of **A**.

How to actually find the appropriate sequence of row operations is best illustrated by example. Consider the 3×3 matrix **A** and the unit matrix:

$$
\mathbf{A} = \begin{bmatrix} 1 & 2 & 1 \\ 4 & 2 & 2 \\ 2 & 4 & 1 \end{bmatrix}; \quad
\mathbf{I} = \begin{bmatrix} 1 & 0 & 0 \\ 0 & 1 & 0 \\ 0 & 0 & 1 \end{bmatrix}.
\tag{5.5a}
$$

We first apply transformations to zero all but the first element in the first column of **A**. Subtracting 4 times the first row from the second yields

$$\mathbf{TA} = \begin{bmatrix} 1 & 2 & 1 \\ 0 & -6 & -2 \\ 2 & 4 & 1 \end{bmatrix}; \ \mathbf{TI} = \begin{bmatrix} 1 & 0 & 0 \\ -4 & 1 & 0 \\ 0 & 0 & 1 \end{bmatrix}, \tag{5.5b}$$

and then subtracting twice the first row from the third yields

$$\mathbf{TA} = \begin{bmatrix} 1 & 2 & 1 \\ 0 & -6 & -2 \\ 0 & 0 & -1 \end{bmatrix}; \ \mathbf{TI} = \begin{bmatrix} 1 & 0 & 0 \\ -4 & 1 & 0 \\ -2 & 0 & 1 \end{bmatrix}. \tag{5.5c}$$

We now go to work on the second column, where, by adding $\frac{1}{3}$ of the second row to the first, we can zero all but the second element:

$$\mathbf{TA} = \begin{bmatrix} 1 & 0 & 1/3 \\ 0 & -6 & -2 \\ 0 & 0 & -1 \end{bmatrix}; \ \mathbf{TI} = \begin{bmatrix} -1/3 & 1/3 & 0 \\ -4 & 1 & 0 \\ -2 & 0 & 1 \end{bmatrix}, \tag{5.5d}$$

and, multiplying the second row by $-\frac{1}{6}$ yields

$$\mathbf{TA} = \begin{bmatrix} 1 & 0 & 1/3 \\ 0 & 1 & 1/3 \\ 0 & 0 & -1 \end{bmatrix}; \ \mathbf{TI} = \begin{bmatrix} -1/3 & 1/3 & 0 \\ 2/3 & -1/6 & 0 \\ -2 & 0 & 1 \end{bmatrix}. \tag{5.5e}$$

Finally, we can reduce the third column of $\mathbf{TA}$ to the required form by adding $\frac{1}{3}$ of the third row to the first and second rows, and then multiplying the third row by -1:

$$\mathbf{TA} = \begin{bmatrix} 1 & 0 & 0 \\ 0 & 1 & 0 \\ 0 & 0 & 1 \end{bmatrix}; \ \mathbf{TI} = \begin{bmatrix} -1 & 1/3 & 1/3 \\ 0 & -1/6 & 1/3 \\ 2 & 0 & -1 \end{bmatrix}. \tag{5.5f}$$

This finds the required inverse.

A moment's thought shows how this algorithm can be generalized to an $N \times N$ matrix and that, when this is done, it requires, for large N, of order N^3 multiplications and additions. Thus, it is computationally tractable, as long as N is not too large.

In practice, several subtleties are important. For example, it might happen that, at some point in the procedure, the diagonal element of **TA** vanishes in the column we are working on. In this case, an interchange of two rows will bring a non-vanishing value to this "pivot" element (if no such row interchange can do this, then the matrix **A** is singular). In fact, numerical accuracy requires that rows be interchanged to bring into the pivot position that element in the column being worked on which has the largest absolute value. Problems associated with numerical round-off can also arise during the inversion if elements of the matrix differ greatly in magnitude. For this reason, it is often useful to scale the rows or columns so that all entries have roughly the same magnitude ("equilibration"). Various special cases (such as when **A** is symmetric or when we are only interested in solving (5.1) for **b** and not for $\mathbf{A}^{-1}$ itself) can result in reductions of the numerical work. For example, in the latter case, we can apply **T** only to the vector **b**, rather than to the whole unit matrix. Finally, if we are interested in computing only the determinant of **A**, successive row transformations, which have a simple and easily calculable effect on the determinant, can be used to bring **TA** to lower or upper diagonal form (all elements vanishing above or below the diagonal, respectively), and then the determinant can be evaluated simply as the product of the diagonal elements.

◻ ◻ ◻   **Exercise 5.1** Use Eq. (5.2) to show that interchanging the rows of a matrix changes the sign of its determinant, that adding a multiple of one row to another leaves its determinant unchanged, and that multiplying one row by a constant multiplies the determinant by that same constant.

◻ ◻ ◻   **Exercise 5.2** Write (or just flowchart) a subroutine that will use the Gauss-Jordan method to find the inverse of a given matrix. Incorporate pivoting as described above to improve the numerical accuracy.

5.2 Eigenvalues of a tri-diagonal matrix

We turn now to the problem of finding the eigenvalues and eigenvectors of an $N \times N$ matrix **A**; that is, the task of finding the N scalars λ_n and their associated N-component vectors $\boldsymbol{\varphi}_n$ which satisfy

$$\mathbf{A}\boldsymbol{\varphi}_n = \lambda_n \boldsymbol{\varphi}_n. \tag{5.6}$$

Equivalently, the eigenvalues are zeros of the N'th degree characteristic polynomial of **A**:

$$P_A(\lambda) \equiv \det(\mathbf{A} - \lambda \mathbf{I}) = \prod_{n=1}^{N} (\lambda_n - \lambda). \qquad (5.7)$$

For simplicity, we will restrict our discussion to matrices $\mathbf{A}$ which are real and symmetric, so that the eigenvalues are always real and the eigenvectors can be chosen to be orthonormal; this is the most common type of matrix arising in modeling physical systems. We will also consider only cases where all or many of the eigenvalues are required, and possibly their associated eigenvectors. If only a few of the largest or smallest eigenvalues of a large matrix are needed, then the iterative methods described in Section 7.4, which involve successive applications of $\mathbf{A}$ or its inverse, can be more efficient.

The general strategy for diagonalizing $\mathbf{A}$ is to reduce the task to the tractable one of finding the eigenvalues and eigenvectors of a symmetric tri-diagonal matrix; that is, one in which all elements are zero except those on or neighboring the diagonal. This can always be accomplished by applying a suitable sequence of orthogonal transformations to the original matrix, as described in the next section. For now, we assume that $\mathbf{A}$ has been brought to a tri-diagonal form and discuss a strategy for finding its eigenvalues.

To find the eigenvalues of a symmetric tri-diagonal matrix, we must find the roots of the characteristic polynomial, (5.7). This polynomial is given in terms of the elements of $\mathbf{A}$ by the determinant

$$P_A(\lambda) = \begin{vmatrix} A_{11}-\lambda & A_{12} & & & & \\ A_{21} & A_{22}-\lambda & A_{23} & & & \\ & A_{32} & A_{33}-\lambda & & & \\ & & A_{43} & & & \\ & & & \cdots & & \\ & & & & A_{N-1N-1}-\lambda & A_{N-1N} \\ & & & & A_{NN-1} & A_{NN}-\lambda \end{vmatrix}, \qquad (5.8)$$

where all elements not shown explicitly are zero and where the symmetry of $\mathbf{A}$ implies that $A_{nm} = A_{mn}$. To find the zeros of P_A, any of the root finding strategies discussed in Section 1.3 can be employed, providing that we can find the numerical value of P_A for a given value of λ. This latter can be done conveniently in a recursive manner. Let $P_n(\lambda)$ be the value of the $n \times n$ sub-determinant of (5.8) formed from the first n rows and columns. Clearly, P_n is a polynomial of degree n, $P_N = P_A$ (the polynomial we are after), and

$$P_1(\lambda) = A_{11}-\lambda; \quad P_2(\lambda) = (A_{22}-\lambda)P_1(\lambda) - A_{12}^2. \qquad (5.9)$$

Moreover, by expanding the determinant for P_n in terms of the minors of the n'th column, it is easy to derive the recursion relation

$$P_n(\lambda) = (A_{nn} - \lambda)P_{n-1}(\lambda) - A_{nn-1}^2 P_{n-2}(\lambda). \qquad (5.10)$$

This, together with the starting values (5.9), allows an efficient evaluation of P_A.

□ □ □   **Exercise 5.3** Prove Eq. (5.10) above.

Several features of the problem above help considerably in finding the roots of P_A. If an algorithm like Newton's method is used, it is quite easy to differentiate (5.10) once or twice with respect to λ and so derive recursion relations allowing the simple evaluation of the first or second derivatives of $P_A(\lambda)$. More importantly, it is possible to show that the number of times the sign changes in the sequence

$$1, \; P_1(\lambda), \; P_2(\lambda), \; \cdots, \; P_N(\lambda)$$

is equal to the number of eigenvalues less than λ. This fact is useful in several ways: it is a means of making sure that no roots are skipped in the search for eigenvalues, it provides a simple way of localizing a root initially, and it can be used with the simple search procedure to locate the root accurately, although perhaps not in the most efficient way possible.

To make a systematic search for all of the eigenvalues (roots of P_A), it is essential to know where to begin looking. If we are working through finding the entire sequence of eigenvalues, a natural guess for the next-highest one is some distance above that eigenvalue just found. Some guidance in how far above and in how to estimate the lowest eigenvalue can be found in Gerschgorin's bounds on the eigenvalues. It is quite simple to show that

$$\lambda_n \geq \min_i \left\{ A_{ii} - \sum_{j \neq i} |A_{ij}| \right\}, \qquad (5.11a)$$

and that

$$\lambda_n \leq \max_i \left\{ A_{ii} + \sum_{j \neq i} |A_{ij}| \right\} \qquad (5.11b)$$

for all n. For the tri-diagonal forms under consideration here, the sums over j in these expressions involve only two terms. The lower bound, Eq. (5.11a), is a convenient place to begin a search for the lowest eigenvalue, and the difference between the upper and lower bounds gives some measure of the average spacing between eigenvalues.

Exercise 5.4 Write a subroutine that finds all of the eigenvalues of ☐ ☐ ☐
a tri-diagonal matrix using the procedure described above,
together with any root-finding algorithm you find convenient. Test
your subroutine on an $N \times N$ tri-diagonal matrix of the form

$$A_{nn} = -2; \quad A_{nn-1} = A_{n-1n} = +1,$$

whose eigenvalues are known analytically to be

$$\lambda_n = -4 \sin^2 \left[\frac{n\pi}{2(N+1)} \right].$$

5.3 Reduction to tri-diagonal form

To apply the method for finding eigenvalues discussed in the
previous section, we must reduce a general real symmetric matrix
A to tri-diagonal form. That is, we must find an orthogonal $N \times N$
matrix, **O**, satisfying

$$O^t O = O O^t = I, \tag{5.12}$$

where O^t is the transpose of **O**, such that $O^t A O$ is tri-diagonal. Elementary considerations of linear algebra imply that the eigenvalues of the transformed matrix are identical to those of the original matrix. The problem, of course, is to find the precise form of
O for any particular matrix **A**. We discuss in this section two
methods for effecting such a transformation.

The Householder method is a common and convenient strategy
for reducing a matrix to tri-diagonal form. It takes the matrix **O**
to be the product of $N-2$ orthogonal matrices,

$$O = O_1 O_2 \cdots O_{N-2}, \tag{5.13}$$

each of which successively transforms one row and column of **A**
into the required form. (Only $N-2$ transformations are required as
the last two rows and columns are already in tri-diagonal form.) A
simple method can be applied to find each of the O_n.

To be more explicit about how the Householder algorithm
works, let us consider finding the first orthogonal transformation,
O_1, which we will choose to annihilate most of the first row and
column of **A**; that is,

$$O_1^t A O_1 = \begin{bmatrix} A_{11} & k^{(1)} & 0 & \cdots & 0 \\ k^{(1)} & A_{22}^{(2)} & A_{23}^{(2)} & \cdots & A_{2N}^{(2)} \\ 0 & A_{32}^{(2)} & A_{33}^{(2)} & \cdots & A_{3N}^{(2)} \\ \vdots & \vdots & \vdots & \vdots & \vdots \\ 0 & A_{N2}^{(2)} & A_{N3}^{(2)} & \cdots & A_{NN}^{(2)} \end{bmatrix}. \tag{5.14}$$

Here, $k^{(1)}$ is a possibly non-vanishing element and the matrix $\mathbf{A}^{(2)}$ is the result of applying the transformation $\mathbf{O}_1$ to the last $N-1$ rows and columns of the original matrix $\mathbf{A}$. Once $\mathbf{O}_1$ is found and applied (by methods discussed below), we can choose $\mathbf{O}_2$ so that it effects the same kind of transformation on the matrix $\mathbf{A}^{(2)}$; that is, it annihilates most of the first row and column of that matrix and transforms its last $N-2$ rows and columns:

$$\mathbf{O}_2^t \mathbf{A}^{(2)} \mathbf{O}_2 = \begin{vmatrix} A_{22}^{(2)} & k^{(2)} & 0 & \cdots & 0 \\ k^{(2)} & A_{33}^{(3)} & A_{34}^{(3)} & \cdots & A_{3N}^{(3)} \\ 0 & A_{43}^{(3)} & A_{44}^{(3)} & \cdots & A_{4N}^{(3)} \\ \vdots & \vdots & \vdots & \vdots & \vdots \\ 0 & A_{N3}^{(3)} & A_{N4}^{(3)} & \cdots & A_{NN}^{(3)} \end{vmatrix}. \tag{5.15}$$

Continued orthogonal transformations of decreasing dimension defined in this way will transform the original matrix into a tri-diagonal one after a total of $N-2$ transformations, the diagonal elements of the transformed matrix being

$$A_{11},\, A_{22}^{(2)},\, A_{33}^{(3)},\, \cdots,\, A_{N-1\,N-1}^{(N-1)},\, A_{NN}^{(N-1)},$$

and the off-diagonal elements being

$$k^{(1)},\, k^{(2)},\, \cdots,\, k^{(N-1)}.$$

It remains, of course, to find the precise form of each of the $\mathbf{O}_n$. We illustrate the procedure with $\mathbf{O}_1$, which we take to be of the form

$$\mathbf{O} = \begin{vmatrix} 1 & \mathbf{0}^t \\ \mathbf{0} & \mathbf{P} \end{vmatrix}, \tag{5.16}$$

where $\mathbf{0}$ is an $N-1$-dimensional column vector of zeros, $\mathbf{0}^t$ is a similar row vector, and $\mathbf{P}$ is an $(N-1)\times(N-1)$ symmetric matrix satisfying

$$\mathbf{P}^2 = \mathbf{I} \tag{5.17}$$

if $\mathbf{O}$ is to be orthogonal. (In these expressions and the following, for simplicity we have dropped all superscripts and subscripts which indicate that it is the first orthogonal transformation of the sequence (5.13) which is being discussed.) A choice for $\mathbf{P}$ that satisfies (5.17) is

$$\mathbf{P} = \mathbf{I} - 2\mathbf{u}\mathbf{u}^t. \tag{5.18}$$

In this expression, I is the $N-1$-dimensional unit matrix, $\mathbf{u}$ is an $N-1$-dimensional unit vector satisfying

$$\mathbf{u^t u}=1, \qquad (5.19)$$

and the last term involves the *outer* product of $\mathbf{u}$ with itself. Each element of (5.18) therefore reads

$$P_{nm}=\delta_{nm}-2u_n u_m,$$

where n and m range from 1 to $N-1$. (It should cause no confusion that the indices on $\mathbf{P}$ and $\mathbf{u}$ range from 1 to $N-1$, while those on the full matrices $\mathbf{O}$ and $\mathbf{A}$ range from 1 to N.)

Under a transformation of the form (5.16), $\mathbf{A}$ becomes

$$\mathbf{O^t A O}=\begin{vmatrix} A_{11} & (\mathbf{P\alpha})^t \\ \\ \mathbf{P\alpha} & \mathbf{A}^{(2)} \end{vmatrix}, \qquad (5.20)$$

where we have defined an $N-1$-dimensional vector $\boldsymbol{\alpha}$ as all elements but the first in the first column of $\mathbf{A}$; that is, $\alpha_i = A_{i+1,1}$ for i ranging from 1 to $N-1$. Upon comparing (5.20) with the desired form (5.14), it is apparent that the action of $\mathbf{P}$ on this vector must yield

$$\mathbf{P\alpha}\equiv\boldsymbol{\alpha}-2\mathbf{u}(\mathbf{u^t}\boldsymbol{\alpha})=\mathbf{k}, \qquad (5.21)$$

where $\mathbf{k}$ is the vector

$$\mathbf{k}=[k,0,0,\cdots,0]^t.$$

Equation (5.21) is what we must solve to determine $\mathbf{u}$, and hence the required transformation, $\mathbf{P}$. To do so, we must first find the scalar k, which is easily done by taking the scalar product of (5.21) with its transpose and using the idempotence of $\mathbf{P}$:

$$\mathbf{k^t k}=k^2=(\mathbf{P\alpha})^t(\mathbf{P\alpha})=\boldsymbol{\alpha^t}\boldsymbol{\alpha}\equiv\alpha^2=\sum_{i=2}^{N} A_{i1}^2, \qquad (5.22)$$

so that $k=\pm\alpha$. Having found $\mathbf{k}$ (we will discuss the choice of the sign in a moment), we can then rearrange (5.21) as

$$\boldsymbol{\alpha}-\mathbf{k}=2\mathbf{u}(\mathbf{u^t}\boldsymbol{\alpha}). \qquad (5.23)$$

Upon taking the scalar product of this equation with itself, using $\boldsymbol{\alpha^t}\mathbf{k}=\pm A_{21}\alpha$, and recalling (5.22), we have

$$2(\mathbf{u^t}\boldsymbol{\alpha})^2=(\alpha^2\mp A_{21}\alpha), \qquad (5.24)$$

so that we can then solve (5.21) for $\mathbf{u}$:

$$\mathbf{u}=\frac{\boldsymbol{\alpha}-\mathbf{k}}{2(\mathbf{u^t}\boldsymbol{\alpha})}. \qquad (5.25)$$

This completes the steps necessary to find **P**. To recapitulate, we first solve (5.22) for k, then calculate the square of the scalar product (5.24), form the vector **u** according to (5.25), and then finally **P** according to (5.18). In evaluating (5.24), considerations of numerical stability recommend taking that sign which makes the right-hand side largest. Note that we need evaluate only the square of the scalar product as the vector **u** enters **P** bilinearly. Note also that a full matrix product need not be performed in evaluating $\mathbf{A}^{(2)}$. Indeed, from (5.18,20) we have

$$\mathbf{A}^{(2)}=(1-2\mathbf{uu}^t)\mathbf{A}(1-2\mathbf{uu}^t)$$

$$=\mathbf{A}-2\mathbf{u}(\mathbf{Au})^t-2(\mathbf{Au})\mathbf{u}^t+4\mathbf{uu}^t(\mathbf{u}^t\mathbf{Au}),\qquad(5.26)$$

where the symbol **A** in this expression stands for the square symmetric matrix formed from the last $N-1$ rows and columns of the original matrix. Thus, to evaluate $\mathbf{A}^{(2)}$, once we have the vector **u** we need only calculate the vector **Au** and the scalar $\mathbf{u}^t\mathbf{Au}$. Finally, we note that numerical stability during the successive transformations is improved if the diagonal element having largest absolute magnitude is in the upper left-hand corner of the sub-matrix being transformed. This can always be arranged by a suitable interchange of rows and columns (which leaves the eigenvalues invariant) after each of the orthogonal transformations is applied.

❏ ❏ ❏   **Exercise 5.5** By writing either a detailed flowchart or an actual subroutine, convince yourself that you understand the Householder algorithm described above.

With an algorithm for transforming **A** to tri-diagonal form, we have specified completely how to find the eigenvalues of a real symmetric matrix. Once these are in hand, the eigenvectors are a relatively simple matter. The method of choice, inverse vector iteration, works as follows. Let $\varphi_n^{(1)}$ be any guess for the eigenvector associated with λ_n. This guess can be refined by evaluating

$$\varphi_n^{(2)}=[\mathbf{A}-(\lambda_n+\varepsilon)\mathbf{I}]^{-1}\varphi_n^{(1)}.\qquad(5.27)$$

Here, ε is a small, non-zero scalar that allows the matrix to be inverted. It is easy to see that this operation enhances that component of $\varphi_n^{(1)}$ along the true eigenvector at the expense of the spurious components. Normalization of $\varphi_n^{(2)}$, followed by repeated refinements according to (5.27) converges quickly to the required eigenvector, often in only two iterations.

An alternative to the Householder method is the Lanczos algorithm [Wh77], which is most suitable when we are interested in many of the lowest eigenvalues of very large matrices. The strategy here is to construct iteratively a set of orthonormal basis vectors, $\{\psi_n\}$, in which $\mathbf{A}$ is explicitly tri-diagonal. To begin the construction, we choose an arbitrary first vector in the basis, ψ_1, normalized so that $\psi_1^t\psi_1=1$. We then form the second vector in the basis as

$$\psi_2=C_2(\mathbf{A}\psi_1-A_{11}\psi_1), \qquad (5.28)$$

where $A_{11}=\psi_1^t\mathbf{A}\psi_1$ (it is *not* the element of $\mathbf{A}$ in the first row and column), and C_2 is a normalization chosen to ensure that $\psi_2^t\psi_2=1$:

$$C_2=[(\mathbf{A}\psi_1)^t(\mathbf{A}\psi_1)-(A_{11})^2]^{-\frac{1}{2}}. \qquad (5.29)$$

It is easy to show that $\psi_2^t\psi_1=0$. Subsequent vectors in the basis are then constructed recursively as

$$\psi_{n+1}=C_{n+1}(\mathbf{A}\psi_n-A_{nn}\psi_n-A_{nn-1}\psi_{n-1}), \qquad (5.30)$$

with

$$C_{n+1}=[(\mathbf{A}\psi_n)^t(\mathbf{A}\psi_n)-(A_{nn})^2-(A_{nn-1})^2]^{-\frac{1}{2}}. \qquad (5.31)$$

Thus, each successive ψ_{n+1} is that unit vector which is coupled to ψ_n by $\mathbf{A}$ and which is orthogonal to both ψ_n and ψ_{n-1}. The matrix $\mathbf{A}$ is explicitly tri-diagonal in this basis since Eq. (5.30) shows that when $\mathbf{A}$ acts on ψ_n, it yields only terms proportional to ψ_n, ψ_{n-1} and ψ_{n+1}. Continuing the recursion (5.30) until ψ_N is generated completes the basis and the representation of $\mathbf{A}$ in it.

The Lanczos method is well-suited to large matrices, as only the ability to apply $\mathbf{A}$ to a vector is required, and only the vectors ψ_n, ψ_{n-1}, and $\mathbf{A}\psi_n$ need be stored at any given time. We must, of course, also be careful to choose ψ_1 so that it is not an eigenvector of $\mathbf{A}$. However, the Lanczos method is not appropriate for finding all of the eigenvalues of a large matrix, as round-off errors in the orthogonalizations of (5.30) will accumulate as the recursive generation of the basis proceeds; i.e., the scalar products of basis vectors with large n with those having small n will not vanish identically.

The real utility of the Lanczos algorithm is in cases where many, but not all, of the eigenvalues are required. Suppose, for example, that we are interested in the 10 lowest eigenvalues of a matrix of dimension 1000. What is done is to generate recursively some number of states in the basis greater than the number of eigenvalues being sought (say 25), and then to find the 10 lowest eigenvalues and eigenvectors of that limited sub-matrix. If an arbitrary linear combination (say the sum) of these eigenvectors is then used as the initial vector in constructing a new basis of dimension 25, it can be shown that iterations of this process (generating a limited basis, diagonalizing the corresponding tri-diagonal matrix, and using the normalized sum of lowest eigenvectors found as the first vector in the next basis) will converge to the required eigenvalues and eigenvectors.

□ □ □ **Exercise 5.6** Write a subroutine that uses the Lanczos method to generate a complete basis in which a symmetric input matrix **A** is tri-diagonal. The program should output the diagonal and off-diagonal elements of **A** in this basis, as well as the basis vectors themselves. Show by explicit example that if the dimension of the matrix is too large, round-off errors cause the basis generated to be inaccurate. Try curing this problem by doing the computation in double-precision arithmetic and observe the results.

□ □ □ **Exercise 5.7** A simple limit of the Lanczos procedure for generating and truncating the basis is when we are interested in only the lowest eigenvalue and retain only ψ_1 and ψ_2 to form a 2×2 matrix to be diagonalized. Show that the lower eigenvalue of this matrix is always less than or equal to A_{11}, so that iterations of the procedure lead to a monotonically decreasing estimate for the lowest eigenvalue of **A**. This procedure is closely related to the time-evolution algorithm discussed in Section 7.4.

5.4 Determining nuclear charge densities

The distribution of electrical charge within the atomic nucleus is one of the most basic aspects of nuclear structure. The interactions of electrons and muons with the nucleus can be used to determine this distribution with great precision, as these particles interact almost exclusively through the well-understood electromagnetic interaction (for a general discussion, see [Fo66]). In this example, we will explore how the experimentally determined cross sections for the elastic scattering of high-energy (several 100 MeV) electrons from nuclei can be analyzed to determine the nuclear charge distribution; the method relies on the solution of a set of linear equations by matrix inversion.

To illustrate the basic idea, we begin by considering the scattering of non-relativistic electrons of momentum k and energy E from a localized charge distribution, $\rho(\mathbf{r})$, which contains Z protons, so that

$$\int d\mathbf{r}\,\rho(\mathbf{r})=Z, \tag{5.32}$$

and which is fixed in space. The electrons interact with this charge distribution through the Coulomb potential it generates, $V(\mathbf{r})$, which satisfies Poisson's equation,

$$\nabla^2 V=4\pi\alpha\rho, \tag{5.33}$$

where $\alpha=1/137.036$ is the fine structure constant. (We henceforth work in units where $\hbar = c = 1$, unless explicitly indicated, and note that $\hbar c = 197.329$ MeV−fm.)

The Born approximation to quantum scattering discussed in the previous chapter illustrates how electron scattering cross sections carry information about the charge distribution, although precision work requires a better approximation to the scattering, as discussed below. Equations (4.18, 4.21) show that the Born cross section for scattering through an angle θ is proportional to the square of the Fourier transform of the potential at the momentum transfer $\mathbf{q}$, where $q = 2k \sin\frac{1}{2}\theta$. The Fourier transform of (5.33) results in

$$V(\mathbf{q}) = -\frac{4\pi\alpha}{q^2}\,\rho(\mathbf{q}), \qquad (5.34)$$

where

$$\rho(\mathbf{q}) = \int d\mathbf{r}\, e^{-i\mathbf{q}\cdot\mathbf{r}}\rho(\mathbf{r}) \qquad (5.35)$$

is the Fourier transform of the charge density. Thus, the differential cross section can be written as (we use σ as a shorthand notation for $d\sigma/d\Omega$):

$$\sigma = \sigma_{ruth}\,|F(\mathbf{q})|^2. \qquad (5.36)$$

Here, the Rutherford cross section for scattering from a point charge of strength Z is

$$\sigma_{ruth} = \frac{4Z^2\alpha^2 m^2}{q^4}, \qquad (5.37)$$

m is the electron mass, and the nuclear "form factor" is $F(\mathbf{q}) = Z^{-1}\rho(\mathbf{q})$. For the spin-0 nuclei we will be considering, the charge density is spherically symmetric, so that $\rho(\mathbf{r}) = \rho(r)$, and

$$F(\mathbf{q}) = F(q) = \frac{4\pi}{Zq}\int_0^\infty dr\, r\,\sin qr\,\rho(r). \qquad (5.38)$$

Equation (5.36) illustrates how electron scattering is used to study nuclear structure. Deviations of the measured cross section from the Rutherford value are a direct measure of the Fourier transform of the nuclear charge density. It also shows that scatterings at high momentum transfers are needed to probe the nucleus on a fine spatial scale. As nuclear sizes are typically several fermis (10^{-13} cm), a spatial resolution better than 1 fm is desirable, implying momentum transfers of several fm^{-1} and so beam energies of several hundred MeV. (Recall that $2k$ is the maximum momentum transfer which can be achieved with a beam of momentum k.)

At energies of several hundred MeV, the electron is highly relativistic (the electron mass is only 0.511 MeV) and so the discussion above must be redone beginning with the Dirac equation rather than the Schroedinger equation. One trivial change in the final expressions that result is that ultra-relativistic kinematics are used for the electron (i.e., its momentum is proportional to its energy), so that the momentum transfer can be written as

$$q = 2E \sin\tfrac{1}{2}\theta. \tag{5.39}$$

The only other modification to Eq. (5.36) is that the Rutherford cross section is replaced by the Mott cross section,

$$\sigma = \sigma_{mott} \, |F(q)|^2; \quad \sigma_{mott} = \frac{4Z^2\alpha^2 E^2 \cos^2 \tfrac{1}{2}\theta}{q^4}, \tag{5.40}$$

where the additional factor of $\cos\tfrac{1}{2}\theta$ in the scattering amplitude arises from the spin-$\tfrac{1}{2}$ nature of the electron.

Let us now consider how measured cross sections are to be analyzed to determine the charge density. Suppose that we have a set of I experimental values of the cross section for elastic scattering of electrons from a particular nucleus at a variety of momentum transfers (e.g., angular distributions at one or several beam energies). Let these values be σ_i^e and their statistical uncertainties by Δ_i. Suppose also that we parametrize the nuclear charge density by a set of N parameters, C_n, so that $\rho(r) = \rho(r;\{C_n\})$. Of course, the C's must be chosen so that the normalization constraint (5.32) is satisfied. That is,

$$Z(\{C_n\}) \equiv 4\pi \int_0^\infty r^2 \rho(r;\{C_n\}) = Z . \tag{5.41}$$

A specific choice for this parametrization will be discussed shortly.

The usual methods of data analysis [Be69] state that the "best" values of the parameters C_n implied by the data are those that minimize

$$\chi^2 \equiv \sum_{i=1}^{I} \frac{(\sigma_i^e - \sigma_i^t)^2}{\Delta_i^2} , \tag{5.42}$$

subject to the normalization constraint (5.41) above. Here, σ_i^t is the "theoretical" cross section calculated from the appropriate Mott cross section and nuclear form factor; it depends parametrically upon the C's. This minimum value of χ^2 measures the quality of the fit, a satisfactory value being about the number of degrees of freedom (the number of data points less the number of parameters).

There are several computational strategies for finding the parameters which minimize χ^2 or, equivalently, which satisfy the N non-linear equations

$$\frac{\partial \chi^2}{\partial C_n} = 0. \tag{5.43}$$

This is a specific example of the commonly encountered problem of minimizing a non-linear function of several parameters; it is often fraught with difficulties [Ac70], not the least of which is finding a local minimum, rather than the global one usually sought. Most strategies are based on an iterative refinement of a "guess" for the optimal C_n, and, if this guess is close to the required solution, there is usually no problem. One commonly used approach is to compute the direction of the N-dimensional gradient (left-hand side of (5.43)) either analytically or numerically at the current guess and then to generate the next guess by stepping some distance in C-space directly away from this direction. Alternatively, a multi-dimensional generalization of the Newton-Rafson method, Eq. (1.14), can be used. The simple approach we will adopt here is based on a local linearization of χ^2 about the current guess, C_n^0. For small variations of the C's about this point,

$$C_n = C_n^0 + \delta C_n,$$

the theoretical cross sections can be expanded as

$$\sigma_i^t = \sigma_i^0 + \sum_{n=1}^{N} W_{in} \delta C_n ; \quad W_{in} \equiv \frac{\partial \sigma_i^t}{\partial C_n},$$

so that χ^2 is given by

$$\chi^2 = \sum_{i=1}^{I} (\sigma_i^e - \sigma_i^0 - \sum_{n=1}^{N} W_{in} \delta C_n)^2 / \Delta_i^2, \tag{5.44}$$

where σ_i^0 is the theoretical cross section corresponding to the current guess, C^0. The charge normalization constraint (5.41) can be linearized similarly as

$$Z(\{C_n\}) = Z + \sum_{n=1}^{N} \frac{\partial Z}{\partial C_n} \delta C_n, \tag{5.45}$$

where we have assumed that the current guess is normalized properly, so that $Z(\{C_n^0\}) = Z$.

Equations for the δC's which make χ^2 stationary can be had by requiring

$$\frac{\partial}{\partial \delta C_n} (\chi^2 - 2\lambda Z) = 0. \tag{5.46}$$

Here, we have used introduced a Lagrange multiplier, 2λ, to ensure the proper charge normalization. Some simple algebra using Eqs. (5.44,45) then results in the linear equations

$$\sum_{m=1}^{N} a_{nm} \delta C_m + \frac{\partial Z}{\partial C_n} \lambda = b_n,$$ (5.47a)

$$\sum_{m=1}^{N} \frac{\partial Z}{\partial C_m} \delta C_m = 0,$$ (5.47b)

with

$$a_{nm} = \sum_{i=1}^{I} \frac{W_{in} W_{im}}{\Delta_i^2}; \quad b_n = \sum_{i=1}^{I} \frac{(\sigma_i^e - \sigma_i^0) W_{in}}{\Delta_i^2}.$$ (5.48)

For a given guess for the C's, Eqs. (5.47) can be solved to determine what change in the C's will reduce χ^2, subject to the validity of the linearization. The new improved values can then be used as the next guess, the process being continued until χ^2 converges to a minimum. To solve Eqs. (5.47), it is convenient to define the $(N+1) \times (N+1)$ matrix

$$A_{nm} = A_{mn} = a_{nm}; \quad A_{n,N+1} = A_{N+1,n} = \frac{\partial Z}{\partial C_n}; \quad A_{N+1,N+1} = 0, \text{ (5.49a)}$$

and the $N+1$-component vector

$$B_n = b_n; \quad B_{N+1} = 0.$$ (5.49b)

(The indices m and n in these expressions range from 1 to N.) Equations (5.47) can therefore be written as

$$\sum_{m=1}^{N+1} A_{nm} \delta C_m = B_n,$$ (5.50)

which is simply solved by inversion of the symmetric matrix A:

$$\delta C_n = \sum_{m=1}^{N+1} A_{nm}^{-1} B_m = \sum_{m=1}^{N} A_{nm}^{-1} b_m.$$ (5.51)

After the process described above converges to the optimal values to the C's, we can then enquire into the precision with which the parameters are determined. This can be related to the matrix A defined above as follows. Let us consider an ensemble of data sets in which each experimental cross section fluctuates independently about its mean value. That is,

$$<\delta \sigma_i^e \delta \sigma_j^e> = \Delta_i^2 \delta_{ij},$$ (5.52)

where $< \cdots >$ denotes ensemble average, and $\delta \sigma_i^e$ is the deviation of σ_i^e from its mean value. Such fluctuations lead to fluctuations of the C's about their optimal value. Since the b_n are linearly related to the experimental cross sections by (5.48), these fluctuations are given by (5.51) as

$$\delta C_n = \sum_{m=1}^{N} A_{nm}^{-1} \delta b_m = \sum_{m=1}^{N} A_{nm}^{-1} \sum_{i=1}^{I} \frac{\delta \sigma_i^e W_{im}}{\Delta_i^2}. \qquad (5.53)$$

Upon using this equation twice (for δC_n and δC_m) and taking the ensemble average, we have, after some algebra,

$$<\delta C_n \delta C_m> = A_{nm}^{-1}. \qquad (5.54)$$

The uncertainties in the density parameters can be translated into a more physical statement about the correlated uncertainties in the density itself by expanding the parametrized density about the optimal parameters. In particular, the correlated uncertainties in the densities at radii r and r' can be written as

$$<\delta\rho(r)\delta\rho(r')> = \sum_{n=1}^{N} \sum_{m=1}^{N} \frac{\partial\rho(r)}{\partial C_n} \frac{\partial\rho(r')}{\partial C_m} <\delta C_n \delta C_m>. \qquad (5.55)$$

Thus, $<\delta\rho(r)\delta\rho(r)>^{1/2}$ can be taken as a measure of the precision with which $\rho(r)$ is determined, although it should be remembered that the uncertainties in the density determined at different spatial points are correlated through (5.55).

We now turn to the parametrization of $\rho(r)$. On general grounds (for example, the validity of treating the nucleus as a sharp-surface, incompressible liquid drop), we expect the charge density to be fairly uniform within the nuclear interior and to fall to zero within a relatively thin surface region. This picture was largely confirmed by early electron scattering experiments [Ho57], although the measured cross sections extended only to momentum transfers less than about 1.5 fm^{-1}, which is barely enough resolution to see the nuclear surface. Such data could therefore yield only limited information, and a suitable form for the density with only three parameters was found to be of the form of a Fermi function:

$$\rho(r) = \frac{\rho_0}{1 + e^{4.4(r-R_0)/t}}, \qquad (5.56)$$

as illustrated in Figure 5.1. Since it turns out that $4.4R_0/t \gg 1$, the parameter ρ_0 is essentially the central density (i.e., $\rho(r=0)$) the parameter R_0 is the radius where the charge density drops to one-half of its central value, and t, the surface thickness, is a measure of the distance over which this drop occurs. These three parameters are, of course, not all independent, as ρ_0 can be related to R_0 and t by the normalization constraint (5.31). Systematic studies of electron scattering from a number of nuclei showed that, for a nucleus with A nucleons, suitable values were $t = 2.4$ fm and $R_0 = 1.07A^{1/3}$ fm.

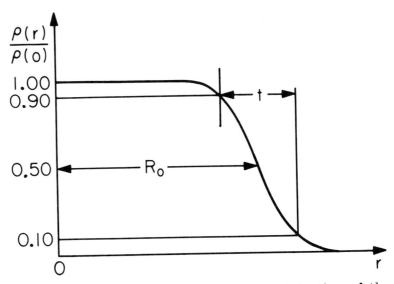

Figure 5.1 The Fermi function parametrization of the nuclear charge density. The density falls to one-half of its central value at the radius R_0; the distance over which the density drops from 90% to 10% of its central value is t. (After [Ho57].)

More modern studies of electron scattering use higher energy beams, and so cover momentum transfers to approximately 4 fm^{-1}. A typical example of the range and quality of the data is shown in Figure 5.2. Such data have led to a number of "model-independent" analyses [Fr75], in which a very flexible parametrization of the charge is used. One of these [Si74] takes the charge density to be the sum of many (of order 10) Gaussian "lumps", and adjusts the location of each lump and the amount of charge it carries to best fit the data. An alternative [Fr73], which we pursue here, assumes that the density vanishes outside a radius R, considerably beyond the nuclear surface, and expands the density as a finite Fourier series of N terms. That is, for $r < R$, we take

$$r\rho(r) = \sum_{n=1}^{N} C_n \sin\left[\frac{n\pi r}{R}\right],$$ (5.57)

and for $r > R$, we take $\rho(r) = 0$. Equations (5.32) and (5.38), together with some straightforward algebra, show that

$$Z(\{C_n\}) = 4R^2 \sum_{n=1}^{N} \frac{(-)^{n+1}}{n} C_n \, ,$$ (5.58)

and that

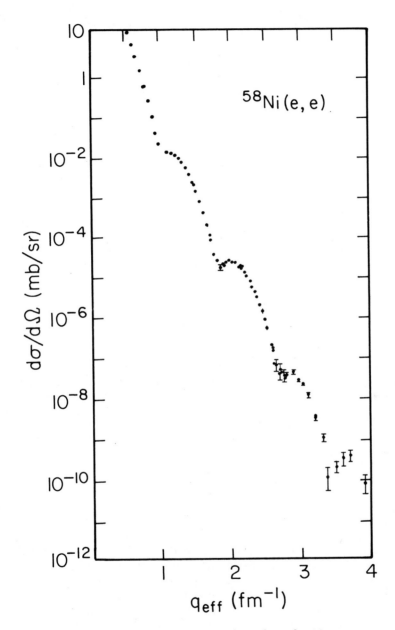

Figure 5.2 Experimental cross sections for the elastic scattering of electrons from ^{58}Ni at an energy of 449.8 MeV. (From [Ca80].)

$$F(q) = \frac{4\pi R^2}{Z} \sum_{n=1}^{N} C_n (-)^n \frac{n\pi}{(qR)^2 - n^2\pi^2} \frac{\sin qR}{qR}. \qquad (5.59)$$

These expressions, together with Eqs. (5.38,40), can be used to derive explicit forms for W_{in}, and hence for the χ^2 minimization procedure described above. It is useful to note that the n'th term in the Fourier expansion of the density gives a contribution to the form factor which is peaked near $q_n \equiv n\pi/R$. Hence, truncation of the series at N implies little control over Fourier components with q larger than $q_N = N\pi/R$. This maximum wavenumber contained in the parametrization must be commensurate with the maximum momentum transfer covered by the experimental data, q_{max}. If $q_N < q_{max}$, then data at larger momentum transfers will not be well described, while if $q_N > q_{max}$, then the C_n's with $n > q_{max}R/\pi$ will not be well-determined by the fit.

We now return to the adequacy of the Born approximation for describing the experimental data. For charge distributions of the shape expected (roughly like those in Figure 5.1), the relatively sharp nuclear surface induces zeros in $F(q)$. That is, there are particular momentum transfers (or equivalently, scattering angles) where the Born scattering amplitude vanishes. At these values of q, corrections associated with the distortion of the plane wave of the incident electron by the nuclear Coulomb potential become relatively more important, so that the experimental data has these zeros largely filled in, often to the extent that they become merely shoulders in a rapidly falling cross section (see Figure 5.2). A precision determination of the nuclear charge density from the measured cross sections therefore requires a more sophisticated description of the scattering of the electrons from the Coulomb potential.

The rigorous solution to this problem involves solving the Dirac equation for the electron in the Coulomb potential generated by the assumed charge distribution [Ye54, Fr73]. This results in a calculation very similar to that done in Project IV: a partial wave decomposition, solution of the radial equation in each partial wave to determine the phase-shift, and then summation over partial waves to determine the scattering amplitude at each scattering angle. While such a calculation is certainly possible, it is too large for a small computer. However, the eikonal approximation to Dirac scattering can be sufficiently accurate and results in a reasonable amount of numerical work.

The eikonal approximation for ultra-relativistic electrons scattering from a charge distribution results in final expressions very similar to those derived in Section 4.3 [Ba64]. The cross section is related to a complex scattering amplitude, f, by

$$\sigma = \cos^2 \tfrac{1}{2}\theta \, | f |^2, \tag{5.60}$$

and the scattering amplitude is given by the Fourier-Bessel transform (compare Eq. (4.30))

$$f = -ik \int\limits_0^\infty J_0(qb)[e^{2i\chi(b)} - 1]b\,db, \tag{5.61}$$

where the profile function is the integral of the potential along the straight-line trajectory at impact parameter b:

$$\chi = -\tfrac{1}{2} \int\limits_{-\infty}^{\infty} V(r)\,dz.$$

For the particular parametrization (5.57), ρ vanishes beyond R, and $V(r)$ is simply $-Z\alpha/r$ in this region. It is therefore convenient to split the impact parameter integral (5.61) into regions $b < R$ and $b > R$, so that the latter integral can be done analytically. This results in

$$f = f_{outer} + f_{inner} = f_{outer} - ik \int\limits_0^R J_0(qb)[e^{2i\chi(b)} - 1]b\,db, \tag{5.62}$$

where

$$f_{outer} = \frac{ik}{q^2}\Big[-2i\,\alpha Z X J_0(X) X^{2i\alpha Z} S_{-2i\alpha Z,-1}(X) \tag{5.63}$$

$$+ X J_1(X) X^{2i\alpha Z} S_{1-2i\alpha Z,0}(X) - X J_1(X) \Big], \tag{5.63}$$

with $X = qR$. In this expression, $S_{\mu\nu}$ is a Lommel function (essentially the incomplete integral of the product of a Bessel function and a power), which has the useful asymptotic expansion for large X

$$S_{\mu\nu}(X) \sim X^{\mu-1}\Big[1 - \frac{(\mu-1)^2 - \nu^2}{X^2} $$
$$+ \frac{\{(\mu-1)^2 - \nu^2\}\{(\mu-3)^2 - \nu^2\}}{X^4} - \cdots \Big]. \tag{5.64}$$

Note that since we will typically have $R \gtrsim 8$ fm and $q \gtrsim 1$ fm^{-1}, X will be 8 or more, and so the number of terms shown should be quite sufficient.

The interior contribution to the scattering amplitude, f_{inner}, involves the potential for $r < R$ and so depends upon the detailed form of the nuclear charge density. Given the simple relation between the potential and density through Poisson's equation (5.33), it should not be too surprising that it is possible to express

$\chi(b < R)$ directly in terms of the density:

$$\chi(b) = -Z\alpha\log\left[\frac{b}{R}\right] - 4\pi\alpha\int_b^R r^2\rho(r)\varphi(b/r)dr, \qquad (5.65)$$

where

$$\varphi(y) \equiv \log\left[\frac{1+(1-y^2)^{\frac{1}{2}}}{y}\right] - (1-y^2)^{\frac{1}{2}}.$$

The parametrization (5.57) then allows the profile function to be written in the form

$$\chi(b) = -Z\alpha\log\left[\frac{b}{R}\right] + \sum_{n=1}^N C_n\chi_n(b); \qquad (5.66a)$$

$$\chi_n(b) = -4\pi\alpha R^2\int_{b/R}^1 zdz\sin(n\pi z)\varphi(b/zR), \qquad (5.66b)$$

with the change of integration variable $z = r/R$.

Equations (5.61-64,66) complete the specification of the eikonal cross section in terms of the nuclear charge density. It is also easy to show that the quantities W_{in} are given by

$$W_{in} = 2\cos^2\tfrac{1}{2}\theta_i\ \mathrm{Re}\left[f_i^*\frac{\partial f_i}{\partial C_n}\right]; \qquad (5.67a)$$

$$\frac{\partial f_i}{\partial C_n} = 2k\int_0^R J_0(q_ib)e^{2i\chi(b)}\chi_n(b)bdb. \qquad (5.67b)$$

Several fine points remain to be discussed before turning to the program. One is that the nucleus is not infinitely heavy, and so recoils when struck by the electron. To correct for this, the cross sections calculated above, which have all been for an electron impinging on a static charge distribution, must be divided by the factor

$$\eta \equiv 1 + 2\frac{E}{M}\sin^2\tfrac{1}{2}\theta,$$

where M is the target mass (roughly $A\times940$ MeV). Note that this correction vanishes as M becomes large compared to E. Second, because of the Coulomb attraction, electrons at the nucleus have a slightly higher momentum than the beam value. In the eikonal treatment, this can be approximately corrected for by replacing q in all of the formulas by

$$q_{eff} = q\left(1 - \frac{\overline{V}}{E}\right), \qquad (5.68)$$

and by multiplying the eikonal scattering amplitude by $(q_{eff}/q)^2$.

Here, $\bar{V}$ is a suitable average of the Coulomb potential over the nuclear volume, conveniently given by

$$\bar{V} = -\frac{4}{3} \frac{Z\alpha}{R_0},$$

where R_0 is the nuclear radius. Similarly, in the Born approximation (5.40), the nuclear form factor is to be evaluated at q_{eff}, but the Mott cross section is still to be evaluated at q.

The BASIC program for Example 5, whose listing can be found in Appendix B and in the file EXAM5.BAS on the *Computational Physics* diskette, uses the method described above to analyze electron scattering cross sections for the nuclei ^{40}Ca ($Z=20$), ^{58}Ni ($Z=28$), and ^{208}Pb ($Z=82$) to determine their charge densities. Many different data sets for each target, taken at a variety of beam energies and angles (see, for example, [Fr77]), have been converted for equivalent cross sections at one beam energy [Ca80]. The integrals (5.62,66b,67b) are evaluated by 20-point Gauss-Legendre quadrature and the matrix inversion required by (5.51) is performed by an efficient implementation of Gauss-Jordan elimination [Ru63].

After requesting the target nucleus, the boundary radius R, and the number of sine functions N, the program iterates the procedure described above, displaying at each iteration plots of the fit, fractional error in the fit, and the charge density of the nucleus, together with the values of the C's and the density. The program also allows for initial iterations with fewer sines than are ultimately required, the number then being increased every few iterations. This technique improves the convergence of the more rapidly oscillating components of the density as it reduces the effective dimension of the χ^2 search by first converging the smoother density components. Note that, at each iteration, only points at an effective momentum transfer smaller than that described accurately by the current number of sines are used in calculating χ^2.

The following exercises will be useful in understanding the physical and numerical principles important in this calculation.

Exercise 5.8 To check that the program is working properly, use □ □ □ the eikonal formulas and Fermi-function charge density described above to generate a set of "pseudo-data" cross sections ranging from q between 0.5 and 4 fm^{-1}. Make sure that these cross sections are given at intervals typical of the experimental data and also assign errors to them typical of the data. Then verify the extent to which the correct (i.e., Fermi-function) charge distribution results when these pseudo-data are used as input to the program. One potential source of error you should investigate is the "completeness error". That is, the ability of a finite Fourier series to reproduce densities of the form expected. This can be studied by changing the number of sines used in the fit.

☐ ☐ ☐ **Exercise 5.9** Run the code with the actual experimental data to determine the charge densities for the three nuclei treated. Study the quality of the fit and the errors in the density extracted as functions of the number of expansion terms used. Also verify that the fitting procedure converges to the same final solution if the initial guess for the density is varied within reasonable limits. Note that the converged solutions do not have a uniform interior density, but rather show "wiggles" due to the specific shell-model structure of each nucleus.

☐ ☐ ☐ **Exercise 5.10** Extend the program to calculate the moments of the density,

$$M_k \equiv \left[\frac{4\pi}{Z} \int_0^\infty r^{2+k} \rho(r) dr \right]^{1/k},$$

and their uncertainties for integer k ranging from 2 to 5. This can be done conveniently by using Eq. (5.57) to express these moments directly in terms of the C_n. Verify that the size of these moments is consistent with an $A^{1/3}$ scaling of the nuclear radius.

☐ ☐ ☐ **Exercise 5.11** A simple model for the doubly-magic nucleus ^{40}Ca is a Slater determinant constructed by putting four nucleons (neutron and proton, spin up and down) in the 10 lowest orbitals of a spherically-symmetric harmonic oscillator potential. Show that the charge density predicted by this model is

$$\rho(r) = \frac{1}{\pi^{3/2} r_0^3} e^{-(r/r_0)^2} \left[5 + 4 \left(\frac{r}{r_0} \right)^4 \right],$$

where $r_0 = (\hbar/m\omega)^{1/2}$, m being the nucleon mass and ω being the harmonic oscillator frequency. Determine the value of r_0 (and hence ω) required to reproduce the experimental value of the root-mean-square radius (M_2), and then compare the detailed form of ρ predicted by this model with that obtained from the data.

☐ ☐ ☐ **Exercise 5.12** Modify the program so that it calculates the Born, rather than eikonal, cross section and show that the former is grossly inadequate to describe the data by comparing the Born cross sections for a Fermi distribution with the experimental data.

Exercise 5.13 Modify the program to calculate the "off-diagonal" ▫ ▫ ▫
uncertainties in the density determined using Eq. (5.55) and show
that the correlation in the uncertainties largely decreases as the
separation of the points increases.

Project V: A schematic shell model

The assumption of particles moving independently in a potential is a common approximation in treating the quantum mechanics of many-particle systems. For example, this picture underlies
the Hartree-Fock method for treating the many electrons in an
atom (see Project III), the shell model for treating the many
nucleons in a nucleus, and many models for treating the quarks in
a hadron. However, it is only a rough approximation in many cases
and a detailed description of the exact quantum eigenstates often
requires consideration of the "residual" interactions between the
particles making up the system. In atoms and hadrons, the residual interactions are not strong enough to induce qualitative
changes in the spectrum expected from simply placing the particles in different orbitals. In nuclei, however, the coherence and
strength of the residual interactions can lead to a "collective"
behavior in which many nucleons participate in the excitations
and the character of the eigenstates is quite different from that
expected naively.

Realistic calculations of the effects of residual interactions are
quite complex and involve the diagonalization of large matrices
expressing the way in which the interaction moves particles
among the orbitals [Mc80]. However, the phenomenon of collec-
tivity can be illustrated by a schematic shell model introduced by
by Lipkin *et al.* [Li65], which we consider in this project. The
model is non-trivial, yet simple enough to be soluble with only a
modest numerical diagonalization. It has therefore served as a
testing ground for approximations to be used in treating many-
body systems and is also the prototype for more sophisticated
group-theoretic models of nuclear spectra [Ar81].

V.1 Definition of the model

The schematic model consists of N distinguishable particles
labeled by $n = 1, 2, , \cdots, N$. Each of these can occupy one of two
orbitals, a lower and an upper, having energies $+\frac{1}{2}$ and $-\frac{1}{2}$ and dis-
tinguished by $s = -1$ and $s = +1$, respectively (see Figure V.1).
There are then 2^N states in the model, each defined by which particles are "up" and which are "down", and each having an

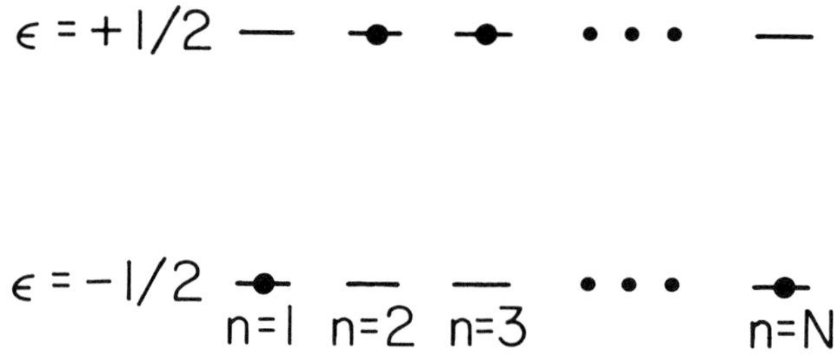

$$\epsilon = +1/2 \ \text{—} \quad \bullet \!\text{—} \quad \bullet \!\text{—} \quad \bullet \bullet \bullet \quad \text{—}$$

$$\epsilon = -1/2 \ \underset{\substack{\text{—}\bullet\\ n=1}}{} \ \underset{\substack{\text{—}\\ n=2}}{} \ \underset{\substack{\text{—}\\ n=3}}{} \quad \bullet\bullet\bullet \quad \underset{\substack{\text{—}\bullet\\ n=N}}{}$$

Figure V.1 Illustration of the orbitals in the schematic shell model

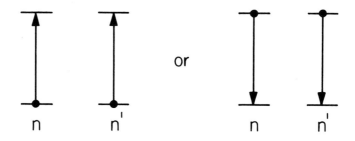

or

Figure V.2 Illustration of the residual interaction in the schematic model

unperturbed energy (in the absence of the residual interaction) equal to one-half of the difference between the number up and the number down.

The residual interaction in the schematic model is one that changes simultaneously the states of any pair of particles; it therefore couples states of the unperturbed system. In particular, it is convenient to take a residual interaction of strength $-V$ that promotes any two particles that are "down" to "up" or that drops any two that are "up" to "down", as illustrated in Figure V.2. An interaction that simultaneously raises one particle and lowers another can also be included, but introduces no qualitatively new features.

To define the model precisely, it is easiest to use the language of second quantization and introduce the operators $a_{ns}^{\dagger}$ and their adjoints, a_{ns}, which respectively create and annihilate a particle in the orbital ns. As the model is constructed so that no more than one particle can occupy each orbital, it is convenient to endow these operators with the usual fermion anti-commutation relations,

$$\{a_{ns}^{\dagger}, a_{n's'}^{\dagger}\}=0,\qquad\qquad\text{(V.1a)}$$

$$\{a_{ns}, a_{n's'}\}=0,\qquad\qquad\text{(V.1b)}$$

$$\{a_{ns}^{\dagger}, a_{n's'}\}=\delta_{nn'}\delta_{ss'}.\qquad\qquad\text{(V.1c)}$$

The operator $a_{ns}^{\dagger}a_{ns}$ then counts the number of particles in the orbital ns (i.e., it has eigenvalues 0 and 1), while the operators $a_{n1}^{\dagger}a_{n-1}$ and $a_{n-1}^{\dagger}a_{n1}$ respectively raise or lower the n'th particle. In terms of these operators, the Hamiltonian for the model can be written as

$$H=\tfrac{1}{2}\sum_{n=1}^{N}(a_{n1}^{\dagger}a_{n1}-a_{n-1}^{\dagger}a_{n-1})$$

$$-\tfrac{1}{2}V\sum_{n=1}^{N}\sum_{n'=1}^{N}(a_{n1}^{\dagger}a_{n-1}a_{n'1}^{\dagger}a_{n'-1}+a_{n-1}^{\dagger}a_{n1}a_{n'-1}^{\dagger}a_{n'1}),\qquad\text{(V.2)}$$

where we need not worry about the unphysical $n'=n$ term in the residual interaction since an attempt to raise or lower the same particle twice yields a vanishing state.

V.2 The exact eigenstates

The Hamiltonian described above can be represented as a matrix coupling the 2^N states among themselves. Apart from symmetry and a vanishing of many of its elements, this matrix is not obviously special, so that its numerical diagonalization to find the exact eigenstates and eigenvalues is a substantial problem for all but the smallest values of N. Fortunately, the relatively simple structure of the problem allows it to be expressed in the familiar language of the group SU(2) (the quantum angular momentum operators), a transformation that carries us a good deal of the way to an exact diagonalization and also affords some insight into the problem.

Let us define the operators

$$J_z=\tfrac{1}{2}\sum_{n=1}^{N}(a_{n1}^{\dagger}a_{n1}-a_{n-1}^{\dagger}a_{n-1}),\qquad\qquad\text{(V.3a)}$$

$$J_+=\sum_{n=1}^{N}a_{n1}^{\dagger}a_{n-1},\quad J_-=(J_+)^{\dagger}=\sum_{n=1}^{N}a_{n-1}^{\dagger}a_{n1}.\qquad\text{(V.3b)}$$

Thus, J_z measures (one-half of) the difference between the number of "up" particles and the number of "down" particles, while J_+ and J_- coherently raise or lower all of the particles. Using these operators, the Hamiltonian (V.2) can be written as

$$H=J_z-\tfrac{1}{2}V(J_+^2+J_-^2)=J_z-V(J_x^2-J_y^2),\qquad\text{(V.4)}$$

where we have introduced the operators

$$J_x = \tfrac{1}{2}(J_+ + J_-), \quad J_y = -\tfrac{1}{2}i(J_+ - J_-). \tag{V.5}$$

Using the fundamental anti-commutation rules (V.1), it is easy to show that the three operators J_z, $J_\pm$ satisfy the commutation rules of a quantum angular momentum:

$$[J_z, J_\pm] = \pm J_\pm, \quad [J_+, J_-] = 2J_z. \tag{V.6}$$

Thus, although these "quasi-spin" operators have nothing to do with a physical angular momentum, all of the techniques and experience in dealing with quantum spin operators can be applied immediately to this problem.

We can begin by realizing that since the total quasi-spin operator

$$J^2 = J_x^2 + J_y^2 + J_z^2 \tag{V.7}$$

commutes with $J_{x,y,z}$, it commutes with the Hamiltonian and thus each eigenstate of H can be labeled by its total quasi-spin, j, where $j(j+1)$ is the eigenvalue of J^2. The eigenstates therefore can be classified into non-degenerate multiplets of $2j+1$ states each, all states in a multiplet having the same j; the Hamiltonian has a non-vanishing matrix element between two states only if they belong to the same multiplet. However, since H does not commute with J_z, its eigenstates will not be simultaneously eigenstates of this latter operator.

To see what values of j are allowed, we can classify the 2^N states of the model (not the eigenstates of H) according to their eigenvalues of J_z. The one state with the largest eigenvalue of J_z has all particles up, corresponding to an eigenvalue $m = \tfrac{1}{2}N$. We therefore expect one multiplet of $2 \cdot \tfrac{1}{2}N + 1 = N+1$ states corresponding to $j = \tfrac{1}{2}N$. Turning next to states with $m = \tfrac{1}{2}N - 1$, we see that there are N states, corresponding to one of the N particles down and all of the rest up. One linear combination of these N states (the totally symmetric one) belongs to the $j = \tfrac{1}{2}N$ multiplet, implying that there are $N-1$ multiplets with $j = \tfrac{1}{2}N - 1$. Continuing in a similar way, there are $\tfrac{1}{2}N(N-1)$ states with $m = \tfrac{1}{2}N - 2$ (two particles down, the rest up), of which one linear combination belongs to the $j = \tfrac{1}{2}N$ multiplet and $N-1$ linear combinations belong to the $j = \tfrac{1}{2}N - 1$ multiplets. There are thus $\tfrac{1}{2}N(N-3)$ multiplets with $j = \tfrac{1}{2}N - 2$. By continuing in this way, we can classify the 2^N states of the model into multiplets with j running from $\tfrac{1}{2}N$ down to 0 or $\tfrac{1}{2}$, if N is even or odd, respectively.

Because H involves only the quasi-spin operators, its action within a multiplet depends only upon the value of j involved. For a given N, the spectrum of one multiplet therefore serves for all with the same j and, in fact, also serves for multiplets with this j in systems of larger N. We can therefore restrict our attention to the multiplet with the largest value of j, $\frac{1}{2}N$.

By these considerations, we have reduced the problem of diagonalizing the full Hamiltonian to one of diagonalizing (V.4) in the $N+1$ eigenstates of J_z belonging to the multiplet with $j=\frac{1}{2}N$. We can label these states as $|m\rangle$, where m runs from $-j$ to j in integer steps. Using the usual formulas for the action of the angular momentum raising and lowering operators,

$$J_{\pm}|m\rangle = C_m^{\pm}|m\pm 1\rangle; \quad C_m^{\pm} = [j(j+1)-m(m\pm 1)]^{\frac{1}{2}}, \qquad (V.8)$$

we can write the elements of H in this basis as

$$\langle m'|H|m\rangle = m\delta_{m'm}$$
$$-\tfrac{1}{2}V[C_m^+ C_{m+1}^+ \delta_{m'm+2} + C_m^- C_{m-1}^- \delta_{m'm-2}]. \qquad (V.9)$$

Thus, H is tri-diagonal in this basis and in fact separates into two uncoupled problems involving the states $m=-j, -j+2, \cdots, +j$ and $m=-j+1, -j+3, \cdots, +j-1$ when N is even (j is integral) and the states $m=-j, -j+2, \cdots, +j-1$ and $m=-j+1, -j+3, \cdots, +j$ when N is odd (j is half-integral). For small values of j, the resulting matrices can be diagonalized analytically, while for larger systems, the numerical methods discussed in this chapter can be employed to find the eigenvalues and eigenvectors.

The quasi-spin method allows us to make one other statement about the exact solution of the model. Let

$$R = e^{i\pi(J_x + J_y)/2^{\frac{1}{2}}}$$

be the unitary operator effecting a rotation in quasi-spin space by an angle of π about the axis $(\hat{x}+\hat{y})/2^{\frac{1}{2}}$. It is easy to see that this rotation transforms the quasi-spin operators as

$$RJ_x R^\dagger = J_y; \quad RJ_y R^\dagger = J_x; \quad RJ_z R^\dagger = -J_z, \qquad (V.10)$$

so that the Hamiltonian (V.4) transforms as

$$RHR^\dagger = -H. \qquad (V.11)$$

Hence, if ψ is an eigenstate of H with energy E, then $R\psi$ is an eigenstate of H with energy $-E$. Thus, the eigenvalues of H come in pairs of equal magnitude and opposite sign (or are 0). This also allows us to see that if N is even, at least one of the eigenvalues will be zero.

V.3 Approximate eigenstates

We turn now to approximate methods for solving the schematic model and consider first ordinary Rayleigh-

Schroedinger perturbation theory. If we treat the J_z term in (V.4) as the unperturbed Hamiltonian and the $J_x^2 - J_y^2$ term as the perturbation, then the unperturbed eigenstates are $|m\rangle$ with unperturbed energies $E_m^{(0)} = m$ and the perturbation series is an expansion in powers of V. Since $\langle m | J_\pm^2 | m \rangle = 0$, the energies are unperturbed in first order and are perturbed in second order as

$$\Delta E_m^{(2)} = \frac{\langle m+2 | \tfrac{1}{2} V J_+^2 | m \rangle^2}{E_m^{(0)} - E_{m+2}^{(0)}} + \frac{\langle m-2 | \tfrac{1}{2} V J_-^2 | m \rangle^2}{E_m^{(0)} - E_{m-2}^{(0)}}. \qquad (V.12)$$

Using Eq. (V.9) for the matrix elements involved, explicit expressions for the second-order energies can be derived; the fourth-order terms can also be done with a bit of patient algebra.

While the weak-coupling (small-V) limit can be treated by straightforward perturbation theory, an approximation for the strong-coupling (large-V) situation is not so obvious. One appealing approach is the semiclassical method, valid in the limit of large N [Ka79, Sh80]. The discussion begins by considering the equations of motion for the time-dependence of the expectation values of the quasi-spin operators:

$$i\frac{d}{dt}\langle J_i \rangle = \langle [J_i, H] \rangle, \qquad (V.13)$$

where J_i is any one of the three components of the quasi-spin. Using the Hamiltonian (V.4) and the commutation rules (V.6), it is easy to write out Eq. (V.13) for each of the three components:

$$\frac{d}{dt}\langle J_x \rangle = -\langle J_y \rangle + V \langle J_z J_y + J_y J_z \rangle, \qquad (V.14a)$$

$$\frac{d}{dt}\langle J_y \rangle = \langle J_x \rangle + V \langle J_z J_x + J_x J_z \rangle, \qquad (V.14b)$$

$$\frac{d}{dt}\langle J_z \rangle = -2V \langle J_y J_x + J_x J_y \rangle. \qquad (V.14c)$$

Unfortunately, these equations do not form a closed set, as the time derivatives of expectation values of single quasi-spin operators depend upon expectation values of bi-linear products of these operators. However, if we ignore the fact that the J's are operators, and put, for example,

$$\langle J_x J_y \rangle = \langle J_x \rangle \langle J_y \rangle,$$

then a closed set of equations results:

$$\frac{d}{dt}\langle J_x \rangle = -\langle J_y \rangle + 2V \langle J_z \rangle \langle J_y \rangle, \qquad (V.15a)$$

$$\frac{d}{dt}\langle J_y \rangle = \langle J_x \rangle + 2V \langle J_z \rangle \langle J_x \rangle, \qquad (V.15b)$$

$$\frac{d}{dt}\langle J_z\rangle=-4V\langle J_y\rangle\langle J_x\rangle.\tag{V.15c}$$

It is easy to show that Eqs. (V.15) conserve

$$J^2\equiv\langle J_x\rangle^2+\langle J_y\rangle^2+\langle J_z\rangle^2,$$

so that a convenient parametrization is in terms of the usual polar angles,

$$\langle J_z\rangle=J\cos\theta,\ \ \langle J_x\rangle=J\sin\theta\cos\varphi,\ \ \langle J_y\rangle=J\sin\theta\sin\varphi,\tag{V.16}$$

where

$$J=[j(j+1)]^{\frac{1}{2}}\approx\tfrac{1}{2}N.$$

If we further define the variables

$$p\equiv\tfrac{1}{2}N\cos\theta,\ \ q\equiv\varphi+\frac{1}{4}\pi,\tag{V.17}$$

so that $|p|\leq\tfrac{1}{2}N$ and $0\leq q\leq2\pi$, then, upon introducing the convenient coupling constant

$$\chi\equiv NV\approx2JV,$$

Eqs. (V.15) can be written as

$$\frac{dp}{dt}=2\frac{\chi}{N}(\frac{N^2}{4}-p^2)\cos2q=-\frac{\partial E(p,q)}{\partial q},\tag{V.18a}$$

$$\frac{dq}{dt}=1+2\frac{\chi}{N}p\sin2q=\frac{\partial E(p,q)}{\partial p}.\tag{V.18b}$$

Here, E is the expectation value of the Hamiltonian

$$E(p,q)\equiv\langle H\rangle=\langle J_z\rangle-V(\langle J_x\rangle^2-\langle J_y\rangle^2)\tag{V.19a}$$

$$=p-\frac{\chi}{N}(\frac{N^2}{4}-p^2)\sin2q.\tag{V.19b}$$

By these manipulations, we have transformed the quasi-spin problem into one of a time-dependent classical system with "momentum" p and "coordinate" q satisfying equations of motion (V.18). As these equations are of the canonical form, p and q are canonically conjugate and (V.19b) can be identified as the Hamiltonian for this classical system.

We can now infer properties of the quantum eigenstates by analyzing this classical system. Using Eq. (V.19b), we can calculate the trajectories in phase-space; i.e., the value of p for a given q and $E=\langle H\rangle$. Equivalently, we can consider contour lines of E in the (p,q) plane. In an analysis of these contours, it is useful to distinguish between weak ($\chi<1$) and strong ($\chi>1$) coupling cases.

Figure V.3 shows a typical weak-coupling case, $N=8$ and $\chi=0.25$. The trajectory with the lowest allowed energy, which we

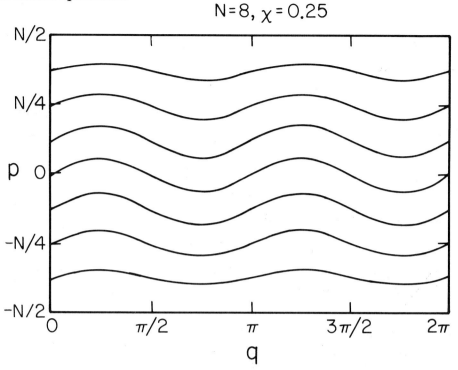

Figure V.3 Contours of the Hamiltonian (V.19b) for the weak coupling case $N=8$ and $\chi=0.25$. The contour passing through $(p=0,\ q=0)$ has zero energy. The upper and lower boundary lines $p=\pm\tfrac{1}{2}N$ are contours at energy $\pm\tfrac{1}{2}N$, respectively. Sucessive contours shown differ in energy by 1. (From [Ka79].)

can identify with the ground state of the quantum system, has $p=-\tfrac{1}{2}N$, $E=-\tfrac{1}{2}N$, and q satisfying the differential equation (V.18b). Note that this trajectory ranges over all values of q.

In the strong coupling case, $\chi>1$, the trajectories look as shown in Figure V.4, which corresponds to $N=8$ and $\chi=2.5$. There are now minima in the energy surface at $p=-\tfrac{1}{2}N/\chi$, $q=\pi/4$, $5\pi/4$, which, using the equations of motion (V.18), can be seen to correspond to stationary points (i.e., time-independent solutions of the equations of motion). The energy at these minima is $E=-N(\chi+\chi^{-1})/4$, which we can identify as an estimate of the ground state energy. There is therefore a "phase transition" at $\chi=1$ where the ground state changes from one where $p=-\tfrac{1}{2}N$ to one where $p>-\tfrac{1}{2}N$, the energy being continuous across this transition. This qualitative change in the nature of the ground state is a direct consequence of the coherence and strength of the

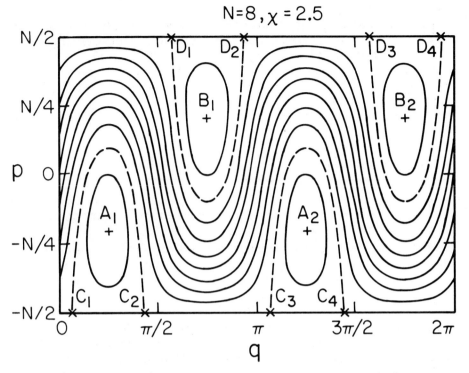

Figure V.4 Similar to Figure V.3 for the strong coupling case $N=8$, $\chi=2.5$. The points $A_{1,2}$, $B_{1,2}$ are minima and maxima, respectively, and C_{1-4}, D_{1-4} are saddle points. Contours through the saddle points, which have energy $\pm\frac{1}{2}N$, are drawn as dashed lines; these separate trajectories that are localized and extended in q. Successive contours differ in energy by 1. (From [Ka79].)

interactions among the particles. It should also be noted that, in direct analogy with the usual quantum double-well problem, the presence of two degenerate minima in Figure V.4 suggests that the ground and first-excited states will be nearly degenerate, the splitting between them caused only by "tunneling" between the wells.

The semiclassical analysis can be extended to predict the energies of excited states as well, essentially by applying the analog of the Bohr-Sommerfeld quantization rule given in Eqs. (1.20,21) [Sh80]. One finds that N controls the quantum nature of the problem, with the system becoming more "classical" as the number of particles increases. In the weak coupling case, all trajectories extend over the full range of q. However, for strong coupling, there are two types of trajectories: those that vibrate around the energy minima or maxima (and are hence confined to limited regions of q) and those that extend throughout all q. It is easy to show from (V.19b) that the transitions between these two behaviors occur at $|E|=J$ (e.g., the dashed lines in Figure V.4). In analogy with the usual quantum double-well, we expect states associated with the confined trajectories to occur in nearly degenerate pairs, the degeneracy being broken more strongly as the energy increases above the minima or decreases below the maxima.

V.4 Solving the model

The simple semiclassical picture we have discussed above can be tested by an analysis of the exact eigenstates of the model. This can be carried out through the following sequence of steps.

□ □ □ **Step 1** Verify the algebra in the discussion above. In particular, show that Eq. (V.4) is a valid representation of the Hamiltonian. Also evaluate the second-order perturbation to the energies, Eq. (V.12).

□ □ □ **Step 2** Verify the correctness of Eqs. (V.14,18,19) and the discussion of the energy contours in the strong and weak coupling cases. In the strong coupling case, linearize the equations of motion about the minima or maxima and show that the frequency of harmonic motion about these points is $[2(\chi^2-1)]^{\frac{1}{2}}$. This is the spacing expected between the pairs of nearly degenerate states in the exact spectrum.

□ □ □ **Step 3** Write a program that finds the eigenvalues of the tridiagonal Hamiltonian matrix (V.9), perhaps by modifying that written for Exercise 5.4. Note that the numerical work can be reduced by treating basis states with even and odd m separately and by using the symmetry of the spectrum about $E=0$.

□ □ □ **Step 4** Use the program written in Step 3 to calculate the spectrum of the model for selected values of N between 8 and 40 and for values of χ between 0.1 and 5. At weak coupling, compare your results with the perturbation estimates (V.12). For strong couplings, compare your ground state energy with the semiclassical estimate and verify the expected pairwise degeneracy of the low-lying states. Also compare the excitation energies of these states with your estimate in Step 2. How does the accuracy of the semiclassical estimate change with N?

□ □ □

Step 5 Write a program that takes the eigenvalues found in Step 2 and solves for the eigenvectors by inverse vector iteration (Eq. (5.27)). For selected values of N and χ, compute the expectation value of J_z for each eigenvector. In a semiclassical interpretation, this expectation value can be identified with the time-average of p over the associated trajectory in phase space. Verify that $\langle J_z \rangle$ changes through the spectrum in accord with what you expect from Figures V.3, V.4.

Chapter 6

Elliptic
Partial Differential
Equations

Partial differential equations are involved in the description of virtually every physical situation where quantities vary in space or in space and time. These include phenomena as diverse as diffusion, electromagnetic waves, hydrodynamics, and quantum mechanics (Schroedinger waves). In all but the simplest cases, these equations cannot be solved analytically and so numerical methods must be employed for quantitative results. In a typical numerical treatment, the dependent variables (such as temperature or electrical potential) are described by their values at discrete points (a lattice) of the independent variables (e.g., space and time) and, by appropriate discretization, the partial differential equation is reduced to a large set of difference equations. Although these difference equations then can be solved, in principle, by the direct matrix methods discussed in Chapter 5, the large size of the matrices involved (dimension comparable to the number of lattice points, often more than several thousand) makes such an approach impractical. Fortunately, the locality of the original equations (i.e., they involve only low-order derivatives of the dependent variables) makes the resulting difference equations "sparse" in the sense that most of the elements of the matrices involved vanish. For such matrices, iterative methods of inversion and diagonalization can be very efficient. These methods are the subject of this and the following chapter.

Most of the physically important partial differential equations are of second order and can be classified into three types: parabolic, elliptic, or hyperbolic. Roughly speaking, parabolic equations involve only a first-order derivative in one variable, but have second order derivatives in the remaining variables. Examples are the diffusion equation and the time-dependent Schroedinger equation, which are first order in time, but second order in space. Elliptic equations involve second order derivatives in each of the independent variables, each derivative having the same sign when all terms in the equation are grouped on one side. This class includes Poisson's equation for the electrostatic potential and the time-independent Schroedinger equation, both in two or more spatial variables. Finally, there are hyperbolic equations, which involve second derivatives of opposite sign, such as the wave equation describing the vibrations of a stretched string. In this chapter, we will discuss some numerical methods appropriate for elliptic equations, reserving the discussion of parabolic equations for Chapter 7. Hyperbolic equations often can be treated by methods similar to those we discuss, although unique difficulties do arise [Ri67].

For concreteness in our discussion, we will consider particular forms of elliptic boundary value and eigenvalue problems for a field φ in two spatial dimensions (x,y). The boundary value problem involves the equation

$$-\left[\frac{\partial^2}{\partial x^2}+\frac{\partial^2}{\partial y^2}\right]\varphi=S(x,y). \tag{6.1}$$

Although this is not the most general elliptic form, it nevertheless covers a wide variety of situations. For example, in an electrostatics problem, φ is the potential and S is related to the charge density (compare Eq. (3.2)), while in a steady-state heat diffusion problem, φ is the temperature, and S is the local rate of heat generation or loss. Our discussion can be generalized straightforwardly to other elliptic cases, such as those involving three spatial dimensions or a spatially varying diffusion or dielectric constant.

Of course, Eq. (6.1) by itself is an ill-posed problem, as some sort of boundary conditions are required. These we will take to be of the Dirichlet type; that is, φ is specified on some large closed curve in the (x,y) plane (conveniently, the unit square) and perhaps on some additional curves within it (see Figure 6.1). The boundary value problem is then to use (6.1) to find φ everywhere within the unit square. Other classes of boundary value problems, such as Neumann (where the normal derivative of φ is specified on the surfaces) and mixed (where a linear combination of φ and its normal derivative is specified), can be handled by very similar methods.

The eigenvalue problems we will be interested in might involve an equation of the form

$$-\left[\frac{\partial^2}{\partial x^2}+\frac{\partial^2}{\partial y^2}\right]\varphi+V(x,y)\varphi=\varepsilon\varphi, \tag{6.2}$$

together with a set of Dirichlet boundary conditions. This might arise, for example, as the time-independent Schroedinger equation, where φ is the wavefunction, V proportional to the potential, and ε is related to the eigenvalue. Such an equation might also be used to describe the fields in an acoustic or electromagnetic

y = |

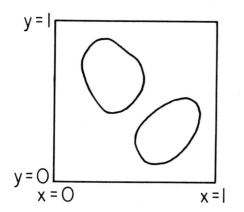
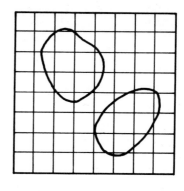

y = 0

x = 0   x = |

Figure 6.1 (Left) A two-dimensional boundary value problem with Dirichlet boundary conditions. Values of φ are specified on the edges of the unit square and on the surfaces within. (Right) Discretization of the problem on a uniform cartesian lattice.

waveguide, where ε is then related to the square of the cut-off frequency. The eigenvalue problem then consists of finding the values ε_λ and the associated eigenfunctions φ_λ for which Eq. (6.2) and the boundary conditions are satisfied. As methods for solving such problems are closely related to those for solving a related parabolic equation, we will defer their discussion to Chapter 7.

6.1 Discretization and the variational principle

Our first step is to cast Eq. (6.1) in a form suitable for numerical treatment. To do so, we define a lattice of points covering the region of interest in the (x,y) plane. For convenience, we take the lattice spacing, h, to be uniform and equal in both directions, so that the unit square in covered by $(N+1)\times(N+1)$ lattice points (see Figure 6.1). These points can be labeled by indices (i,j), each of which runs from 0 to N, so that the coordinates of the point (i,j) are $(x_i=ih, y_j=jh)$. If we then define $\varphi_{ij}=\varphi(x_i,y_j)$, and similarly for S_{ij}, it is then straightforward to apply the three-point difference approximation (1.7) for the second derivative in each direction and so approximate (6.1) as

$$-\left[\frac{\varphi_{i+1j}+\varphi_{i-1j}-2\varphi_{ij}}{h^2}+\frac{\varphi_{ij+1}+\varphi_{ij-1}-2\varphi_{ij}}{h^2}\right]=S_{ij}, \qquad (6.3a)$$

or, in a more convenient notation,

$$-\left[(\delta_i^2\varphi)_{ij}+(\delta_j^2\varphi)_{ij}\right]=h^2 S_{ij}. \qquad (6.3b)$$

Here, δ_i^2 is the second-difference operator in the i index,

$$(\delta_i^2\varphi)_{ij} \equiv \varphi_{i+1j} + \varphi_{i-1j} - 2\varphi_{ij},$$

and δ_j^2 is defined similarly.

Although Eq. (6.3) is the equation we will be solving, it is useful to derive it in a different way, based on a variational principle. Such an approach is handy in cases where the coordinates are not cartesian, as discussed in Section 6.3 below, or when more accurate difference formulas are required. It also is guaranteed to lead to symmetric (or hermitian) difference equations, an often useful property. The variational method also affords some insight into how the solution algorithm works. A good review of this approach can be found in [Ad84].

Consider the quantity E, defined to be a functional of the field φ of the form

$$E = \int_0^1 dx \int_0^1 dy \left[\tfrac{1}{2}(\nabla\varphi)^2 - S\varphi \right]. \tag{6.4}$$

In some situations, E has a physical interpretation. For example, in electrostatics, $-\nabla\varphi$ is the electric field and S is the charge density, so that E is indeed the total energy of the system. However, in other situations, such as the steady-state diffusion equation, E should be viewed simply as a useful quantity.

It is easy to show that, at a solution to (6.1), E is stationary under all variations $\delta\varphi$ that respect the Dirichlet boundary conditions imposed. Indeed, the variation is

$$\delta E = \int_0^1 dx \int_0^1 dy \left[\nabla\varphi\cdot\nabla\delta\varphi - S\delta\varphi \right], \tag{6.5}$$

which, upon integrating the second derivative term by parts, becomes

$$\delta E = \int_C dl\, \delta\varphi\, \mathbf{n}\cdot\nabla\varphi + \int_0^1 dx \int_0^1 dy\, \delta\varphi[-\nabla^2\varphi - S], \tag{6.6}$$

where the line integral is over the boundary of the region of interest (C) and $\mathbf{n}$ is the unit vector normal to the boundary. Since we consider only variations that respect the boundary conditions, $\delta\varphi$ vanishes on C, so that the line integral does as well. Demanding that δE be zero for all such variations then implies that φ satisfies (6.1). This then furnishes a variational principle for our boundary value problem.

To derive a discrete approximation to the partial differential equation based on this variational principle, we first approximate E in terms of the values of the field at the lattice points and then vary with respect to them. The simplest approximation to E is to employ the two-point difference formula to approximate each first derivative in $(\nabla\varphi)^2$ at the points halfway between the lattice points and to use the trapezoidal rule for the integrals. This leads to

$$E = \tfrac{1}{2} \sum_{i=1}^{N} \sum_{j=1}^{N} \left[(\varphi_{ij} - \varphi_{i-1\,j})^2 + (\varphi_{ij} - \varphi_{ij-1})^2 \right] - h^2 \sum_{i=1}^{N-1} \sum_{j=1}^{N-1} S_{ij}\varphi_{ij}. \quad (6.7)$$

Putting

$$\frac{\partial E}{\partial \varphi_{ij}} = 0$$

for all ij then leads to the difference equation derived previously, (6.3). Of course, a more accurate discretization can be obtained by using better approximations for the first derivatives and for the integrals, taking care that the accuracies of both are commensurate. It is also easy to show from (6.7) that not only is E stationary at the solution, but that it is a minimum as well.

Exercise 6.1 Show that the vanishing of the derivatives of Eq. (6.7) with respect to the values of φ at the lattice points lead to the difference Eq. (6.3). Prove, or argue heuristically, that E is a minimum when φ is the correct solution.

Exercise 6.2 Use the differentiation formulas given in Table 1.2 to derive discretizations that are more accurate than (6.3). Can you see how the boundary conditions are to be incorporated in these higher-order formulas? What are the corresponding discretizations of E analogous to (6.7)? (These are not trivial; see [Fl78].)

We must now discuss where the boundary conditions enter the set of linear equations (6.3). Unless the coordinate system is well adapted to the geometry of the surfaces on which the boundary conditions are imposed (e.g., the surfaces are straight lines in cartesian coordinates or arcs in cylindrical or spherical coordinates), the lattice points will only roughly describe the geometry (see Fig. 6.1). One can always improve the accuracy by using a non-uniform lattice spacing and placing more points in the regions near the surfaces or by transforming to a coordinate system in which the boundary conditions are expressed more naturally. In any event, the boundary conditions will then provide the values of the φ_{ij} at some subset of lattice points. At a point far away from one of the boundaries, the boundary conditions do not enter (6.3) directly. However, consider (6.3) at a point just next to a boundary, say $(i, N-1)$. Since φ_{iN} is specified as part of the boundary conditions (as it is on the whole border of the unit square), we can rewrite (6.3b) as

$$4\varphi_{iN-1}-\varphi_{i+1N-1}-\varphi_{i-1N-1}-\varphi_{iN-2}=h^2 S_{iN-1}+\varphi_{iN}\ ; \qquad (6.8a)$$

that is, φ_{iN} enters not as an unknown, but rather as an inhomogeneous, known term. Similarly, if a Neumann boundary condition were imposed at a surface, say $\partial\varphi/\partial y=g(x)$ at $y=1$ or, equivalently, $j=N$, then this could be approximated by the discrete boundary condition

$$\varphi_{iN}-\varphi_{iN-1}=hg_i,$$

which means that at $j=N-1$, Eq. (6.3) would become

$$3\varphi_{iN-1}-\varphi_{i+1N-1}-\varphi_{i-1N-1}-\varphi_{iN-2}=h^2 S_{iN-1}+hg_i\ . \qquad (6.8b)$$

These considerations, and a bit more thought, show that the discrete approximation to the differential equation (6.1) is equivalent to a system of linear equations for the unknown values of φ at the interior points. In a matrix notation, this can be written as

$$\mathbf{M}\varphi=\mathbf{s}, \qquad (6.9)$$

where $\mathbf{M}$ is the matrix appearing in the linear system (6.3) and the inhomogeneous term $\mathbf{s}$ is proportional to S at the interior points and is linearly related to the specified values of φ or its derivatives on the boundaries. In any sort of practical situation there are a very large number of these equations (some 2500 if $N=50$, say), so that solution by direct inversion of $\mathbf{M}$ is impractical. Fortunately, since the discrete approximation to the Laplacian involves only neighboring points, most of the elements of $\mathbf{M}$ vanish (it is sparse) and there are then efficient iterative techniques for solving (6.9). We begin their discussion by considering an analogous, but simpler, one-dimensional boundary value problem, and then return to the two-dimensional case.

6.2 An iterative method for boundary value problems

The one-dimensional boundary value problem analogous to the two-dimensional problem we have been discussing can be written as

$$-\frac{d^2\varphi}{dx^2}=S(x), \qquad (6.10)$$

with $\varphi(0)$ and $\varphi(1)$ specified. The related variational principle involves

$$E=\int_0^1 dx\left[\frac{1}{2}\left(\frac{d\varphi}{dx}\right)^2-S\varphi\right], \qquad (6.11)$$

which can be discretized on a uniform lattice of spacing $h = 1/N$ as

$$E = \frac{1}{2h} \sum_{i=1}^{N} (\varphi_i - \varphi_{i-1})^2 - h \sum_{i=1}^{N-1} S_i \varphi_i . \qquad (6.12)$$

When varied with respect to φ_i, this yields the difference equation

$$2\varphi_i - \varphi_{i+1} - \varphi_{i-1} = h^2 S_i, \qquad (6.13)$$

which is, of course, just the naive discretization of Eq. (6.10).

Methods of solving the boundary value problem by integrating forward and backward in x were discussed in Chapter 3, but we can also consider (6.13), together with the known values of φ_0 and φ_N, as a set of linear equations. For a modest number of points (say $N \lesssim 100$), the linear system above can be solved by the direct methods discussed in Chapter 5 and, in fact, a very efficient special direct method exists for such "tri-diagonal" systems, as discussed in Chapter 7. However, to illustrate the iterative methods appropriate for the large sparse matrices of elliptic partial differential equations in two or more dimensions, we begin by rewriting (6.13) in a "solved" form for φ_i:

$$\varphi_i = \frac{1}{2}[\varphi_{i+1} + \varphi_{i-1} + h^2 S_i]. \qquad (6.14)$$

Although this equation is not manifestly useful, since we don't know the φ's on the right-hand side, it can be interpreted as giving us an "improved" value for φ_i based on the values of φ at the neighboring points. Hence the strategy (Gauss-Seidel iteration) is to guess some initial solution and then to sweep systematically through the lattice (say from left to right), successively replacing φ at each point by an improved value. Note that the most "current" values of the $\varphi_{i \pm 1}$ are to be used in the right-hand side of Eq. (6.14). By repeating this sweep many times, an initial guess for φ can be "relaxed" to the correct solution.

To investigate the convergence of this procedure, we generalize Eq. (6.14) so that at each step of the relaxation φ_i is replaced by a linear mixture of its old value and the "improved" one given by (6.14):

$$\varphi_i \rightarrow \varphi'_i = (1-\omega)\varphi_i + \omega \frac{1}{2}[\varphi_{i+1} + \varphi_{i-1} + h^2 S_i]. \qquad (6.15)$$

Here, ω is a parameter that can be adjusted to control the rate of relaxation: "over-relaxation" corresponds to $\omega > 1$, while "under-relaxation" means $\omega < 1$. The optimal value of ω that maximizes the rate of relaxation will be discussed below. To see that (6.15) results in an "improvement" in the solution, we calculate the change in the energy functional (6.12), remembering that all φ's except φ_i are to be held fixed. After some algebra, one finds

$$E' - E = -\frac{\omega(2-\omega)}{2h}[\frac{1}{2}(\varphi_{i+1} + \varphi_{i-1} + h^2 S_i) - \varphi_i]^2 \leq 0, \qquad (6.16)$$

so that, as long as $0<\omega<2$, the energy never increases, and should thus converge to the required minimum value as the sweeps proceed. (The existence of other, spurious minima of the energy would imply that the linear system (6.13) is not well posed.)

□ □ □   **Exercise 6.3** Use Eqs. (6.12) and (6.15) to prove Eq. (6.16).

As an example of this relaxation method, let us consider the one-dimensional boundary-value problem of the form (6.10) with

$$S(x)=12x^2; \quad \varphi(0)=\varphi(1)=0.$$

The exact solution is

$$\varphi(x)=x(1-x^3)$$

and the energy is

$$E=-\frac{9}{14}=-0.64286.$$

The following BASIC code implements the relaxation algorithm and prints out the energy after each sweep of the 21-point lattice. An initial guess of $\varphi_i=0$ is used, which is clearly quite far from the truth.

```
10 N%=20: H=1/N%                              'lattice size and spacing
15 OMEGA=1                                    'relaxation parameter
20 DIM PHI(N%)                                'function to be relaxed
25 DIM S(N%)                                  'array for h^2*source
30 '
35 FOR I%=0 TO N%                             'initialization
40    X=I%*H
45    S(I%)=H*H*12*X*X                        'establish source array
50    PHI(I%)=0                               'initial guess is phi=0
55 NEXT I%
60 '
65 FOR IT%=1 TO 500                           'iteration loop
70    FOR I%=1 TO N%-1                        'sweep through the lattice
75       PHIP=(PHI(I%-1)+PHI(I%+1)+S(I%))/2
80       PHI(I%)=(1-OMEGA)*PHI(I%)+OMEGA*PHIP 'relaxation
85    NEXT I%
90    IF (IT%-1) MOD 10 <> 0 THEN GOTO 130    'sometimes calculate the energy
95       PRINT USING "iteration=### ";IT%
100      E=0:
105      FOR I%=1 TO N%:                      'loop over the lattice
110         E=E+((PHI(I%)-PHI(I%-1))/H)^2/2   'add gradient term to the energy
115         E=E-S(I%)/H/H*PHI(I%)             'add source  term to the energy
120      NEXT I%
125      PRINT USING "energy=+##.####^^^^";E*H
130 NEXT IT%
```

Results for the energy as a function of iteration number are shown in Table 6.1 for three different values of ω. Despite the rather poor

Table 6.1 Convergence of the energy functional during relaxation of a 1-D boundary value problem

Iteration	$\omega=0.5$	$\omega=1.0$	$\omega=1.5$
1	-0.01943	-0.04959	-0.09459
21	-0.24267	-0.44024	-0.60688
41	-0.36297	-0.56343	-0.63700
61	-0.44207	-0.61036	-0.63831
81	-0.49732	-0.62795	-0.63836
101	-0.53678	-0.63450	-0.63837
121	-0.56517	-0.63693	-0.63837
141	-0.58563	-0.63783	-0.63837
161	-0.60037	-0.63817	-0.63837
181	-0.61100	-0.63829	-0.63837
201	-0.61866	-0.63834	-0.63837
221	-0.62418	-0.63836	-0.63837
241	-0.62815	-0.63836	-0.63837
. . . .			
381	-0.63734	-0.63837	-0.63837
401	-0.63763	-0.63837	-0.63837

initial guess for φ, the iterations converge and the converged energy is independent of the relaxation parameter, but differs somewhat from the exact answer due to discretization errors (i.e., h not vanishingly small); the discrepancy can be reduced, of course, by increasing N. A detailed examination of the solution indicates good agreement with the analytical result. Note that the rate of convergence clearly depends upon ω. A general analysis [Wa66] shows that the best choice for the relaxation parameter depends upon the lattice size and on the geometry of the problem; it is usually greater than 1. The optimal value can be determined empirically by examining the convergence of the solution for only a few iterations before choosing a value to be used for many iterations.

Exercise 6.4 Use the code above to verify that the energy approaches the analytical value as the lattice is made finer. Investigate the accuracy of the solution at each of the lattice points and note that the energy can be considerably more accurate than the solution itself; this is a natural consequence of the minimization of E at the correct solution. Use one of the higher-order discretizations you derived in Exercise 6.2 to solve the problem.

The application of the relaxation scheme described above to two- (or even three-) dimensional problems is now straightforward. Upon solving (6.3a) for φ_{ij}, we can generate the analogue of (6.15):

$$\varphi_{ij} \to \varphi'_{ij} = (1-\omega)\varphi_{ij} + \frac{\omega}{4}[\varphi_{i+1j} + \varphi_{i-1j} + \varphi_{ij+1} + \varphi_{ij-1} + h^2 S_{ij}]. \quad (6.17)$$

If this algorithm is applied successively to each point in the lattice, say sweeping the rows in order from top to bottom and each row from left to right, one can show that the energy functional (6.7) always decreases (if ω is within the proper range) and that there will be convergence to the required solution.

Several considerations can serve to enhance this convergence in practice. First, starting from a good guess at the solution (perhaps one with similar, but simpler, boundary conditions) will reduce the number of iterations required. Second, an optimal value of the relaxation parameter should be used, either estimated analytically or determined empirically, as described above. Third, it may sometimes be more efficient to concentrate the relaxation process, for several iterations, in some sub-area of the lattice where the trial solution is known to be particularly poor, thus not wasting effort on already-relaxed parts of the solution. Finally, one can always do a calculation on a relatively coarse lattice that relaxes with a small amount of numerical work, and then interpolate the solution found onto a finer lattice to be used as the starting guess for further iterations.

6.3 More on discretization

It is often the case that the energy functional defining a physical problem has a form more complicated than the simple "integral of the square of the derivative" that we have been considering so far. For example, in an electrostatics problem with spatially-varying dielectric properties or in a diffusion problem with a spatially-varying diffusion coefficient, the boundary-value problem (6.1) is modified to

$$-\nabla \cdot D \nabla \varphi = S(x,y), \quad (6.18)$$

where $D(x,y)$ is the dielectric constant or diffusion coefficient, and the corresponding energy functional is (compare Eq. (6.4))

$$E = \int_0^1 dx \int_0^1 dy \left[\tfrac{1}{2} D (\nabla \varphi)^2 - S \varphi \right]. \quad (6.19)$$

Although it is possible to discretize Eq. (6.18) directly, it should be evident from the previous discussion that a far better procedure is to discretize (6.19) first and then to differentiate with respect to the field variables to obtain the difference equations to be solved.

To see how this works out in detail, consider the analog of the one-dimensional problem defined by (6.11),

$$E = \int_0^1 dx \left[\tfrac{1}{2} D(x) \left(\frac{d\varphi}{dx} \right)^2 - S\varphi \right], \tag{6.20}$$

The discretization analogous to (6.12) is

$$E = \frac{1}{2h} \sum_{i=1}^{N} D_{i-\frac{1}{2}} (\varphi_i - \varphi_{i-1})^2 - h \sum_{i=1}^{N-1} S_i \varphi_i, \tag{6.21}$$

where $D_{i-\frac{1}{2}}$ is the diffusion constant at the half-lattice points. This might be known directly if we have an explicit formula for $D(x)$, or it might be approximated with appropriate accuracy by $\tfrac{1}{2}(D_i + D_{i-1})$. Note that, in either case, we have taken care to center the differencing properly. Variation of this equation then leads directly to the corresponding difference equations (compare Eq. (6.13)),

$$(D_{i+\frac{1}{2}} + D_{i-\frac{1}{2}})\varphi_i - D_{i+\frac{1}{2}}\varphi_{i+1} - D_{i-\frac{1}{2}}\varphi_{i-1} = h^2 S_i. \tag{6.22}$$

These can then be solved straightforwardly by the relaxation technique described above.

A problem treated in cylindrical or spherical coordinates presents very much the same kind of situation. For example, when the diffusion or dielectric properties are independent of space, the energy functional in cylindrical coordinates will involve

$$E = \int_0^\infty dr \, r \left[\tfrac{1}{2} \left(\frac{d\varphi}{dr} \right)^2 - S\varphi \right], \tag{6.23}$$

where r is the cylindrical radius. (We suppress here the integrations over the other coordinates.) This is of the form (6.20), with $D(r) = r$ and an additional factor of r appearing in the source integral. Discretization on a lattice $r_i = hi$ in analogy to (6.21) then leads to the analog of (6.22),

$$2r_i\varphi_i - r_{i+\frac{1}{2}}\varphi_{i+1} - r_{i-\frac{1}{2}}\varphi_{i-1} = h^2 r_i S_i. \tag{6.24}$$

At $i=0$, this equation just tells us that $\varphi_1 = \varphi_{-1}$, or equivalently, that $\partial\varphi/\partial r = 0$ at $r=0$. This is to be expected, as, in the electrostatics language, Gauss' law allows no radial electric field at $r=0$. At $i=1$, Eq. (6.24) gives an equation involving three unknowns, φ_0, φ_1, and φ_2, but putting $\varphi_0 = \varphi_1$ as a rough approximation to the zero-derivative boundary condition gives an equation involving only two unknowns, which is what we expect at a boundary on the basis of our experience with the cartesian problems discussed above.

A more elegant discritization of problems with cylindrical symmetry naturally incorporates the zero-derivative boundary condition at $r=0$ by working on a lattice defined by $r_i=(i-\tfrac{1}{2})h$. In this case, Eq. (6.24) is still valid, but for $i=1$ the coefficient of the term involving φ_{i-1} vanishes, giving directly an equation with only two unknowns, φ_1 and φ_2.

□ □ □   **Exercise 6.5** Verify that variation of Eq. (6.21) leads to Eq. (6.22). Write down explicitly the discretizations of Eq. (6.23) on the $r_i=ih$ and $r_i=(i-\tfrac{1}{2})h$ lattices and verify their variation leads to (6.24) in either case. Why can the contribution to the energy functional from the region between $r=0$ and $r=\tfrac{1}{2}h$ be safely neglected on the $r_i=(i-\tfrac{1}{2})$ lattice to the same order as the accuracy of approximation used for the derivative?

6.4 Elliptic equations in two dimensions

The application of relaxation methods to elliptic boundary value problems is illustrated by the program for Example 6, which solves Laplace's equation, $\nabla^2\varphi=0$, on a uniform rectangular lattice of unit spacing with Dirichlet boundary conditions specified on the lattice borders and on selected points within the lattice. This situation might describe the steady-state temperature distribution within a plate whose edges and certain interior regions are held at specified temperatures, or it might describe an electrostatics problem specified by a number of equipotential surfaces. During the iterations, the potential is displayed, as is the energy functional (6.7) to monitor convergence. The code's source listing can be found in Appendix B and in the file EXAM6.BAS on the *Computational Physics* diskette.

The following exercises, phrased in the language of electrostatics, might help you to understand better the physical and numerical principles illustrated by this example.

□ □ □   **Exercise 6.6** Verify that the solutions corresponding to particular boundary conditions in the interior of the lattice agree with your intuition. You might try fixing a single interior point to a potential different from that of the boundary, fixing two symmetrically located points to different potentials, or fixing a whole line to a given potential. Other possibilities include constructing a "Faraday cage" (a closed region bounded by a surface at fixed potential), studying a quadrupole pattern of boundary conditions, or calculating the capacitance of various configurations of conductors. Study of what happens when you increase or decrease the size of the lattice is also interesting.

Exercise 6.7 For a given set of boundary conditions, investigate the effects of under-relaxation and over-relaxation on the convergence of the relaxation process. Change the discretization of the Laplacian to a higher-order formula as per Exercise 6.2 and observe any the changes in the solution or the efficiency with which it is approached.

Exercise 6.8 Modify the code to solve Poisson's equation; i.e., allow for a charge density to be specified throughout the lattice. Use this to study the solutions for certain simple charge distributions. For example, you might try computing the potential between two (line) charges as a function of their separation and comparing it with your analytical expectations.

Exercise 6.9 Modify the code to use Neumann boundary conditions on selected borders and interior regions of the lattice. Study the solution for simple geometries and compare it with what you expect.

Exercise 6.10 Modify the code to allow for a spatially-varying dielectric constant. (Note that you must change both the relaxation formula and the energy functional.) Study the solutions for selected simple geometries of the dielectric constant (e.g., a half-space filled with dielectric) and simple boundary conditions.

Exercise 6.11 An alternative to the Dirichlet boundary conditions are periodic boundary conditions, in which the potentials on the left and right and top and bottom borders of the rectangle are constrained to be equal, but otherwise arbitrary. That is,

$$\varphi_{i1}=\varphi_{iN}; \; \varphi_{1j}=\varphi_{Nj}$$

for all i and j. This might correspond to the situation in a crystal, where the charge density is periodic in space. Modify the code to incorporate these boundary conditions into Poisson's equation and verify that the solutions look as expected for simple charge distributions.

Exercise 6.12 Change the relaxation formula and energy functional to treat situations with an azimuthal symmetry by re-interpreting one of the coordinates as the cylindrical radius while retaining the other as a cartesian coordinate. Use the resulting program to model a capacitor made of two circular disks, and in particular calculate the capacitance and potential field for varying separations between the disks. Compare your results with your expectations for very large or very small separations.

Project VI: Steady-state hydrodynamics in two dimensions

The description of the flow of fluids is one of the richest and most challenging problems that can be treated on a computer. The non-linearity of the equations and the complexity of phenomena they describe (e.g., turbulence) sometimes make computational fluid dynamics more of an art than a science, and several book-length treatments are required to cover the field adequately (see, for example, [Ro76]). In this project, we will consider one relatively simple situation that can be treated by the relaxation methods for elliptic equations described in this chapter and that will serve to give some idea of the problems involved. This situation is the time-independent flow of a viscous, incompressible fluid past an object. For simplicity, we will take the object to be translationally invariant in one direction transverse to the flow, so that the fluid has a non-trivial motion only in two-coordinates, (x,y); this might describe a rod or beam placed in a steady flow of water with incident speed V_0. We will also consider only the case where the cross-section of this rod is a rectangle with dimensions $2W$ transverse to the flow and T along the flow (see Figure VI.1). This will greatly simplify the coding needed to treat the boundary conditions, while still allowing the physics to be apparent. We begin with an exposition of the basic equations and their discretization, follow with a brief discussion of the boundary conditions, and then give some guidance in writing the program and in extracting some understanding from it.

VI.1 The equations and their discretization

In describing the flow of a fluid through space, at least two fields are important: ρ, the mass density, and $\mathbf{V}$, the velocity, of the fluid element at each point in space. These are related through two fundamental equations of hydrodynamics [La59]:

$$\frac{\partial \rho}{\partial t} + \nabla \cdot \rho \mathbf{V} = 0; \tag{VI.1a}$$

$$\frac{\partial \mathbf{V}}{\partial t} = -(\mathbf{V} \cdot \nabla)\mathbf{V} - \frac{1}{\rho}\nabla P + \nu \nabla^2 \mathbf{V}. \tag{VI.1b}$$

The first of these (the continuity equation) expresses the conservation of mass, and states that the density can change at a point in space only due to a net in- or out-flow of matter. The second equation (Navier-Stokes) expresses the conservation of momentum, and states that the velocity changes in response to convection, $(\mathbf{V} \cdot \nabla)\mathbf{V}$, spatial variations in the pressure, ∇P, and viscous forces $\nu \nabla^2 \mathbf{V}$, where ν is the kinematic viscosity, assumed constant in our discussion. In general, the pressure is given in terms of the density and temperature through an "equation of state", and when

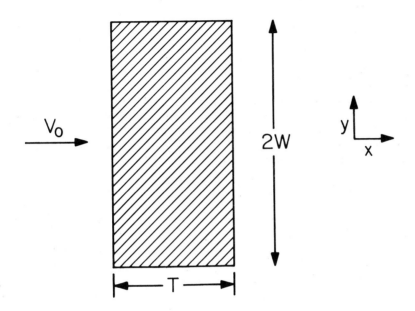

Figure VI.1 Geometry of the two-dimensional flow past a plate to be treated in this project.

the temperature varies as well, an additional equation embodying the conservation of energy is also required. We will assume that the temperature is constant throughout the fluid.

We will be interested in studying time-independent flows, so that all time derivatives can be set to zero in these equations. Furthermore, we will assume that the fluid is incompressible, so that the density is constant (this is a good approximation for water under many conditions). Equations (VI.1) then become

$$\nabla \cdot \mathbf{V} = 0; \tag{VI.2a}$$

$$(\mathbf{V} \cdot \nabla) \mathbf{V} = -\frac{1}{\rho} \nabla P + \nu \nabla^2 \mathbf{V}. \tag{VI.2b}$$

For two-dimensional flow, these equations can be written explicitly in terms of the x and y components of the velocity field, denoted by u and v, respectively:

$$\frac{\partial u}{\partial x} + \frac{\partial v}{\partial y} = 0; \tag{VI.3a}$$

$$u \frac{\partial u}{\partial x} + v \frac{\partial u}{\partial y} = -\frac{1}{\rho} \frac{\partial P}{\partial x} + \nu \nabla^2 u; \tag{VI.3b}$$

$$u\frac{\partial v}{\partial x}+v\frac{\partial v}{\partial y}=-\frac{1}{\rho}\frac{\partial P}{\partial y}+\nu\nabla^2 v. \qquad\text{(VI.3c)}$$

Here, as usual,

$$\nabla^2=\frac{\partial^2}{\partial x^2}+\frac{\partial^2}{\partial y^2}.$$

Equations (VI.3) are three scalar equations for the fields u, v, and P. While these equations could be solved directly, it is more convenient for two-dimensional problems to replace the velocity fields by two equivalent scalar fields: the stream function, $\psi(x,y)$, and the vorticity, $\zeta(x,y)$. The first of these is introduced as a convenient way of satisfying the continuity equation (VI.3a). The stream function is defined so that

$$u=\frac{\partial\psi}{\partial y}; \quad v=-\frac{\partial\psi}{\partial x}. \qquad\text{(VI.4)}$$

It is easily verified that this definition satisfies the continuity equations (VI.3a) for any function ψ and that such a ψ exists for all flows that satisfy the continuity condition (VI.2a). Furthermore, one can see that $(\mathbf{V}\cdot\nabla)\psi=0$, so that $\mathbf{V}$ is tangent to contour lines of constant ψ, the "stream lines".

The vorticity is defined as

$$\zeta=\frac{\partial u}{\partial y}-\frac{\partial v}{\partial x}, \qquad\text{(VI.5)}$$

which is seen to be (the negative of) the curl of the velocity field. From the definitions (VI.4), it follows that ζ is related to the stream function ψ by

$$\nabla^2\psi=\zeta. \qquad\text{(VI.6)}$$

An equation for ζ can be derived by differentiating (VI.3b) with respect to y and (VI.3c) with respect to x. Upon subtracting one from the other and invoking the continuity equation (VI.3a) and the definitions (VI.4), one finds, after some algebra,

$$\nu\nabla^2\zeta=\left[\frac{\partial\psi}{\partial y}\frac{\partial\zeta}{\partial x}-\frac{\partial\psi}{\partial x}\frac{\partial\zeta}{\partial y}\right]. \qquad\text{(VI.7)}$$

Finally, an equation for the pressure can be derived by differentiating (VI.3b) with respect to x and adding it to the derivative of (VI.3c) with respect to y. Upon expressing all velocity fields in terms of the stream function, one finds, after a bit of re-arranging,

$$\nabla^2 P=2\rho\left[\left(\frac{\partial^2\psi}{\partial x^2}\right)\left(\frac{\partial^2\psi}{\partial y^2}\right)-\left(\frac{\partial^2\psi}{\partial x\,\partial y}\right)^2\right]. \qquad\text{(VI.8)}$$

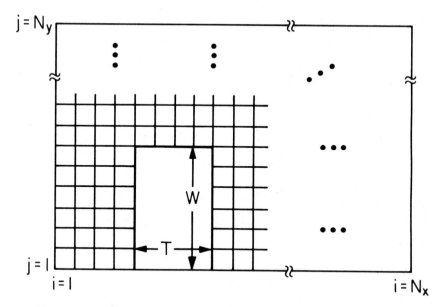

Figure VI.2 The lattice to be used in calculating the fluid flow past the plate illustrated in Figure VI.1.

Equations (VI.6-8) are a set of non-linear elliptic equations equivalent to the original hydrodynamic equations (VI.3). They are particularly convenient in that if it is just the velocity field we want, only the two equations (VI.6,7) need be solved simultaneously, since neither involves the pressure. If we do happen to want the pressure, it can be obtained by solving (VI.8) for P after we have found ψ and ζ.

To solve Eqs. (VI.6,7) numerically, we introduce a two-dimensional lattice of uniform spacing h having N_x and N_y points in the x and y directions, respectively, and use the indices i and j to measure these coordinates. (See Figure VI.2.) Note that since the centerline of the rectangle is a line of symmetry, we need only treat the upper half-plane, $y > 0$. Moreover, it is convenient to place the lattice so that the edges of the plate are on lattice points. The location of the plate relative to the up- and down-stream edges of the lattice is arbitrary, although it is wise to place the plate far enough forward to allow the "wake" to develop behind it, yet not too close to the upstream edge, whose boundary conditions can influence the flow pattern spuriously. It is also convenient to scale the equations by measuring all lengths in units of h and all velocities in units of the incident fluid velocity, V_0. The stream function is then measured in units of $V_0 h$, while the vorticity is in units of V_0/h, and the pressure is conveniently scaled by ρV_0^2. Upon differencing (VI.6) in the usual way, we have

$$(\delta_i^2 \psi)_{ij} + (\delta_j^2 \psi)_{ij} = \zeta_{ij}, \qquad (VI.9)$$

where δ^2 is the symmetric second-difference, as in Eq. (6.3b). Similarly, (VI.7) can be differenced as

$$(\delta_i^2 \zeta)_{ij} + (\delta_j^2 \zeta)_{ij} = \frac{R}{4}[(\delta_j \psi)_{ij}(\delta_i \zeta)_{ij} - (\delta_i \psi)_{ij}(\delta_j \zeta)_{ij}], \qquad (VI.10)$$

where the symmetric first-difference operator is

$$(\delta_i \psi)_{ij} \equiv \psi_{i+1j} - \psi_{i-1j}$$

and similarly for δ_j. The lattice Reynolds number, $R = V_0 h / \nu$, is a dimensionless measure of the strength of the viscous forces or, equivalently, of the speed of the incident stream. It is related to the physical Reynolds number Re by replacing the mesh spacing by the width of the rectangle: $Re = 2WV_0/\nu$. Finally, the pressure equation (VI.8) can be differenced as

$$[(\delta_i^2 P)_{ij} + (\delta_j^2 P)_{ij}] = 2[(\delta_i^2 \psi)_{ij}(\delta_j^2 \psi)_{ij} - \frac{1}{16}(\delta_i \delta_j \psi)_{ij}^2]. \qquad (VI.11)$$

VI.2 Boundary conditions

In order for the elliptic problems (VI.6-8) to be well posed, we must specify either the values or the normal derivatives of the stream function, the vorticity, and the pressure at all boundaries of the lattice shown in Figure VI.2. These boundaries can be classified into three groups, in the notation of Figure VI.3:

i)   the centerline boundaries (A and E);
ii) the boundaries contiguous with the rest of the fluid (F, G, and H);
iii) the boundaries of the plate itself (B, C, and D).

We treat ψ and ζ on each of these in turn, followed by the pressure boundary conditions. Throughout this discussion, we use unscaled (i.e., physical) quantities. It also helps to recall that in the freely flowing fluid (i.e., no obstruction), the solution is $u = V_0$, $v = 0$, so that $\psi = V_0 y$ and $\zeta = 0$.

The boundary conditions on the centerline surfaces A and E are determined by symmetry. The y component of the velocity, v, must vanish here, so that the x derivative of ψ vanishes. It follows that A and E are each stream lines. Moreover, since the normal velocity (and hence the tangential derivative of ψ) also vanishes on B, C, and D, the entire surface ABCDE is a single stream line, and

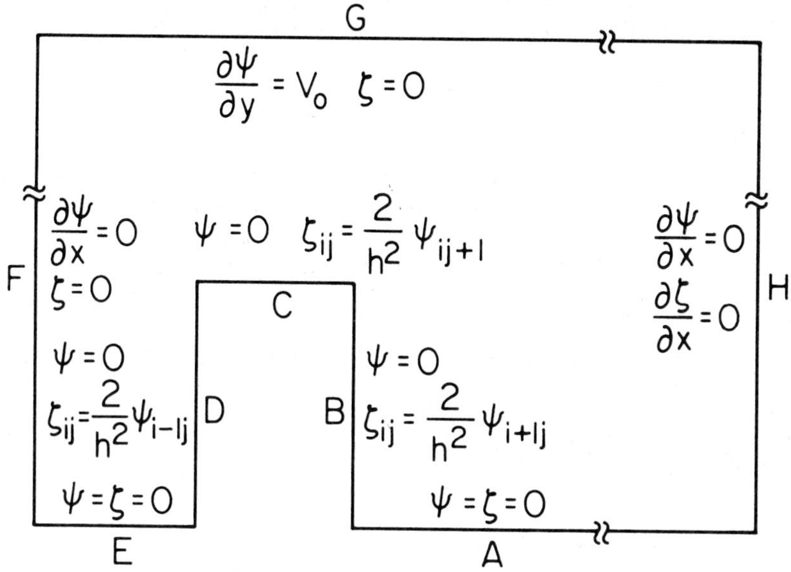

Figure VI.3 Boundary conditions to be imposed on the stream function and vorticity

we may arbitrarily put $\psi=0$ on it. Note that the velocities depend only upon the derivatives of ψ, so that the physical description is invariant to adding a spatially independent constant to the stream function; the choice of ψ on this streamline fixes the constant. From symmetry, we can also conclude that the vorticity vanishes on A and E.

The boundary conditions on the upstream surface F are also fairly straightforward. This surface is contiguous with the smoothly flowing incident fluid, so that specifying

$$v = -\frac{\partial \psi}{\partial x} = 0; \ \zeta = 0 \qquad \text{on F,}$$

as is the case far upstream, seems reasonable. Similarly, if the lattice is large enough, we may expect the upper boundary G to be in free flow, so that fixing

$$u = \frac{\partial \psi}{\partial y} = V_0; \ \zeta = 0 \qquad \text{on G,}$$

is one appropriate choice. The downstream boundary H is much more ambiguous and, as long as it is sufficiently far from the plate, there should be many plausible choices. However, the boundary conditions on a boundary that approaches the plate can influence the shape of the solution found. One convenient choice is to say that nothing changes beyond the lattice boundary, so that

$$\frac{\partial \psi}{\partial x} = \frac{\partial \zeta}{\partial x} = 0 \quad \text{on H.}$$

At the walls of the plate, (B, C, and D) one of the correct boundary conditions is that the normal velocity of the fluid is zero. This we have used already by requiring that this surface be a stream line. However, the other boundary condition appropriate for viscous flow is that the tangential velocity be zero. Implementing this by setting the normal derivative of ψ to zero would be an overspecification of the elliptic problem for ψ. Instead, the "no-slip" boundary condition is imposed on the vorticity. Consider a point ij on the upper surface of the plate, C. We can write the stream function at the point on lattice spacing above this one, $ij+1$, as a Taylor series in y:

$$\psi_{ij+1} = \psi_{ij} + h\left.\frac{\partial \psi}{\partial y}\right|_{ij} + \frac{h^2}{2}\left.\frac{\partial^2 \psi}{\partial y^2}\right|_{ij} + \cdots . \tag{VI.12}$$

Since, at the wall,

$$\frac{\partial \psi}{\partial y} = u = 0$$

and

$$\frac{\partial^2 \psi}{\partial y^2} = \frac{\partial u}{\partial y} = \zeta,$$

because $\partial v / \partial x = 0$, (VI.12) can be reduced to

$$\zeta_{ij} = 2\frac{\psi_{ij+1} - \psi_{ij}}{h^2} \quad \text{on C}, \tag{VI.13a}$$

which provides a Dirichlet boundary condition for ζ. (This is the general form of the boundary condition; recall that we had previously specified $\psi_{ij} = 0$ on the plate boundaries.) The same kind of arguments can be applied to the surfaces B and D to yield

$$\zeta_{ij} = 2\frac{\psi_{i+1j} - \psi_{ij}}{h^2} \quad \text{on B}; \tag{VI.13b}$$

$$\zeta_{ij} = 2\frac{\psi_{i-1j} - \psi_{ij}}{h^2} \quad \text{on D}. \tag{VI.13c}$$

Note that there is an ambiguity at the "corners", where surfaces B and C and D and C intersect, as here the vorticity can be computed in two ways (horizontal or vertical difference of ψ). In practice, there are several ways of resolving this: use one form and verify that the other gives similar values (a check on the accuracy of the calculation), use the average of the two methods, or use the horizontal value when relaxing the point just to the right or left of the corner and the vertical value when relaxing the point just above the corner.

The boundary conditions for the pressure on all surfaces are of the Neumann type, and follow from Eqs. (VI.3). We leave their explicit finite-difference form in terms of ψ and ζ as an exercise. Note that from symmetry, $\partial P/\partial y = 0$ on the centerlines A and E.

VI.3 Solving the equations

Our general strategy will be to solve the coupled non-linear elliptic equations (VI.9,10) for the stream function and vorticity using the relaxation methods discussed in this chapter. One possible iteration scheme goes as follows. We begin by choosing trial values corresponding to the free-flowing solution: $\psi = y$ and $\zeta = 0$. We then perform one relaxation sweep of (VI.9) to get an improved value of ψ. The Dirichlet boundary conditions for ζ on the walls of the plate are then computed from (VI.13) and then one relaxation sweep of (VI.10) is performed. With the new value of ζ so obtained, we go back to a sweep of (VI.9) and repeat the cycle many times until convergence. If required, we can then solve the pressure equation (VI.11) with Neumann boundary conditions determined from (VI.3). This iteration scheme can be implemented in the following sequence of steps:

Step 1 Write a section of code to execute one relaxation sweep of (VI.9) for ψ given ζ and subject to the boundary conditions discussed above. Be sure to allow for an arbitrary relaxation parameter. This code can be adapted from that for Example 6; in particular, the technique of using letters on the screen as a crude contour plot can be useful in displaying ψ.

Step 2 Write a section of code that calculates the boundary conditions for ζ on the plate walls if ψ is known.

Step 3 Write a section of code that does one relaxation sweep of (VI.10) for ζ if ψ is given. Build in the boundary conditions discussed above and be sure to allow for an arbitrary relaxation parameter. If your computer has two displays, it is useful to display ψ on one screen and ζ on the other.

Step 4 Combine the sections of code you wrote in the previous three steps into a program that executes a number of iterations of the coupled ψ-ζ problem. Test the convergence of this scheme on several coarse lattices for several different Reynolds numbers and choices of relaxation parameters.

□ □ □   **Step 5** Calculate the flow past several plates by running your program to convergence on some large lattices (say 24×70) for several increasing values of the lattice Reynolds number. A typical size of the plate might be $W=8h$ and $T=8h$, while lattice Reynolds numbers might run from 0.1 to 8. For the larger Reynolds numbers, you might find instabilities in the relaxation procedure due to the nonlinearity of the equations. These can be suppressed by using relaxation parameters as small as 0.1 and by using as trial solutions the flow patterns obtained at smaller Reynolds numbers. Verify that the flow around the plate is smooth at small Reynolds numbers but that at larger velocities the flow separates from the back edge of the plate and a small vortex (eddy) develops behind it. Check also that your solutions are accurate by running two cases with different lattice Reynolds numbers but the same physical Reynolds number.

□ □ □   **Step 6** Two physically interesting quantities that can be computed from the flow pattern are the net pressure and viscous forces per unit area of the plate, F_P and F_v. By symmetry, these act in the x direction and are measured conveniently in terms of the flow of momentum incident on the face of the plate per unit area, $2W\rho V_0^2$. The pressure force is given by:

$$F_P = \int_D P dy - \int_B P dy, \qquad (VI.14)$$

where the integrals are over the entire front and back surfaces of the plate. This shows clearly that it is only the relative values of the pressure over the plate surfaces that matter. These can be obtained from the flow solution using Eqs. (VI.3b,c). Consider, for example, the front face of the plate. Using (VI.3c), together with (VI.4,6) and the fact that the velocities vanish at the plate surface, we can write the derivative of the pressure along the front face as:

$$\frac{\partial P}{\partial y} = -\nu\rho \frac{\partial \zeta}{\partial x}. \qquad (VI.15)$$

Hence, ζ from the flow solution can be integrated to find the (relative) pressure along the front face. Similar expressions can be derived for the top and back faces, so that the pressure (apart from an irrelevant additive constant) can be computed on all faces by integrating up the front face, across the top, and down the back face. The net viscous force per unit area of the plate is due only to the flow past the top (and bottom) surfaces and is given by

$$F_v = 2\rho\nu \int_C \frac{\partial u}{\partial y} dx. \qquad \text{(VI.16a)}$$

However, since the boundary conditions on the top surface of the plate require $\partial v / \partial x = 0$, this can be written as

$$F_v = 2\rho\nu \int_C \zeta dx, \qquad \text{(VI.16b)}$$

which is conveniently evaluated from the flow solution. Investigate the variation of the viscous and pressure forces with Reynolds number and compare your results with what you expect.

Step 7 Three other geometries can be investigated with only minor □ □ □ modifications of the code you've constructed. One of these is a "jump", in which the thickness of the plate is increased until its downstream edge meets the downstream border of the lattice (H). Another is a pair of rectangular plates placed symmetrically about a centerline, so that the fluid can flow between them, as well around them (a crude nozzle). Finally, two plates, one behind the other, can also be calculated. Modify your code to treat each of these cases and explore the flow patterns at various Reynolds numbers.

Chapter 7

Parabolic Partial Differential Equations

Typical parabolic partial differential equations one encounters in physical situations are the diffusion equation

$$\frac{\partial \varphi}{\partial t} = \nabla \cdot (D \nabla \varphi) + S, \qquad (7.1)$$

where D is the (possibly space dependent) diffusion constant and S is a source function, and the time-dependent Schroedinger equation

$$i\hbar \frac{\partial \varphi}{\partial t} = -\frac{\hbar^2}{2m} \nabla^2 \varphi + V\varphi, \qquad (7.2)$$

where V is the potential. In contrast to the boundary value problems encountered in the previous chapter, these problems are generally of the initial value type. That is, we are given the field φ at an initial time and seek to find it at a later time, the evolution being subject to certain spatial boundary conditions (e.g., the Schroedinger wavefunction vanishes at very large distances or the temperature or heat flux is specified on some surfaces). The methods by which such problems are solved on the computer are basically straightforward, although a few subtleties are involved. We will also see that they provide a natural way of solving elliptic eigenvalue problems, particularly if it is only the few lowest or highest eigenvalues that are required.

7.1 Naive discretization and instabilities

We begin our discussion by treating diffusion in one dimension with a uniform diffusion constant. We take x to vary between 0 and 1 and assume Dirichlet boundary conditions that specify the value of the field at the end points of this interval. After appropriate scaling, the analogue of (7.1) is

$$\frac{\partial \varphi}{\partial t} = \frac{\partial^2 \varphi}{\partial x^2} + S(x,t). \qquad (7.3)$$

As usual, we approximate the spatial derivatives by finite differences on a uniform lattice of $N+1$ points having spacing

$h = 1/N$, while the time derivative is approximated by the simplest first-order difference formula assuming a time step Δt. Using the superscript n to label the time step (that is, $\varphi^n \equiv \varphi(t_n)$, $t_n = n \Delta t$), we can approximate (7.3) as

$$\frac{\varphi_i^{n+1} - \varphi_i^n}{\Delta t} = \frac{1}{h^2} (\delta^2 \varphi^n)_i + S_i^n, \tag{7.4}$$

At $i=1$ and $i=N-1$, this equation involves the Dirichlet boundary conditions specifying φ_0 and φ_N.

Equation (7.4) is an *explicit* differencing scheme since, given φ at one time, it is straightforward to solve for φ at the next time. Indeed, in an obvious matrix notation, we have:

$$\varphi^{n+1} = (1 - H \Delta t) \varphi^n + S^n \Delta t, \tag{7.5}$$

where the action of the operator H is defined by

$$(H\varphi)_i \equiv -\frac{1}{h^2} (\delta^2 \varphi)_i.$$

As an example of how this explicit scheme might be applied, consider the case where $S=0$ with the boundary conditions $\varphi(0) = \varphi(1) = 0$. Suppose that we are given the initial condition of a gaussian centered about $x = \frac{1}{2}$,

$$\varphi(x, t=0) = e^{-20(x - \frac{1}{2})^2} - e^{-20(x - \frac{3}{2})^2} - e^{-20(x + \frac{1}{2})^2},$$

where the latter two "image" gaussians approximately ensure the boundary conditions at $x=1$ and $x=0$, respectively, and we seek to find φ at later times. The following BASIC code applies (7.5) to do this on a lattice with $N=25$.

```
5  N%=25: H=1/N%                                          'lattice parameters
10 DT=.001: DTH=DT/H^2                                    'time step, useful cnst
15 DIM PHI(N%)                                            'array for the solution
20 DEF FNEXACT(X)=FNGAUSS(X)-FNGAUSS(X-1)-FNGAUSS(X+1)    'functions for the
25 DEF FNGAUSS(X)=EXP(-20*(X-.5)^2/(1+80*T))/SQR(1+80*T)  'analytical solution
30 '
35 T=0                                                    'define initial conds.
40 PHI(0)=0: PHI(N%)=0
45 FOR I%=1 TO N%-1
50     PHI(I%)=FNEXACT(I%*H)
55 NEXT I%
60 '
65 FOR IT%=1 TO 50                                        'loop over time steps
70     PT=0                                               'old PHI at last point
75     FOR I%=1 TO N%-1                                   'loop over the lattice
80         PS=PHI(I%)+DTH*(PT+PHI(I%+1)-2*PHI(I%))        'new PHI at this point
85         PT=PHI(I%)                                     'old PHI for next point
90         PHI(I%)=PS                                     'store new PHI
95     NEXT I%
100 '
```

```
105    IF IT% MOD 5<>0 THEN GOTO 140                    'output every 5 steps
110        T=IT%*DT
115        PRINT USING "time=#.######";T
120        FOR I%=1 TO N%-1
125            EXACT=FNEXACT(I%*H)
130            PRINT USING "phi(##)=+#.###^^^^ exact=+#.###^^^^";I%,PHI(I%),EXACT
135        NEXT I%
140 NEXT IT%
```

Typical results from this code are shown in Figure 7.1. Things seem to be working fine for a time step of 0.00075 and the results agree reasonably well with the analytical solution of a spreading gaussian,

$$\varphi(x,t)=\tau^{-\frac{1}{2}}\left[e^{-20(x-\frac{1}{2})^2/\tau}-e^{-20(x-\frac{3}{2})^2/\tau}-e^{-20(x+\frac{1}{2})^2/\tau}\right]; \quad \tau=1+80t.$$

This time step is, however, quite small compared to the natural time scale of the solution, $t\approx 0.01$, so that many steps are required before anything interesting happens. If we try to increase the time step, even to only 0.001, things go very wrong: an unphysical instability develops in the numerical solution, which quickly acquires violent oscillations from one lattice point to another soon after $t=0.02$.

It is easy to understand what is happening here. Let the set of states ψ_λ be the eigenfunctions of the discrete operator H with eigenvalues ε_λ. Since H is an hermitian operator, the eigenvalues are real and the eigenvectors can be chosen to be orthonormal. We can expand the solution at any time in this basis as

$$\varphi^n=\sum_\lambda \varphi_\lambda^n\psi_\lambda.$$

The exact time evolution is given by

$$\varphi^n=e^{-nH\Delta t}\varphi^0,$$

so that each component of the solution should evolve as

$$\varphi_\lambda^n=e^{-n\varepsilon_\lambda\Delta t}\varphi_\lambda^0.$$

This corresponds to the correct behavior of the diffusion equation, where short-wavelength components (with larger eigenvalues) disappear more rapidly as the solution "smooths out". However, (7.5) shows that the explicit scheme will evolve the expansion coefficients as

$$\varphi_\lambda^n=(1-\varepsilon_\lambda\Delta t)^n\varphi_\lambda^0. \tag{7.6}$$

As long as Δt is chosen to be small enough, the factor in (7.6) approximates $e^{-n\varepsilon_\lambda\Delta t}$ and the short-wavelength components damp with time. However, if the time step is too large, one or more of the quantities $1-\varepsilon_\lambda\Delta t$ has an absolute value larger than unity. The

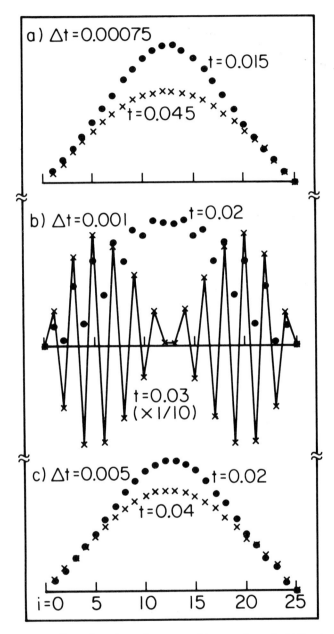

Figure 7.1 Results for the one-dimensional diffusion of a gaussian; a fixed lattice spacing $h = 0.04$ is used in all calculations. a) and b) result from the explicit algorithm (7.5) while c) is the implicit algorithm (7.8).

corresponding components of the initial solution, even if present only due to very small numerical round-off errors, are then amplified with each time step, and soon grow to dominate.

To quantify this limit on Δt, we have some guidance in that the eigenvalues of H are known analytically in this simple model problem. It is easily verified that the functions

$$(\psi_\lambda)_i = \sin \frac{\lambda \pi i}{N}$$

are (un-normalized) eigenfunctions of H with the correct boundary conditions on a lattice of $N+1$ points for $\lambda = 1, 2, \cdots, N-1$ and that the associated eigenvalues are

$$\varepsilon_\lambda = \frac{4}{h^2} \sin^2 \frac{\lambda \pi}{2N}.$$

The largest eigenvalue of H is $\varepsilon_{N-1} \approx 4h^{-2}$, which corresponds to an eigenvector that alternates in sign from one lattice point to the next. Requiring $|1 - \varepsilon_{N-1} \Delta t| < 1$ then restricts Δt to be less than $\frac{1}{2}h^2$, which is 0.0008 in the example we are considering.

The question of stability is quite distinct from that of accuracy, as the limit imposed on the time step is set by the spatial step used and not by the characteristic time scale of the solution, which is much larger. The explicit scheme we have discussed is unsatisfactory, as the instability forces us to use a much smaller time step than is required to describe the evolution adequately. Indeed, the situation gets even worse if we try to use a finer spatial lattice to obtain a more accurate solution. Although the restriction on Δt that we have derived is rigorous only for the simple case we have considered, it does provide a useful guide for more complicated situations, as the eigenvector of H with the largest eigenvalue will always oscillate from one lattice point to the next; its eigenvalue is therefore quite insensitive to the global features of the problem.

Exercise 7.1 Use the code above to verify that the instability sets in for smaller time steps if the spatial lattice is made finer and that the largest possible stable time step is roughly $\frac{1}{2}h^2$. Show that the instability is present for other initial conditions, and that round-off causes even the exact lowest eigenfunction of H to be unstable under numerical time evolution.

7.2 Implicit schemes and the inversion of tri-diagonal matrices

One way around the instability of the explicit algorithm described above is to retain the general form of (7.4), but to replace the second space derivative by that of the solution at the new time. That is, (7.4) is modified to

$$\frac{\varphi_i^{n+1}-\varphi_i^n}{\Delta t} = \frac{1}{h^2}(\delta^2\varphi^{n+1})_i + S_i^n. \qquad (7.7)$$

This is an *implicit* scheme, since the unknown, φ^{n+1}, appears on both side of the equation. We can, of course, solve for φ^{n+1} as

$$\varphi^{n+1}= \frac{1}{1+H\Delta t}[\varphi^n + S^n \Delta t]. \qquad (7.8)$$

This scheme is equivalent to (7.5) to lowest order in Δt. However, it is much better in that larger time steps can be used, as the operator $(1+H\Delta t)^{-1}$ has eigenvalues $(1+\varepsilon_\lambda\Delta t)^{-1}$, all of whose moduli are less than 1 for any Δt. All components of the solution therefore decrease with each time step, as they should. Although this decrease is inaccurate (i.e., not exponential) for the most rapidly oscillating components, such components should not be large in the initial conditions if the spatial discretization is accurate. In any event, there is no amplification, which implies stability. For the slowly varying components of the solution corresponding to small eigenvalues, the evolution closely approximates the exponential at each time step.

Note that if we had tried to be more accurate than (7.7) and had used the average of the second space derivatives at the two time steps involved,

$$\frac{\varphi_i^{n+1}-\varphi_i^n}{\Delta t} = \frac{1}{h^2}(\delta^2\tfrac{1}{2}[\varphi^{n+1}+\varphi^n])_i + S_i^n, \qquad (7.9)$$

so that the time evolution is effected by

$$\varphi^{n+1}= \frac{1}{1+\tfrac{1}{2}H\Delta t}[(1-\tfrac{1}{2}H\Delta t)\varphi^n + S^n \Delta t], \qquad (7.10)$$

this would have been almost as good as the implicit scheme, because the components of the solution are diminished by factors whose absolute values are less than one.

A potential drawback of the implicit scheme (7.8) is that it requires the solution of a set of linear equations (albeit tridiagonal) at each time step to find φ^{n+1}; this is equivalent to the application of the inverse of the matrix $1+H\Delta t$ to the vector appearing in brackets. Since the inverse itself is time-independent, it might be found only once at the beginning of the calculation and then used for all times, but application still requires of order N^2 operations if done directly. Fortunately, the following algorithm (Gaussian elimination and back-substitution, [Va62]) provides a very efficient solution (of order N operations) of a tri-diagonal system of equations such as that posed by Eq. (7.8).

Let us consider trying to solve the tri-diagonal linear system of equations $\mathbf{A}\varphi = \mathbf{b}$ for the unknowns φ_i:

$$A_i^- \varphi_{i-1} + A_i^0 \varphi_i + A_i^+ \varphi_{i+1} = b_i. \tag{7.11}$$

Here, the $A_i^{\pm,0}$ are the only non-vanishing elements of $\mathbf{A}$ and the b_i are known quantities. This is the form of the problem posed by the evaluation of (7.8) for φ^{n+1}, where φ_0 and φ_N are given by the Dirichlet boundary conditions. In particular,

$$b_i = \varphi_i^n + S_i^n \Delta t, \quad A_i^0 = 1 + \frac{2\Delta t}{h^2}, \quad A_i^{\pm} = -\frac{\Delta t}{h^2}.$$

To solve this system of equations, we assume that the solution satisfies a one-term forward recursion relation of the form

$$\varphi_{i+1} = \alpha_i \varphi_i + \beta_i, \tag{7.12}$$

where the α_i and β_i are coefficients to be determined. Substituting this into (7.11), we have

$$A_i^- \varphi_{i-1} + A_i^0 \varphi_i + A_i^+ (\alpha_i \varphi_i + \beta_i) = b_i, \tag{7.13}$$

which can solved for φ_i to yield

$$\varphi_i = \gamma_i A_i^- \varphi_{i-1} + \gamma_i (A_i^+ \beta_i - b_i), \tag{7.14}$$

with

$$\gamma_i = -\frac{1}{A_i^0 + A_i^+ \alpha_i}. \tag{7.15}$$

Upon comparing Eq. (7.14) with (7.12), we can identify the following backward recursion relations for the α's and β's:

$$\alpha_{i-1} = \gamma_i A_i^-; \tag{7.16a}$$

$$\beta_{i-1} = \gamma_i (A_i^+ \beta_i - b_i). \tag{7.16b}$$

The strategy to solve the system should now be clear. We will use the recursion relations (7.15,16) in a backwards sweep of the lattice to determine the α_i and β_i for i running from $N-2$ down to 0. The starting values to be used are

$$\alpha_{N-1} = 0, \quad \beta_{N-1} = \varphi_N,$$

which will guarantee the correct value of φ at the last lattice point. Having determined these coefficients, we can then use the recursion relation (7.12) in a forward sweep from $i=0$ to $N-1$ to determine the solution, with the starting value φ_0 known from the boundary conditions. We have then determined the solution in only two sweeps of the lattice, involving of order N arithmetic operations.

The following code implements the algorithm (7.8) for the model diffusion problem we have been considering. Results are shown in Figure 7.1 c), where it is clear that accurate solutions can be obtained with a much larger time step than can be used with the explicit scheme; the increase in numerical effort per time step is only about a factor of two. Note that the α_i and γ_i are independent of **b**, so that, as the inversion must be done at every time step, it is more efficient to compute these coefficients only once and store them at the beginning of the calculation (lines 70-95); only the β_i then need be computed for each inversion (lines 115-130).

```
5 N%=25: H=1/N%                                      'lattice parameters
10 DT=.005: DTH=DT/H^2                               'time step, useful cnst
15 DIM PHI(N%)                                       'array for the solution
20 DIM ALPHA(N%),BETA(N%),GAMMA(N%)                  'coefficient arrays
25 DEF FNEXACT(X)=FNGAUSS(X)-FNGAUSS(X-1)-FNGAUSS(X+1) 'functions for the
30 DEF FNGAUSS(X)=EXP(-20*(X-.5)^2/(1+80*T))/SQR(1+80*T)' analytical solution
35 '
40 T=0                                               'define initial conds.
45 PHI(0)=0: PHI(N%)=0
50 FOR I%=1 TO N%-1
55    PHI(I%)=FNEXACT(I%*H)
60 NEXT I%
65 '                                                 'find ALPHA and GAMMA
70 AP=-DTH: AZ=1+2*DTH                               'A coefficients
75 ALPHA(N%-1)=0: GAMMA(N%-1)=-1/AZ                  'starting values
80 FOR I%=N%-1 TO 1 STEP -1                          'backward sweep
85    ALPHA(I%-1)=GAMMA(I%)*AP
90    GAMMA(I%-1)=-1/(AZ+AP*ALPHA(I%-1))
95 NEXT I%
100 '
105 FOR IT%=1 TO 50                                  'loop over time steps
110 '                                                'find BETA
115    BETA(N%-1)=0                                  'starting value
120    FOR I%=N%-1 TO 1 STEP -1                      'backward sweep
125       BETA(I%-1)=GAMMA(I%)*(AP*BETA(I%)-PHI(I%))
130    NEXT I%
135 '                                                'find new PHI
140    PHI(0)=0                                      'starting value
145    FOR I%=0 TO N%-1                              'forward sweep
150       PHI(I%+1)=ALPHA(I%)*PHI(I%)+BETA(I%)
155    NEXT I%
160 '
165    T=IT%*DT                                      'output every time step
170    PRINT USING "time=#.######";T
175    FOR I%=1 TO N%-1
180       EXACT=FNEXACT(I%*H)
185       PRINT USING "phi(##)=+#.###^^^^ exact=+#.###^^^^";I%,PHI(I%),EXACT
190    NEXT I%
195 NEXT IT%
```

Exercise 7.2 Use the code above to investigate the accuracy and stability of the implicit algorithm for various values of Δt and for various lattice spacings. Study the evolution of various initial conditions and verify that they correspond to your intuition. Incorporate sources or sinks along the lattice and study the solutions that arise when φ vanishes everywhere at $t=0$.

Exercise 7.3 Modify the code above to apply the algorithm (7.10) and study its stability properties. Explore the effects of taking different linear combinations of φ^n and φ^{n+1} on the right-hand side of Eq. (7.9) (e.g., $\frac{3}{4}\varphi^n + \frac{1}{4}\varphi^{n+1}$).

Exercise 7.4 A simple way to impose the Neumann boundary condition

$$\left.\frac{\partial \varphi}{\partial x}\right|_{x=0} = g$$

is to require that $\varphi_1 = \varphi_0 + hg$. Show that this implies that the forward recursion of (7.12) is to be started with

$$\varphi_0 = \left[\frac{hg - \beta_0}{\alpha_0 - 1}\right].$$

What is the analogous expression implied by the more accurate constraint $\varphi_1 = \varphi_{-1} + 2hg$? What are the initial conditions for the backward recursion of the α and β coefficients if Neumann boundary conditions are imposed at the right-hand edge of the lattice? Modify the code above to incorporate the boundary condition that φ have vanishing derivative at $x=0$ (as is appropriate if an insulator is present in a heat conduction problem) and observe its effect on the solutions. Show that the inversion scheme for tri-diagonal matrices discussed above cannot be applied if periodic boundary conditions are required (i.e., $\varphi_N = \varphi_0$).

Exercise 7.5 Solve the boundary value problem posed in Exercise 3.4 by discretization and inversion of the resulting tri-diagonal matrix. Compare your solution with those obtained by the Green's function method and with the analytical result.

7.3 Diffusion and boundary value problems in two dimensions

The discussion above shows that diffusion in one dimension is best handled by an implicit method and that the required inversion of a tri-diagonal matrix is a relatively simple task. It is therefore natural to attempt to extend this approach to two or more spatial dimensions. For the two-dimensional diffusion equation,

$$\frac{\partial \varphi}{\partial t} = \nabla^2 \varphi,$$

the discretization is straightforward and, following our development for the one-dimensional problem, the time evolution should be effected by

$$\varphi^{n+1} = \frac{1}{1 + H \Delta t} \varphi^n, \tag{7.17}$$

where

$$(H\varphi)_{ij} \equiv -\frac{1}{h^2} [(\delta_i^2 \varphi)_{ij} + (\delta_j^2 \varphi)_{ij}].$$

Unfortunately, while H is very sparse, it is not tri-diagonal, so that the algorithm that worked so well in one dimension does not apply; some thought shows that H cannot be put into a tri-diagonal form by any permutation of its rows or columns. However, the fact that H can be written as a sum of operators that separately involve differences only in the i or j indices:

$$H = H_i + H_j ; \quad H_{i,j} = -\frac{1}{h^2} \delta_{i,j}^2, \tag{7.18}$$

means that an expression equivalent to (7.17) through order Δt is

$$\varphi^{n+1} = \frac{1}{1 + H_i \Delta t} \frac{1}{1 + H_j \Delta t} \varphi^n. \tag{7.19}$$

This can now be evaluated exactly, as each of the required inversions involves a tri-diagonal matrix. In particular, if we define the auxilliary function $\varphi^{n+\frac{1}{2}}$, we can write

$$\varphi^{n+\frac{1}{2}} = \frac{1}{1 + H_j \Delta t} \varphi^n ; \quad \varphi^{n+1} = \frac{1}{1 + H_i \Delta t} \varphi^{n+\frac{1}{2}}.$$

Thus, $(1 + H_j \Delta t)^{-1}$ is applied by forward and backward sweeps of the lattice in the j direction, independently for each value of i. The application of $(1 + H_i \Delta t)^{-1}$ is then carried out by forward and backward sweeps in the i direction, independently for each value of j. This "alternating-direction" scheme is easily seen to be stable for all values of the time step and is generalized straightforwardly to three dimensions.

The ability to invert a tri-diagonal matrix exactly and the idea of treating separately each of the second derivatives of the Laplacian also provides a class of iterative alternating-direction methods [Wa66] for solving the elliptic boundary value problems discussed in Chapter 6. The matrix involved is written as the sum of several parts, each containing second differences in only one of the lattice indices. In two dimensions, for example, we seek to solve (see Eq. (6.3b))

$$(H_i + H_j)\varphi = s. \tag{7.20}$$

If we add a term $\omega\varphi$ to both sides of this equation, where ω is a constant discussed below, the resulting expression can be solved for φ in two different ways:

$$\varphi = \frac{1}{\omega + H_i}[s - (H_j - \omega)\varphi]; \tag{7.21a}$$

$$\varphi = \frac{1}{\omega + H_j}[s - (H_i - \omega)\varphi]. \tag{7.21b}$$

This pair of equations forms the basis for an iterative method of solution: they are solved in turn, φ on the right-hand sides being taken as the previous iterate; the solution involves only the inversion of tri-diagonal matrices. The optimal choice of the "acceleration parameter", ω, depends on the matrices involved and can even be taken to vary from one iteration to the next to improve convergence. A rather complicated analysis is required to find the optimal values [Wa66]. However, a good rule of thumb is to take

$$\omega = (\alpha\beta)^{\frac{1}{2}}, \tag{7.22}$$

where α and β are, respectively, lower and upper bounds to the eigenvalues of H_i and H_j. In general, these alternating direction methods are much more efficient than the simple relaxation scheme discussed in Chapter 6, although they are slightly more complicated to program. Note that there is a slight complication when boundary conditions are also specified in the interior of the lattice.

7.4 Iterative methods for eigenvalue problems

Our analysis of the diffusion equation (7.1) shows that the net result of time evolution is to enhance those components of the solution with smaller eigenvalues of H relative to those with larger eigenvalues. Indeed, for very long times it is only that component with the lowest eigenvalue that is significant, although it has very small amplitude. This situation suggests a scheme for finding the lowest eigenvalue of an elliptic operator, as defined by (6.2): guess a trial eigenvector and subject it to a fictitious time evolution that will "filter" it to the eigenvector having lowest eigenvalue. Since we are dealing with a linear problem, the relentlessly decreasing or increasing magnitude of the solution can be avoided by renormalizing continuously as time proceeds.

To make the discussion concrete, consider the time-independent Schroedinger equation in one dimension with $\hbar = 2m = 1$. The eigenvalue problem is

$$\left[-\frac{d^2}{dx^2} + V(x)\right]\varphi = \varepsilon\varphi, \tag{7.23}$$

with the normalization condition

$$\int dx\,\varphi^2=1$$

(φ can always be chosen to be real if V is). The corresponding energy functional is

$$E=\int dx\left[\frac{1}{2}\left(\frac{d\varphi}{dx}\right)^2+V(x)\varphi^2(x)\right].\tag{7.24}$$

On general grounds, we know that E is stationary at an eigenfunction with respect to variations of φ that respect the normalization condition and that the value of E at this eigenfunction is the associated eigenvalue.

To derive a discrete approximation to this problem, we discretize E as

$$E=\sum_i h\left[\frac{1}{2}\frac{(\varphi_i-\varphi_{i-1})^2}{h^2}+V_i\varphi_i^2\right],\tag{7.25}$$

and the normalization constraint takes the form

$$\sum_i h\,\varphi_i^2=1.\tag{7.26}$$

Variation with respect to φ_i gives the eigenvalue problem

$$(H\varphi)_i\equiv-\frac{1}{h^2}(\delta^2\varphi)_i+V_i\varphi_i=\varepsilon\varphi_i,\tag{7.27}$$

with ε entering as a Lagrange multiplier ensuring the normalization. (Compare with the derivation of the Hartree-Fock equations given in Project III.)

We can interpret (7.27) as defining the problem of finding the real eigenvalues and eigenvectors of a (large) symmetric tridiagonal matrix representing the operator H. Although direct methods for solving this problem were discussed in Chapter 5, they cannot be applied to the very large banded matrices that arise in two- and three-dimensional problems. However, in such cases, one is usually interested in the few highest or lowest eigenvalues of the problem, and for these the diffusion analogy is appropriate. Thus, we consider the problem

$$\frac{\partial\varphi}{\partial\tau}=-H\varphi,$$

where τ is a "fake" time. For convenience, we suppose that things have been arranged so that the lowest eigenvalue of H is positive definite. (This can be guaranteed by shifting H by a spatially-independent constant chosen so that the resultant V_i is positive for all i.)

To solve this "fake" diffusion problem, any of the algorithms discussed above can be employed. The simplest is the explicit scheme analogous to (7.5):

$$\varphi^{n+1} \sim (1 - H\Delta\tau)\varphi^n, \qquad (7.28)$$

where $\Delta\tau$ is a small, positive parameter. Here, the symbol $\sim$ is used to indicate that φ^{n+1} is to be normalized to unity according to (7.26). For an initial guess, we can choose φ^0 to be anything not orthogonal to the exact eigenfunction, although guessing something close to the solution will speed the convergence. At each step in this refining process, computation of the energy from (7.25) furnishes an estimate of the true eigenvalue.

As an example, consider the problem where

$$V=0, \; \varphi(0) = \varphi(1) = 0;$$

this corresponds to a free particle in hard-walled box of unit length. The analytical solutions for the normalized eigenfunctions are

$$\psi_\lambda = 2^{\frac{1}{2}} \sin \lambda\pi x,$$

and the associated eigenvalues are

$$\varepsilon_\lambda = \lambda^2\pi^2.$$

Here, λ is a non-zero integer. The following BASIC program implements the scheme (7.28) on a lattice of 21 points and calculates the energy (7.25) at each iteration. The initial trial function $\varphi \sim x(1-x)$ is used, which roughly approximates the shape of the exact ground state.

```
5   N%=20: H=1/N%                               'define lattice parameters
10  DT=.0005: DTH=DT/H^2                         'time step, useful cnst.
15  DIM PHI(N%)                                  'array for the solution
20  '
25  FOR I%=0 TO N%                               'initial guess for PHI
30      X=I%*H: PHI(I%)=X*(1-X)
35  NEXT I%
40  GOSUB 1000                                   'normalize guess for PHI
45  '
50  FOR IT%=1 TO 60                              'iteration loop
55      PT=0                                     'apply (1-H*DT)
60      FOR I%=1 TO N%-1
65          PS=PHI(I%)+DTH*(PT+PHI(I%+1)-2*PHI(I%))
70          PT=PHI(I%): PHI(I%)=PS
75      NEXT I%
80      GOSUB 1000                               'normalize PHI
85  '
90      E=0                                      'calculate the energy
95      FOR I%=1 TO N%
100         E=E+(PHI(I%)-PHI(I%-1))^2
105     NEXT I%
```

```
110  '                                              'output
115    PRINT USING "iteration=## energy=+## ####^^^^";IT%,E/H
120  NEXT IT%
125  END
1000 '*****************************************************************
1005 'subroutine to normalize phi to unity
1010 '*****************************************************************
1015 NORM=0
1020 FOR I%=1 TO N%
1025     NORM=NORM+PHI(I%)^2
1030 NEXT I%
1035 NORM=1/SQR(H*NORM)
1040 FOR I%=1 TO N%
1045     PHI(I%)=PHI(I%)*NORM
1050 NEXT I%
1055 RETURN
```

The results generated by the code above for several different values of $\Delta\tau$ are shown in Table 7.1. Note that for values of $\Delta\tau$ smaller than the stability limit, $\frac{1}{2}h^2 \approx 0.00125$, the energy converges to the expected answer and does so more rapidly for larger $\Delta\tau$. However, if $\Delta\tau$ becomes too large, the large-eigenvalue components of trial eigenfunction are amplified rather than diminished, and the energy obtained finally corresponds to the *largest* eigenvalue of H, the exact finite difference value of which is (see the discussion following Eq. (7.6))

$$1600 \sin^2\left[\frac{19\pi}{40}\right] = 1590.15 \, .$$

This phenomenon then suggests how to find the eigenvalue of an operator with the largest absolute value: simply apply H to a trial function many times.

Although the procedure outlined above works, it is unsatisfactory in that the limitation on the size of $\Delta\tau$ caused by the lattice spacing often results in having to iterate many times to refine a poor trial function. This can be alleviated to some extent by choosing a good trial function. Even better is to use an implicit scheme (such as (7.8)) that does not amplify the large-eigenvalue components for any value of $\Delta\tau$. Another possibility is to use $\exp(-H\Delta\tau)$ to refine the trial function. While exact calculation of this matrix can be difficult if large dimensions are involved, it can be well-represented by its series expansion through a finite number of terms. This series is easy to evaluate since it involves only applying H many times to a trial state and generally a larger $\Delta\tau$ can be used than if only the first order approximation to the exponential, (7.28), is employed.

We have shown so far how the method we have discussed can

Table 7.1 Convergence of the lowest eigenvalue of the square-well problem. The analytic result is 9.86960; the exact finite-difference value is 9.84933.

Iteration	$\Delta\tau=.0005$	$\Delta\tau=.0010$	$\Delta\tau=.0015$
4	9.93254	9.90648	9.88909
8	9.90772	9.87841	9.86569
12	9.89109	9.86441	9.87311
16	9.87946	9.85719	10.05357
20	9.87116	9.85342	12.34984
24	9.86519	9.85146	42.96950
28	9.86086	9.85044	379.69290
32	9.85773	9.84991	1301.501
36	9.85544	9.84963	1565.746
40	9.85378	9.84949	1587.891
44	9.85257	9.84941	1589.690
48	9.85169	9.84937	1589.966
52	9.85105	9.84935	1590.061

be used to find the lowest or highest eigenvalue of an operator or matrix. To see how to find other eigenvalues and their associated eigenfunctions, consider the problem of finding the second-lowest eigenvalue. We first find the lowest eigenvalue and eigenfunction by the method described above. A trial function for the second eigenfunction is then guessed and refined in the same way, taking care, however, that at each stage of the refinement the solution remain orthogonal to the lowest eigenfunction already found. This can be done by continuously projecting out that component of the solution not orthogonal to the lowest eigenfunction. (This projection is not required when there is some symmetry, such as reflection symmetry, that distinguishes the two lowest solutions and that is preserved by the refinement algorithm.) Having found, in this way, the second-lowest eigenfunction, the third lowest can be found similarly, taking care that during *its* refinement, it remains orthogonal to both of the eigenfunctions with lower eigenvalues. This process cannot be applied to find more than the few lowest (or highest) eigenvectors, however, as numerical round-off errors in the orthogonalizations to the many previously-found eigenvectors soon grow to dominate.

Although the methods described above have been illustrated by a one-dimensional example, but it should be clear that they can be applied directly to find the eigenvalues and eigenfunctions of elliptic operators in two or more dimensions, for example via Eq. (7.19).

□ □ □   **Exercise 7.6** Extend the code given above to find the eigenfunctions and eigenvalues of the first two excited states of each parity in the one-dimensional square-well. Compare your results with the exact solutions and with the analytical finite-difference values.

□ □ □   **Exercise 7.7** Write a program (or modify that given above) to find the few lowest and few highest eigenvalues and eigenfunctions of the Laplacian operator in a two-dimensional region consisting of a square with a square hole cut out of its center. Investigate how your results vary as functions of the size of the hole.

7.5 The time-dependent Schroedinger equation

The numerical treatment of the time-dependent Schroedinger equation for a particle moving in one dimension provides a good illustration of the power of the techniques discussed above and some striking examples of the operation of quantum mechanics [Go67]. We consider the problem of finding the evolution of the (complex) wavefunction φ under Eq. (7.2), given its value at some initial time. After spatial discretization in the usual way, we have the parabolic problem

$$\frac{\partial \varphi_i}{\partial t} = -i(H\varphi)_i, \tag{7.29}$$

where H is the operator defined in (7.27) and we have put $\hbar=1$. To discretize the time evolution, we have the possibility of employing the analogue of any of the three methods discussed above for the diffusion equation. The explicit method related to (7.5), in which the evolution is effected by $(1-iH\Delta t)$, is unstable for *any* value of Δt because the eigenvalues of this operator,

$$(1-i\varepsilon_\lambda\Delta t),$$

have moduli

$$(1+\varepsilon_\lambda^2\Delta t^2)^{\frac{1}{2}}$$

greater than unity. The implicit scheme analogous to (7.8), in contrast, is stable for all Δt, as the moduli of the eigenvalues of the evolution operator,

$$(1+\varepsilon_\lambda^2\Delta t^2)^{-\frac{1}{2}},$$

are always less than one. This is still unsatisfactory, though, as the numerical evolution then does not have the important unitarity property of the exact evolution; the norm of the wavefunction continually decreases with time. Fortunately, the analogue of (7.10) turns out to be very suitable. It is

$$\varphi^{n+1} = \left(\frac{1-i\frac{1}{2}H\Delta t}{1+i\frac{1}{2}H\Delta t}\right)\varphi^n. \tag{7.30}$$

This evolution operator is manifestly unitary (recall that H is hermitian) with eigenvalues of unit modulus, so that the norm of φ computed according to (7.26) is the same from one time to the next. (Of course, the square of the real wavefunction in (7.26) is to be replaced by the modulus squared of the complex wavefunction.) The algorithm (7.30) also has the desirable feature that it approximates the exact exponential evolution operator, $\exp(-iH\Delta t)$, through second order in Δt, which is one more power than would have been supposed naively.

For actual numerical computation, it is efficient to rewrite (7.30) as

$$\varphi^{n+1} = \left[\frac{2}{1+i\frac{1}{2}H\Delta t} - 1\right]\varphi^n \equiv \chi - \varphi^n. \tag{7.31}$$

This form eliminates the sweep of the lattice required to apply the numerator of (7.30). To find, at each time step, the intermediate function χ defined by

$$(1+i\tfrac{1}{2}H\Delta t)\chi = 2\varphi^n,$$

we write this equation explicitly as

$$-\frac{i\Delta t}{2h^2}\chi_{j+1} + \left[1 + \frac{i\Delta t}{h^2} + \frac{i\Delta t}{2}V_j\right]\chi_j - \frac{i\Delta t}{2h^2}\chi_{j-1} = 2\varphi_j^n, \tag{7.32}$$

which, upon dividing by $-i\Delta t / 2h^2$ becomes

$$\chi_{j+1} + \left[-2 + \frac{2ih^2}{\Delta t} - h^2 V_j\right]\chi_j + \chi_{j-1} = \frac{4ih^2}{\Delta t}\varphi_j^n. \tag{7.33}$$

This has the form of (7.11), and can therefore be solved by the two-sweep method discussed in connection with that equation.

The BASIC program for Example 7, whose source code is contained in Appendix B and in the file EXAM7.BAS on the *Computational Physics* diskette, uses the method described above to solve the time-dependent Schroedinger equation. Several analytical forms of the potential (square well or barrier, gaussian well or barrier, potential step, or parabolic well) can be defined on a 160-point lattice and then altered pointwise using the cursor keys on the keyboard. An initial Gaussian or Lorentzian wavepacket of specified average position and momentum and spatial width can then be set up on the lattice and evolved under the boundary condition that φ vanish at the lattice boundaries. The probability density, $|\varphi|^2$, and potential are displayed at each time step, together with the total probabilities and average positions of those portions of the wavepacket to the left and right of a specified point. When the time-evolution is terminated by pressing "e" , the wavepacket or potential can be altered to study another situation. Since BASIC does not support complex arithmetic, the real and imaginary parts of all expressions have been coded separately.

The following exercises will be useful in improving your understanding of this example. A convenient base-line situation is to use a square-well potential of height or depth 0.1 and half-width 20, an initial wavepacket with average momentum ≈ 0.4 and width ≈ 15, and a time-step of 1 or 2.

□ □ □ **Exercise 7.8** Test the accuracy of the integration by integrating forward for some time interval, changing the sign of the time step, and then continuing the integration for an equal time interval to see if you return to the initial wavepacket. Write a subroutine that calculates the average energy of the wavepacket at each time step and verify that this energy is conserved exactly, independent of the time-step.

□ □ □ **Exercise 7.9** Verify that wavepackets in the absence of any potential and wavepackets in a parabolic potential well behave as you expect them to.

□ □ □ **Exercise 7.10** Send wavepackets of varying widths and incident energies at barriers and wells of various sizes and shapes. Interpret all features that you observe during the time evolution. For square-well and step potentials, compare the fractions of the initial probability transmitted and reflected with the analytical values of the usual transmission and reflection coefficients. Set up a resonance situation by considering scattering from potentials with a "pocket" in them and observe the separation of the initial wavepacket into prompt and delayed components. Set up a "double-well" tunneling situation and observe the evolution.

□ □ □ **Exercise 7.11** Replace the evolution algorithm (7.30) by the unstable explicit method or the non-unitary implicit method and observe the effects.

□ □ □ **Exercise 7.12** Replace the vanishing Dirichlet boundary condition at one lattice end by a zero-derivative Neumann boundary condition and observe what happens when a wavepacket approaches.

□ □ □ **Exercise 7.13** In two- and three-dimensional calculations, the large number of lattice points forces the use of as large a lattice spacing as is possible, so that higher-order approximations to the spatial derivative become necessary [Fl78]. Replace the "three-point" formula for the spatial second derivative by the more accurate "five-point" one listed in Table 1.2. Develop an algorithm to invert the penta-diagonal matrix involved in the time evolution. Implement this in the code and observe any changes in the results or the computational efficiency.

Project VII: Self-organization in chemical reactions

Recent work in several branches of physics has shown that the solutions of non-linear equations can display a rich variety of phenomena. Among these is "pattern selection", in which stable, non-trivial patterns in space and/or time emerge spontaneously from structureless initial conditions. In this project, we will investigate analytically and numerically a model of chemical reactions, the "Brusselator", whose solutions exhibit behavior that is very similar to the striking phenomena observed in actual chemical systems [Wi74]. Our discussion follows that of [Ni77] and [Bo76].

VII.1 Description of the model

We consider a network of chemical reactions in which reagent species A and B are converted into product species D and E through intermediates X and Y:

$$A \to X \,; \tag{VII.1a}$$

$$B + X \to Y + D \,; \tag{VII.1b}$$

$$2X + Y \to 3X \,; \tag{VII.1c}$$

$$X \to E \,. \tag{VII.1d}$$

We assume that the concentrations of A and B are fixed and are spatially uniform and that the species D and E are "dead" in the sense of being chemically inert or being removed from the reaction volume. We also assume that the processes (VII.1a-d) are sufficiently exoergic to make the reverse reactions negligible. Under these conditions, we can write the following equations for the evolution of the concentrations of X and Y:

$$\frac{\partial X}{\partial t} = k_a A - k_b B X + k_c X^2 Y - k_d X + D_X \nabla^2 X \,; \tag{VII.2a}$$

$$\frac{\partial Y}{\partial t} = k_b B X - k_c X^2 Y + D_Y \nabla^2 Y \,. \tag{VII.2b}$$

Here, k_{a-d} are the rate constants for the reactions (VII.1a-d), respectively, and $D_{X,Y}$ are the diffusion constants of the species X, Y.

It is convenient to scale the non-linear diffusion equations (VII.2) to a dimensionless form. If we measure time in units of k_d^{-1}, space in units of l, a characteristic size of the reaction volume, X and Y in units of $(k_d/k_c)^{1/2}$, A in units of $(k_d^3/k_a^2 k_c)^{1/2}$, B in units of k_d/k_b, and $D_{X,Y}$ in units of $k_d l^2$, then Eqs. (VII.2) become

$$\frac{\partial X}{\partial t} = A - (B+1)X + X^2 Y + D_X \nabla^2 X; \tag{VII.3a}$$

$$\frac{\partial Y}{\partial t} = BX - X^2 Y + D_Y \nabla^2 Y. \tag{VII.3b}$$

From this scaling, it is clear that the constants A and B can be of order unity, although $D_{X,Y}$ might be considerably smaller, depending upon the value of l.

One trivial solution to these equations can be found by assuming that X and Y are independent of space and time. Setting all derivatives to zero yields a set of algebraic equations that can be solved for the equilibrium point. After a bit of algebra, we find that equilibrium is at

$$X = X_0 = A, \quad Y = Y_0 = \frac{B}{A}.$$

To completely specify the model, we must give the spatial boundary conditions. Of the various possible choices, two are of particular physical interest. These are the "no-flux" boundary conditions, in which the normal derivatives of X and Y are required to vanish on the surface of the reaction volume (as might the case if the chemistry takes place in a closed vessel) and the "fixed" boundary conditions, where X and Y are required to have their equilibrium values on the surface of the reaction volume.

VII.2 Linear stability analysis

To get some insight into the behavior of this system before beginning to compute, it is useful to analyze the behavior of small perturbations about the equilibrium state. To do so, we put

$$X = X_0 + \delta X(r,t); \tag{VII.4a}$$

$$Y = Y_0 + \delta Y(r,t), \tag{VII.4b}$$

where r represents the spatial variables and δX, δY are small quantities dependent upon space and time. Inserting this into (VII.3) and linearizing in the small quantities, we have

$$\frac{\partial \delta X}{\partial t} = (2X_0 Y_0 - B - 1 + D_X \nabla^2)\delta X + X_0^2 \delta Y; \tag{VII.5a}$$

$$\frac{\partial \delta Y}{\partial t} = (B - 2X_0 Y_0)\delta X + (D_Y \nabla^2 - X_0^2)\delta Y. \tag{VII.5b}$$

We now specialize to the no-flux boundary condition in one dimension. We can then expand δX as

$$\delta X(r,t)= \sum_{m=0}^{\infty} \delta X_m e^{\omega_m t} \cos m\pi r , \qquad (\text{VII.6})$$

and similarly for δY. Here, $0 \le r \le 1$ is the spatial coordinate and the ω_m indicate the stability of each normal mode. In particular, $\text{Re}\,\omega_m < 0$ indicates a stable mode that damps in time, while $\text{Re}\,\omega_m > 0$ indicates an unstable perturbation that grows; a complex ω_m indicates a mode that also oscillates as it grows or damps. For the fixed boundary conditions, the expansion analogous to (VII.6) involves $\sin m\pi r$ rather than $\cos m\pi r$.

Upon introducing the Fourier expansion into the linearized equations (VII.5a,b) and equating the coefficients of each spatial mode, we obtain the following homogeneous eigenvalue equations:

$$\omega_m \delta X_m = (2X_0 Y_0 - B - 1 - D_X m^2 \pi^2)\delta X_m + X_0^2 \delta Y_m ;$$

$$\omega_m \delta Y_m = (B - 2X_0 Y_0)\delta X_m - (X_0^2 + D_Y m^2 \pi^2)\delta Y_m . \qquad (\text{VII.7})$$

It is easy to see that these hold for both types of boundary conditions and that the eigenvalues then satisfy the characteristic equation

$$\omega_m^2 + (\beta_m - \alpha_m)\omega_m + A^2 B - \alpha_m \beta_m = 0, \qquad (\text{VII.8})$$

with

$$\alpha_m = B - 1 - m^2 \pi^2 D_X; \quad \beta_m = A^2 + m^2 \pi^2 D_Y. \qquad (\text{VII.9})$$

(We have here written X_0 and Y_0 explicitly in terms of A and B). The roots of this equation are

$$\omega_m^{\pm} = \tfrac{1}{2}\{\alpha_m - \beta_m \pm [(\alpha_m + \beta_m)^2 - 4A^2 B]^{1/2}\}. \qquad (\text{VII.10})$$

A detailed analysis of Eqs. (VII.9,10) reveals several interesting aspects about the stability of the uniform equilibrium state. It is easy to see that if m becomes large for fixed values of A, B, and $D_{X,Y}$, then the $\omega_m^{\pm}$ become very negative; modes with large wavenumbers are therefore stable. Let us now imagine that A and $D_{X,Y}$ are fixed and B is allowed to vary. Then, since there are complex roots only if the discriminant in (VII.10) is negative, we can conclude that there will be oscillating modes only if

$$(A - \Delta_m^{1/2})^2 < B < (A + \Delta_m^{1/2})^2, \qquad (\text{VII.11})$$

where

$$\Delta_m = 1 + m^2 \pi^2 (D_X - D_Y).$$

Since we must have $\Delta_m > 0$, this implies that there are oscillations

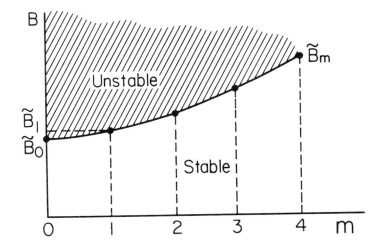

B

$\widetilde{B}_m$

Unstable

$\widetilde{B}_1$
$\widetilde{B}_0$

Stable

0   1   2   3   4   m

Figure VII.1 Stability of the uniform equilibrium state with respect to oscillatory behavior. The $m=0$ mode is absent for fixed boundary conditions. (From [Ni77].)

only when

$$D_Y-D_X<\frac{1}{m^2\pi^2}.$$

Furthermore, if there is a complex eigenvalue, its real part will be positive (an unstable mode) only if

$$B>\widetilde{B}_m\equiv1+A^2+m^2\pi^2(D_X+D_Y).$$

The situation is summarized in Figure VII.1.

. Modes with real, positive frequencies are present only if $\alpha_m\beta_m-A^2B>0$, which implies that

$$B>B_m\equiv1+A^2\left[\frac{D_X}{D_Y}+\frac{1}{D_Ym^2\pi^2}\right]+D_Xm^2\pi^2. \qquad \text{(VII.12)}$$

If we imagine m to be a continuous variable, then it is easy to see that B_m has a minimum value of

$$B_\mu=\left[1+A\left(\frac{D_X}{D_Y}\right)^{\frac{1}{2}}\right]^2$$

at

$$m=\mu=\left[\frac{A}{\pi^2(D_XD_Y)^{\frac{1}{2}}}\right]^{\frac{1}{2}}.$$

Of course, since the physical values of m are restricted to

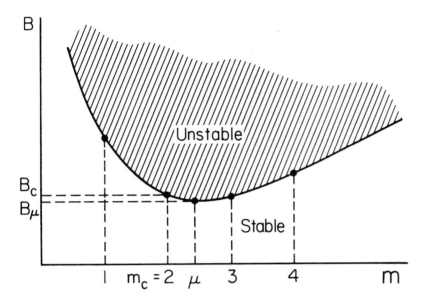

Figure VII.2 Stability of the uniform equilibrium state with respect of exponentially growing perturbations (From [Ni77].)

integers, as B is increased from 0, the non-oscillatory mode m_c that first becomes unstable at $B=B_c$ is that which is closest to the minimum, as shown in Figure VII.2.

To summarize this discussion, the uniform equilibrium state is unstable with respect to perturbations that oscillate in time when $B>\tilde{B}_0$ or $\tilde{B}_1$ for no-flux and fixed boundary conditions, respectively. Similarly, it is unstable with respect to perturbations that grow exponentially in time when $B>B_c$.

VII.3 Numerical solution of the model

Although the stability analysis presented above gives some hint at the behavior of the system for various ranges of the parameters, the full richness of the model in the non-linear regime can be revealed only by numerical experiments. These can be carried out through the following steps.

Step 1 Verify the algebra in the stability analysis discussed above. Assume a fixed value of $A=2$ in a one-dimensional situation and calculate the $\tilde{B}_m$ and B_c for a set of values of the diffusion constants of order 10^{-3}. How do your results change if the reaction takes place in the unit square in two space dimensions?

□ □ □   **Step 2** Write a program that solves the non-linear diffusion equations (VII.3a,b) in one dimension on the interval $[0,1]$ for the case of no-flux boundary conditions; a reasonable number of spatial points might be between 25 and 100. The diffusion terms should be treated implicitly to prevent unphysical numerical instabilities, while the reaction terms can be treated explicitly. Have your program plot out X and Y at each time step.

□ □ □   **Step 3** Use your one-dimensional program to investigate the behavior of the solutions for different values of B, $D_{X,Y}$. A reasonable place to start might be $D_X = 1 \times 10^{-3}$, $D_Y = 4 \times 10^{-3}$; the linear stability analysis should then give some guidance as to what values of B are interesting. Investigate initial conditions corresponding to smooth sinusoidal and point-wise random perturbations of the uniform equilibrium configuration (the latter can be generated with the help of BASIC's RND function). Verify that you can find cases in which the system relaxes back to the uniform state, in which it asymptotically approaches a time-independent solution with a non-trivial spatial variation (dissipative structure), and in which it approaches a space- and time-dependent oscillating solution. Throughout, make sure that your time step is small enough to allow an accurate integration of the equations.

□ □ □   **Step 4** Extend your one-dimensional code to solve the Brusselator with no-flux boundary conditions in two space dimensions using, for example, an alternating-direction algorithm like Eq. (7.19). Plot your results for X at each time. (The technique of displaying an array of characters, as in Example 6, can be useful.) Investigate parameter ranges and initial conditions as in Step 3.

Chapter 8

Monte Carlo
Methods

Systems with a large number of degrees of freedom are often of interest in physics. Among these are the many atoms in a chunk of condensed matter, the many electrons in an atom, or the infinitely many values of a quantum field at all points in a region of space-time. The description of such systems often involves (or can be reduced to) the evaluation of integrals of very high dimension. For example, the classical partition function for a gas of A atoms at a temperature $1/\beta$ interacting through a pair-wise potential v is proportional to the $3A$-dimensional integral

$$Z = \int d^3r_1 \cdots d^3r_A e^{-\beta \sum_{i<j} v(r_{ij})} . \tag{8.1}$$

The straightforward evaluation of an integral like this by one of the quadrature formulas discussed in Chapters 1 or 4 is completely out of the question except for the very smallest values of A. To see why, suppose that the quadrature allows each coordinate to take on 10 different values (not a very fine discretization), so that the integrand must be evaluated at 10^{3A} points. For a modest value of $A=20$ and a very fast computer capable of some 10^7 evaluations per second, this would take some 10^{53} seconds, more than 10^{34} times the age of the universe! Of course, tricks like exploiting the permutation symmetry of the integrand can reduce this estimate considerably, but it still should be clear that direct quadrature is hopeless.

The Monte Carlo methods discussed in this chapter are ways of efficiently evaluating integrals of high dimension. The name "Monte Carlo" arises from the random or "chance" character of the method and the famous casino in Monaco. The essential idea is not to evaluate the integrand at every one of a large number of quadrature points, but rather at only a representative random sampling of abscissae. This is analogous to predicting the results of an election on the basis of a poll of a small number of voters. Although it is by no means obvious that anything sensible can come out of random numbers in a computer, the Monte Carlo strategy turns out to be very appropriate for a broad class of problems in statistical and quantum mechanics. More detailed presentations of the method than that given here can be found in [Ha64] and [Ka85].

8.1 The basic Monte Carlo strategy

Even though the real power of Monte Carlo methods is in evaluating multi-dimensional integrals, it is easiest to illustrate the basic ideas in a one-dimensional situation. Suppose that we have to evaluate the integral

$$I = \int_0^1 f(x)\,dx$$

for some particular function f. Chapters 1 and 4 discussed several different quadrature formulas that employed values of f at very particular values of x (e.g., equally spaced). However, an alternative way of evaluating I is to think about it as the average of f over the interval $[0,1]$. In this light, a plausible quadrature formula is

$$I \approx \frac{1}{N}\sum_{i=1}^{N} f(x_i). \qquad (8.2)$$

Here, the average of f is evaluated by considering its values at N abscissae, $\{x_i\}$, chosen at random with equal probability anywhere within the interval $[0,1]$. We discuss below methods for generating such "random" numbers, but for now it is sufficient to suppose that there is a computer function (e.g., RND in BASIC) that provides as many of them as are required, one after the other.

To estimate the uncertainty associated with this quadrature formula, we can consider $f_i \equiv f(x_i)$ as a random variable and invoke the central limit theorem for large N. From the usual laws of statistics, we have

$$\sigma_I^2 \approx \frac{1}{N}\sigma_f^2 = \frac{1}{N}\left[\frac{1}{N}\sum_{i=1}^{N} f_i^2 - \left(\frac{1}{N}\sum_{i=1}^{N} f_i\right)^2\right], \qquad (8.3)$$

where σ_f^2 is the variance in f; i.e., a measure of the extent to which f deviates from its average value over the region of integration.

Equation (8.3) reveals two very important aspects of Monte Carlo quadrature. First, the uncertainty in the estimate of the integral, σ_I, decreases as $N^{-\frac{1}{2}}$. Hence, if more points are used, we will get a more accurate answer, although the error decreases very slowly with the number of points (a factor of four more numerical work is required to halve the uncertainty in the answer). This is to be contrasted with a method like the trapezoidal rule, where Eqs. (1.8,1.9) show that the error scales like N^{-2}, which affords a much greater accuracy for a given amount of numerical work. (This advantage vanishes in multi-dimensional cases, as we discuss below.) The second important point to realize from (8.3) is that the precision is greater if σ_f is smaller; that is, if f is as smooth as possible. One limit of this is when f is a constant, in which case we need its value at only one point to define its average. To see the other limit, consider a situation in which f is zero except for a very narrow peak about some value of x. If the x_i have an equal probability to lie anywhere between 0 and 1, it is probable that all but a few of them will lie outside the peak of f, and that only these few of the f_i will be non-zero; this will lead to a poorly defined estimate of I.

As an example of a Monte Carlo evaluation of an integral, consider

$$\int_0^1 \frac{dx}{1+x^2} = \frac{\pi}{4} = 0.78540. \tag{8.4}$$

The following BASIC program calculates this integral for the value of N input, together with an estimate of the precision of the quadrature.

```
10 DEF FNF(X)=1/(1+X^2)              'function to integrate
20 INPUT "enter value of N";N%
30 SUMF=0: SUMF2=0                    'zero sums
40 FOR I%=1 TO N%                     'loop over samples
50     X=RND: FX=FNF(X)
60     SUMF=SUMF+FX: SUMF2=SUMF2+FX^2 'add contributions to sums
70 NEXT I%
80 FAVG=SUMF/N%                       'final results
90 SIGMA=SQR((SUMF2/N%-FAVG^2)/N%)
100 PRINT USING "integral=#.##### +- #.#####";FAVG,SIGMA
```

Results of running this program for various values of N are given in the first three columns of Table 8.1. The calculated result is equal to the exact value within a few (usually less than one) standard deviations and the quadrature becomes more precise as N increases.

Since Eq. (8.3) shows that the uncertainty in a Monte Carlo quadrature is proportional to the variance of the integrand, it is easy to devise a general scheme for reducing the variance and improving the efficiency of the method. Let us imagine multiplying and dividing the integrand by a positive weight function $w(x)$, normalized so that

$$\int_0^1 dx\, w(x) = 1.$$

The integral can then be written as

Table 8.1 Monte Carlo evaluation of the integral (8.4) using two different weight functions, $w(x)$. The exact value is 0.78540.

N	$w(x)=1$		$w(x)=\frac{1}{3}(4-2x)$	
	I	σ_I	I	σ_I
10	0.81491	0.04638	0.79982	0.00418
20	0.73535	0.03392	0.79071	0.00392
50	0.79606	0.02259	0.78472	0.00258
100	0.79513	0.01632	0.78838	0.00194
200	0.78677	0.01108	0.78529	0.00140
500	0.78242	0.00719	0.78428	0.00091
1000	0.78809	0.00508	0.78524	0.00064
2000	0.78790	0.00363	0.78648	0.00045
5000	0.78963	0.00227	0.78530	0.00028

$$I=\int_0^1 dx\; w(x)\frac{f(x)}{w(x)}\;.\tag{8.5}$$

If we now make a change of variable from x to

$$y(x)=\int_0^x dx'\, w(x'),\tag{8.6}$$

so that

$$\frac{dy}{dx}=w(x);\;\; y(x=0)=0;\;\; y(x=1)=1,$$

then the integral becomes

$$I=\int_0^1 dy\;\frac{f(x(y))}{w(x(y))}\;.\tag{8.7}$$

The Monte Carlo evaluation of this integral proceeds as above, namely averaging the values of f/w at a random sampling of points uniformly distributed in y over the interval $[0,1]$:

$$I\approx\frac{1}{N}\sum_{i=1}^N \frac{f(x(y_i))}{w(x(y_i))}\;.\tag{8.8}$$

The potential benefit of the change of variable should now be clear. If we choose a w that behaves approximately as f does (i.e., it is large where f is large and small where f is small), then the integrand in (8.7), f/w, can be made very smooth, with a consequent reduction in the variance of the Monte Carlo estimate (8.8). This benefit is, of course, contingent upon being able to find an appropriate w and upon being able to invert the relation (8.6) to obtain $x(y)$.

A more general way of understanding why a change of variable is potentially useful is to realize that the uniform distribution of points in y implies that the distribution of points in x is $dy/dx=w(x)$. This means that points are concentrated about the most "important" values of x where w (and hopefully f) is large, and that little computing power is spent on calculating the integrand for "unimportant" values of x where w and f are small.

As an example of how a change of variable can improve the efficiency of Monte Carlo quadrature, we consider again the integral (8.4). A good choice for a weight function is

$$w(x)=\frac{1}{3}(4-2x),$$

which is positive definite, decreases monotonically over the range of integration (as does f), and is normalized correctly. Moreover, since $f/w=3/4$ at both $x=0$ and $x=1$, w well approximates the behavior of f. According to (8.6), the new integration variable is

$$y=\frac{1}{3}x(4-x),$$

which can be inverted to give

$$x=2-(4-3y)^{\frac{1}{2}}.$$

The following BASIC code then evaluates I according to (8.8).

```
10  DEF FNF(X)=1/(1+X^2): FNW(X)=(4-2*X)/3        'define functions
15  DEF FNX(Y)=2-SQR(4-3*Y)
20  INPUT "enter value of N";N%
30  SUMFW=0: SUMFW2=0                             'zero sums
40  FOR I%=1 TO N%                                'loop over samples
50     Y=RND: X=FNX(Y)
55     FOVERW=FNF(X)/FNW(X)
60     SUMFW=SUMFW+FOVERW: SUMFW2=SUMFW2+FOVERW^2  'add terms to sums
70  NEXT I%
80  FWAVG=SUMFW/N%                                'final results
90  SIGMA=SQR((SUMFW2/N%-FWAVG^2)/N%)
100 PRINT USING "integral=#.##### +- #.#####";FWAVG,SIGMA
```

Results of running this code for various values of N are shown in the last two columns of Table 8.1, where the improvement over the $w=1$ case treated previously is evident.

The one-dimensional discussion above can be readily generalized to d-dimensional integrals of the form $I=\int d^d x\, f(\mathbf{x})$. The analog of (8.2) is

$$I\approx\frac{1}{N}\sum_{i=1}^{N}f(\mathbf{x}_i),\qquad\qquad(8.9)$$

with the several components of the random points $\mathbf{x}_i$ to be chosen independently. Thus, the following BASIC program calculates

$$\pi = 4 \int_0^1 dx_1 \int_0^1 dx_2 \, \theta(1 - x_1^2 - x_2^2);$$

that is, it compares the area of a quadrant of the unit circle to that of the unit square (θ is the unit step function):

```
10 N%=5000: COUNT%=0
20 FOR I%=1 TO N%
30    X=RND: Y=RND:
40    IF X^2+Y^2<1 THEN COUNT%=COUNT%+1
50 NEXT I%
60 PI4=COUNT%/N%: SIGMA=SQR(PI4*(1-PI4)/N%)
70 PRINT USING "PI=#.#### +- #.####";4*PI4,4*SIGMA
```

When run, this program generated the satisfactory result of 3.1424 ± 0.0232.

□ □ □   **Exercise 8.1** Verify that the error estimate used in line 60 of this program is that given by Eq. (8.3).

The change of variable discussed above can also be generalized to many dimensions. For a weight function $w(\mathbf{x})$ normalized so that its integral over the region of integration is unity, the appropriate new variable of integration is $\mathbf{y}$, where the Jacobian is $|\partial \mathbf{y} / \partial \mathbf{x}| = w(\mathbf{x})$. It is generally very difficult (if not impossible) to construct $\mathbf{x}(\mathbf{y})$ explicitly, so that it is more convenient to think about the change of variable in the multi-dimensional case in the sense discussed above; i.e., it distributes the points $\mathbf{x}_i(\mathbf{y}_i)$ with distribution w. Various practical methods for doing this for arbitrary w will be discussed below.

Although the results were satisfactory in the examples given above, Monte Carlo quadrature does not appear to be particularly efficient. Even with the "good" choice of w, the results in Table 8.1 show a precision of only about 10^{-4} for $N=5000$, whereas the conventional trapezoidal formula with 5000 points is accurate to better than 10^{-5}. However, consider evaluating a multi-dimensional integral such as (8.1). Suppose that we are willing to invest a given amount of numerical work (say to evaluate the integrand N times), and wish to compare the efficiencies of conventional and Monte Carlo quadratures. In a conventional quadrature, say a multi-dimensional analog of the trapezoidal rule, if there are a total of N points, then each dimension of a d-dimensional integral is broken up into $\sim N^{1/d}$ intervals of spacing $h \sim N^{-1/d}$. The analog of (1.9) shows that the error in the integral over each cell of volume h^d in the integration region is $O(h^{d+2})$, so that the total error in the conventional quadrature is

$$NO(h^{d+2})=O(N^{-2/d});$$

for large d, this decreases very slowly with increasing N. On the other hand, Eq. (8.3) above shows that the uncertainty of a Monte Carlo quadrature decreases as $N^{-\frac{1}{2}}$, independent of d. Assuming that the prefactors in these estimates are all of order unity, we see that Monte Carlo quadrature is more efficient when $d \gtrsim 4$. Of course, this estimate depends in detail upon the conventional quadrature scheme we use or how good a weight function is used in the Monte Carlo scheme, but the basic point is the very different way in which the two errors scale with increasing N for large d.

8.2 Generating random variables with a specified distribution

The discussion above shows that Monte Carlo quadrature involves two basic operations: generating abscissae randomly distributed over the integration volume with a specified distribution $w(\mathbf{x})$ (which may perhaps be unity) and then evaluating the function f/w at these abscissae. The second operation is straightforward, but it is not obvious how to generate "random" numbers on a deterministic computer. In this section, we will cover a few of the standard methods used.

The generation of uniformly distributed random numbers is the computer operation that underlies any treatment of a more complicated distribution. There are numerous methods for performing this basic task and for checking that the results do indeed correspond to the uniform random numbers required. One of the most common algorithms, and in fact that used in the BASIC RND function, is a "linear congruential" method, which "grows" a whole sequence of random numbers from a "seed" number. The current member of the sequence is multiplied by a first "magic" number, incremented by a second magic number, and then the sum is taken modulo a third magic number. Thus,

$$x_{i+1}=(ax_i+c)\bmod m, \tag{8.10}$$

where i is the sequence label and a, c, and m are the magic numbers. The latter are often very large, with their precise values depending upon the word length of the computer being used. Note that the x_i cannot be truly "random" as they arise from the seed in a well-defined, deterministic algorithm; indeed, two sequences grown from the the same seed are identical. For this reason, they are often termed "pseudo-random". Nevertheless, for many practical purposes pseudo-random numbers generated in this way can be used as if they were truly random. A good discussion of uniform random number generators and of the tests that can be applied to determine if they are working properly can be found in [Kn69].

We have already seen how to choose one-dimensional random variables distributed as a specified weight function $w(x)$. According to the discussion in the previous section, the procedure is to choose y, the incomplete integral of w, uniformly, and then to find x by inverting Eq. (8.6). Thus, if we are faced with evaluating the integral

$$I = \int\limits_0^\infty dx\, e^{-x} g(x),$$

with g a relatively smooth function, it is sensible to generate x between 0 and ∞ with distribution e^{-x}, and then to average g over these values. According to (8.5), we have

$$y = 1 - e^{-x}, \quad x = -\log(1-y).$$

The following line of BASIC code is therefore a subroutine that generates values of X with the required distribution.

```
110 X=-LOG (1-RND): RETURN
```

□ □ □   **Exercise 8.2** Verify that the code above generates X with the required distribution, for example by generating and histogramming a large number of values of X. Use these values to evaluate I and its uncertainty for $g(x)=x$, x^2, and x^3 and compare your results with the analytical values. Can you understand the trend in the uncertainties calculated?

While the method of generating the incomplete integral is infallible, it requires that we be able to find $x(y)$, which can be done analytically only for a relatively small class of functions. For example, if we had wanted to used

$$w(x) = \frac{6}{5}(1 - \tfrac{1}{2}x^2)$$

in our efforts to evaluate Eq. (8.4), we would have been faced with solving a cubic equation to find $y(x)$. While this is certainly possible, choices of w that might follow more closely the behavior of f generally will lead to more complicated integrals that cannot be inverted analytically. However, it is possible to do the integral and inversion numerically. Let us imagine tabulating the values $x^{(j)}$ for which the incomplete integral of w takes on a series of uniformly spaced values $y^{(j)} \equiv j/M$, $j = 0, 1, ..., M$ that span the interval $[0,1]$. Thus,

$$y^{(j)} \equiv \frac{j}{M} = \int\limits_0^{x^{(j)}} dx'\, w(x'). \tag{8.11}$$

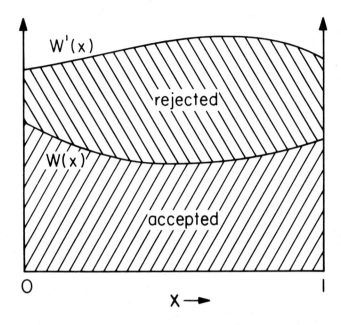

Figure 8.1 Illustration of the von Neumann rejection method for generating random numbers distributed according to a given distribution $w(x)$.

Values of $x^{(j)}$ with j an integer chosen from the set $0, 1, ,\ldots, M$ with equal probability will then approximate the required distribution. (Some special treatment is required to handle the end-points $j=0$ and $j=M$ properly.) To generate the x^j, is, of course, the problem. This can be done by integrating the differential equation $dy/dx=w(x)$ through the simple discretization

$$\frac{y^{(j+1)}-y^{(j)}}{x^{(j+1)}-x^{(j)}}=w(x^{(j)}).$$

Since $y^{(j+1)}-y^{(j)}=1/M$, we have the convenient recursion relation

$$x^{(j+1)}=x^{(j)}+\frac{1}{Mw(x^{(j)})}, \tag{8.12}$$

which can be used with the starting value $x^{(0)}=0$.

Another convenient method for generating one- (or multi-) dimensional random variables is von Neumann rejection, whose geometrical basis is illustrated in Figure 8.1. Suppose we are interested in generating x between 0 and 1 with distribution $w(x)$ and let $w'(x)$ be a positive function such that $w'(x)>w(x)$ over the region of integration. Note that this means that the definite integral of w' is greater than 1. A convenient, but not always useful, choice for w' is any constant greater than the maximum value of w in the region of integration. If we generate points in two dimensions that uniformly fill the area under the curve $w'(x)$ and then "accept" for use only those points that are under $w(x)$, then the accepted points will be distributed according to w. Practically, what is done is to choose two random variables, x_i and η, the former distributed proportional to w' and the latter distributed uniformly between 0 and 1. The value x_i is then accepted if η is less than the ratio $w(x_i)/w'(x_i)$; if a point is rejected, we simply go on and generate another pair of x_i and η. This technique is clearly efficient only if w' is close to w throughout the entire range of integration; otherwise, much time is wasted rejecting useless points.

□ □ □   **Exercise 8.3** Use von Neumann rejection to sample points in the interval $[0,1]$ distributed as $w(x)=\frac{6}{5}(1-\tfrac{1}{2}x^2)$ and evaluate the integral (8.4) and its uncertainty with these points. Compare the efficiency of your calculation for various choices of w',

The Gaussian (or normal) distribution with zero mean and unit variance,

$$w(x)=(2\pi)^{-\frac{1}{2}}e^{-\frac{1}{2}x^2},$$

plays a central role in probability theory and so is often needed in Monte Carlo calculations. It is possible, but not too efficient, to generate this distribution by inverting its incomplete integral using polynomial approximations to the error function. A more "clever" method is based on the central limit theorem, which states that the sum of a large number of uniformly distributed random numbers will approach a Gaussian distribution. Since the mean and variance of the uniform distribution on the interval $[0,1]$ are $1/2$ and $1/12$, respectively, the sum of 12 uniformly distributed numbers will have a mean value of 6 and a variance of 1 and will very closely approximate a Gaussian distribution. Hence, the following BASIC subroutine returns a Gaussian random variable, GAUSS, with zero mean and unit variance:

```
100 'subroutine to generate the normally distributed number GAUSS
110 GAUSS=0
120 FOR I%=1 TO 12
130    GAUSS=GAUSS+RND
140 NEXT I%
150 GAUSS=GAUSS-6
160 RETURN
```

Exercise 8.4 By histogramming a large number of the values of □ □ □
GAUSS produced by the subroutine above, convince yourself that
their distribution is very close to normal. Can you derive a quantitative estimate of the extent to which this distribution deviates
from a Gaussian? Compare with the results of summing 6 or 24
uniformly distributed random numbers. (Note that these latter
sums have means of 3 and 12 and variances of $\frac{1}{2}$ and 2, respectively.)

Another efficient method for generating normally distributed
variables is to consider a Gaussian distribution in two dimensions
(x_1, x_2), for which the number of points in a differential area is
proportional to

$$e^{-\frac{1}{2}(x_1^2+x_2^2)}dx_1 dx_2.$$

In terms of the usual polar coordinates

$$r=(x_1^2+x_2^2)^{\frac{1}{2}}, \quad \theta=\tan^{-1}\frac{x_2}{x_1},$$

the distribution is

$$e^{-\frac{1}{2}r^2}r dr d\theta,$$

or, if $u=\frac{1}{2}r^2$, the distribution is

$$e^{-u}du\, d\theta.$$

Hence, if we generate u between 0 and ∞ with an exponential distribution and θ uniformly between 0 and 2π, then the corresponding values of

$$x_1=(2u)^{\frac{1}{2}}\cos\theta, \quad x_2=(2u)^{\frac{1}{2}}\sin\theta$$

will be distributed normally. Thus, the following BASIC subroutine
returns two normally distributed random variables, GAUSS1 and
GAUSS2:

```
100 'subroutine to return 2 normally distributed numbers, GAUSS1 and GAUSS2
110 TWOU=-2*LOG(1-RND)
120 RADIUS=SQR(TWOU)
130 THETA=2*3.14159*RND
140 GAUSS1=RADIUS*COS(THETA)
150 GAUSS2=RADIUS*SIN(THETA)
160 RETURN
```

To generate a gaussian distribution with mean $\bar{x}$ and variance σ^2,

$$w(x)=(2\pi\sigma)^{-\frac{1}{2}}e^{-\frac{1}{2}(x-\bar{x})^2/\sigma^2},$$

we need only take the values generated by this subroutine, multiply them by σ, and then increment them by $\bar{x}$.

8.3 The algorithm of Metropolis *et al.*

Although the methods we have discussed above for generating random numbers according to a specified distribution can be very efficient, it is difficult or impossible to generalize them to sample a complicated weight function in many dimensions, and so an alternative approach is required. One very general way to produce random variables with a given probability distribution of arbitrary form is known as the Metropolis, Rosenbluth, Rosenbluth, Teller, and Teller algorithm [Me53]. As it requires only the ability to calculate the weight function for a given value of the integration variables, the algorithm has been applied widely in statistical mechanics problems, where the weight function of the canonical ensemble can be a very complicated function of the coordinates of the system (see Eq. (8.1)) and so cannot be sampled conveniently by other methods. However, it is not without its drawbacks.

Although the algorithm of Metropolis *et al.* can be implemented in a variety of ways, we begin by describing one simple realization. Suppose that we want to generate a set of points in a (possibly multi-dimensional) space of variables $\mathbf{X}$ distributed with probability density $w(\mathbf{X})$. The Metropolis algorithm generates a sequence of points, $\mathbf{X}_0$, $\mathbf{X}_1$, $\cdots$, as those visited successively by a random walker moving through $\mathbf{X}$ space; as the walk becomes longer and longer, the points it connects approximate more closely the desired distribution.

The rules by which the random walk proceeds through configuration space are as follows. Suppose that the walker is at a point $\mathbf{X}_n$ in the sequence. To generate $\mathbf{X}_{n+1}$, it makes a trial step to a new point $\mathbf{X}_t$. This new point can be chosen in any convenient manner, for example uniformly at random within a multi-dimensional cube of small side δ about $\mathbf{X}_n$. This trial step is then "accepted" or "rejected" according to the ratio

$$r = \frac{w(\mathbf{X}_t)}{w(\mathbf{X}_n)}.$$

If r is larger than one, then the step is accepted (i.e., we put $\mathbf{X}_{n+1} = \mathbf{X}_t$), while if r is less than one, the step is accepted with probability r. This latter is conveniently accomplished by comparing r with a random number η uniformly distributed in the interval $[0,1]$ and accepting the step if $\eta < r$. If the trial step is not accepted, then it is rejected, and we put $\mathbf{X}_{n+1} = \mathbf{X}_n$. This generates $\mathbf{X}_{n+1}$, and we may then proceed to generate $\mathbf{X}_{n+2}$ by the same process, making a trial step from $\mathbf{X}_{n+1}$. Any arbitrary point, $\mathbf{X}_0$, can be used as the starting point for the random walk.

The following subroutine illustrates the application of the Metropolis algorithm to sample a two-dimensional distribution in the variables X1 and X2. Each call to the subroutine executes another step of the random walk and returns the next values of X1 and X2; the main program must initialize these variables, as well as set the value of DELTA and define the distribution FNW(X1,X2).

```
1000 'subroutine to take a step in the Metropolis algorithm
1010 X1T=X1+DELTA*(2*RND-1)          'take a trial step in square about (X1,X2)
1020 X2T=X2+DELTA*(2*RND-1)
1030 R=FNW(X1T,X2T)/FNW(X1,X2)       'compute the ratio
1040 IF RND>R THEN GOTO 1060         'step rejected
1050    X1=X1T: X2=X2T               'step accepted
1060 RETURN
```

This code could be made more efficient by saving the weight function at the current point of the random walk, so that it need not be computed again when deciding whether or not to accept the trial step; the evaluation of w is often the most time-consuming part of a Monte Carlo calculation using the Metropolis algorithm.

To prove that the algorithm described above does indeed generate a sequence of points distributed according to w, let us consider a large number of walkers starting from different initial points and moving independently through $\mathbf{X}$ space. If $N_n(\mathbf{X})$ is the density of these walkers at $\mathbf{X}$ after n steps, then the net number of walkers moving from point $\mathbf{X}$ to point $\mathbf{Y}$ in the next step is

$$\Delta N(\mathbf{X}) = N_n(\mathbf{X})P(\mathbf{X} \to \mathbf{Y}) - N_n(\mathbf{Y})P(\mathbf{Y} \to \mathbf{X})$$

$$= N_n(\mathbf{Y})P(\mathbf{X} \to \mathbf{Y})\left[\frac{N_n(\mathbf{X})}{N_n(\mathbf{Y})} - \frac{P(\mathbf{Y} \to \mathbf{X})}{P(\mathbf{X} \to \mathbf{Y})}\right]. \tag{8.13}$$

Here, $P(\mathbf{X} \to \mathbf{Y})$ is the probability that a walker will make a transition to $\mathbf{Y}$ if it is at $\mathbf{X}$. This equation shows that there is equilibrium (no net change in population) when

$$\frac{N_n(\mathbf{X})}{N_n(\mathbf{Y})} = \frac{N_e(\mathbf{X})}{N_e(\mathbf{Y})} \equiv \frac{P(\mathbf{Y} \to \mathbf{X})}{P(\mathbf{X} \to \mathbf{Y})} , \tag{8.14}$$

and that changes in $N(\mathbf{X})$ when the system is not in equilibrium tend to drive it toward equilibrium (i.e., $\Delta N(\mathbf{X})$ is positive if there are "too many" walkers at $\mathbf{X}$, or if $N_n(\mathbf{X})/N_n(\mathbf{Y})$ is greater than its equilibrium value). Hence it is plausible (and can be proved) that, after a large number of steps, the population of the walkers will settle down to its equilibrium distribution, N_e.

It remains to show that the the transition probabilities of the Metropolis algorithm lead to an equilibrium distribution of walkers $N_e(\mathbf{X}) \sim w(\mathbf{X})$. The probability of making a step from $\mathbf{X}$ to $\mathbf{Y}$ is

$$P(\mathbf{X} \to \mathbf{Y}) = T(\mathbf{X} \to \mathbf{Y})A(\mathbf{X} \to \mathbf{Y}),$$

where T is the probability of making a trial step from $\mathbf{X}$ to $\mathbf{Y}$ and A

is the probability of accepting that step. If **Y** can be reached from **X** in a single step (i.e., if it is within a cube of side δ centered about **X**), then

$$T(\mathbf{X} \rightarrow \mathbf{Y}) = T(\mathbf{Y} \rightarrow \mathbf{X}),$$

so that the equilibrium distribution of the Metropolis random walkers satisfies

$$\frac{N_e(\mathbf{X})}{N_e(\mathbf{Y})} = \frac{A(\mathbf{Y} \rightarrow \mathbf{X})}{A(\mathbf{X} \rightarrow \mathbf{Y})}. \tag{8.15}$$

If $w(\mathbf{X}) > w(\mathbf{Y})$, then $A(\mathbf{Y} \rightarrow \mathbf{X}) = 1$ and

$$A(\mathbf{X} \rightarrow \mathbf{Y}) = \frac{w(\mathbf{Y})}{w(\mathbf{X})},$$

while if $w(\mathbf{X}) < w(\mathbf{Y})$ then

$$A(\mathbf{Y} \rightarrow \mathbf{X}) = \frac{w(\mathbf{X})}{w(\mathbf{Y})}$$

and $A(\mathbf{X} \rightarrow \mathbf{Y}) = 1$. Hence, in either case, the equilibrium population of Metropolis walkers satisfies

$$\frac{N_e(\mathbf{X})}{N_e(\mathbf{Y})} = \frac{w(\mathbf{X})}{w(\mathbf{Y})},$$

so that the walkers are indeed distributed with the correct distribution.

Note that although we made the discussion concrete by choosing $\mathbf{X}_t$ in the neighborhood of $\mathbf{X}_n$, we can use any transition and acceptance rules that satisfy

$$\frac{w(\mathbf{X})}{w(\mathbf{Y})} = \frac{T(\mathbf{Y} \rightarrow \mathbf{X}) A(\mathbf{Y} \rightarrow \mathbf{X})}{T(\mathbf{X} \rightarrow \mathbf{Y}) A(\mathbf{Y} \rightarrow \mathbf{X})}. \tag{8.16}$$

Indeed, one limiting choice is $T(\mathbf{X} \rightarrow \mathbf{Y}) = w(\mathbf{Y})$, independent of **X**, and $A = 1$. This is the most efficient choice, as no trial steps are "wasted" through rejection. However, this choice is somewhat impractical, because if we knew how to sample w to take the trial step, we wouldn't need to use the algorithm to begin with.

An obvious question is "If trial steps are to be taken within a neighborhood of $\mathbf{X}_n$, how do we choose the step size, δ?". To answer this, suppose that $\mathbf{X}_n$ is at a maximum of w, the most likely place for it to be. If δ is large, then $w(\mathbf{X}_t)$ will likely be very much smaller than $w(\mathbf{X}_n)$ and most trial steps will be rejected, leading to an inefficient sampling of w. If δ is very small, most trial steps will be accepted, but the random walker will never move very far, and so also lead to a poor sampling of the distribution. A good rule of thumb is that the size of the trial step should be chosen so that about half of the trial steps are accepted.

One bane of applying the Metropolis algorithm to sample a distribution is that the points that make up the random walk, $\mathbf{X}_0, \mathbf{X}_1, \cdots$, are not independent of one another, simply from the way in which they were generated; that is, $\mathbf{X}_{n+1}$ is likely to be in the neighborhood of $\mathbf{X}_n$. Thus, while the points might be distributed properly as the walk becomes very long, they are not statistically independent of one another, and some care must be taken in using them to calculate integrals. For example, if we calculate

$$I = \frac{\int d\mathbf{X} w(\mathbf{X}) f(\mathbf{X})}{\int d\mathbf{X} w(\mathbf{X})}$$

by averaging the values of f over the points of the random walk, the usual estimate of the variance, Eq. (8.3), is invalid because the $f(\mathbf{X}_i)$ are not statistically independent. This can be quantified by calculating the auto-correlation function

$$C(k) = \frac{\langle f_i f_{i+k} \rangle - \langle f_i \rangle^2}{\langle f_i^2 \rangle - \langle f_i \rangle^2}. \tag{8.17}$$

Here, $\langle ... \rangle$ indicates average over the random walk; e.g.,

$$\langle f_i f_{i+k} \rangle = \frac{1}{N-k} \sum_{i=1}^{N-k} f(\mathbf{X}_i) f(\mathbf{X}_{i+k}).$$

Of course, $C(0)=1$, but the non-vanishing of C for $k \neq 0$ means that the f's are not independent. What can be done in practice is to compute the integral and its variance using points along the random walk separated by a fixed interval, the interval being chosen so that there is effectively no correlation between the points used. An appropriate sampling interval can be estimated from the value of k for which C becomes small (say $\lesssim 0.1$).

Another issue in applying the Metropolis algorithm is where to start the random walk; i.e., what to take for $\mathbf{X}_0$. In principle, any location is suitable and the results will be independent of this choice, as the walker will "thermalize" after some number of steps. In practice, an appropriate starting point is a probable one, where w is large. Some number of thermalization steps then can be taken before actual sampling begins to remove any dependence on the starting point.

Exercise 8.5 Use the algorithm of Metropolis *et al.* to sample the normal distribution in one dimension. For various trial step sizes, study the acceptance ratio (fraction of trial steps accepted), the correlation function (and hence the appropriate sampling frequency), and the overall computational efficiency. Use the random variables you generate to calculate

$$\int\limits_{-\infty}^{\infty} dx\ x^2 e^{-\frac{1}{2}x^2}$$

and estimate the uncertainty in your answer. Study how your results depend upon where the random walker is started and on how many thermalization steps you take before beginning the sampling. Compare the efficiency of the Metropolis algorithm with that of a calculation that uses one the methods discussed in Section 8.2 to generate the normal distribution directly.

8.4 The Ising model in two dimensions

Models in which the degrees of freedom reside on a lattice and interact locally arise in several areas of condensed matter physics and field theory. The simplest of these is the Ising model [Hu63], which can be taken as a crude description of a magnetic material or a binary alloy. In this example, we will use Monte Carlo methods to calculate the thermodynamic properties of this model.

If we speak in the magnetic language, the Ising model consists of a set of spin degrees of freedom interacting with each other and with an external magnetic field. These might represent the magnetic moments of the atoms in a solid. We will consider in particular a model in two spatial dimensions, where the spin variables are located on the sites of an $N_x \times N_y$ square lattice. The spins can therefore be labeled as S_{ij}, where i, j are the indices for the two spatial directions, or as S_α, where α is a generic site label. Each of these spin variables can be either "up" ($S_\alpha = +1$) or "down" ($S_\alpha = -1$). This mimics the spin-$\frac{1}{2}$ situation, although note that we take the spins to be classical degrees of freedom and do not impose the angular momentum commutation rules characteristic of a quantum description. (Doing so would correspond to the Heisenberg model.)

The Hamiltonian for the system is conventionally written as

$$H = -J \sum_{<\alpha\beta>} S_\alpha S_\beta - B\sum_\alpha S_\alpha. \qquad (8.18)$$

Here, the notation $<\alpha\beta>$ means that the sum is over nearest-neighbor *pairs* of spins; these interact with a strength J (see Figure 8.2). Thus, the spin at site ij interacts with the spins at $i\pm1j$ and $ij\pm1$. (We assume periodic boundary conditions on the lattice, so that, for example, the lower neighbors of the spins with $i = N_x$ are those with $i = 1$ and the left-hand neighbors of those with $j = 1$ are those with $j = N_y$; the lattice therefore has the topology of a torus.) When J is positive, the energy is lower if a spin is in the same direction as its neighbors (ferromagnetism), while when J is

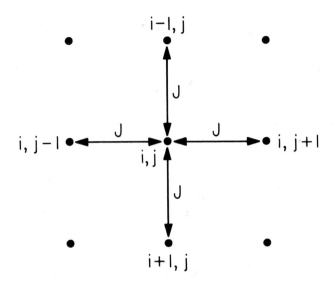

Figure 8.2 Schematic illustration of the two-dimensional Ising model

negative, a spin will tend to be anti-aligned with its neighbors (anti-ferromagnetism). The term involving B represents the interaction of the spins with an external magnetic field, which tends to align all spins in the same direction.

We will be interested in the thermodynamics of this system. In this case, it is convenient to measure the coupling energies J and B in units of the temperature, so that heating the system corresponds to decreasing these couplings. Configurations of the system are specified by giving the values of all $N_x \times N_y \equiv N_s$ spin variables and the weighting of any one of the 2^{N_s} spin configurations, $\mathbf{S}$, in the canonical ensemble is

$$w(\mathbf{S}) = \frac{e^{-H(\mathbf{S})}}{Z}, \tag{8.19}$$

where the partition function is

$$Z(J,B) = \sum_{\mathbf{S}} e^{-H(\mathbf{S})}. \tag{8.20}$$

The thermodynamic quantities we will be interested are the magnetization

$$M = \frac{\partial \log Z}{\partial B} = \sum_{\mathbf{S}} w(\mathbf{S})(\sum_{\alpha} S_{\alpha}), \tag{8.21a}$$

the susceptibility

$$\chi=\frac{\partial M}{\partial B}=\sum_{S}w(S)(\sum_{\alpha}S_{\alpha})^2-M^2,\qquad(8.21\text{b})$$

the energy

$$E=\sum_{S}w(S)H(S),\qquad(8.21\text{c})$$

and the specific heat at constant field,

$$C_B=\sum_{S}w(S)H^2(S)-E^2.\qquad(8.21\text{d})$$

In the limit of an infinitely large lattice ($N_{x,y}\to\infty$), it is possible to solve the Ising model exactly; discussions of the solution, originally due to Onsager, can be found in [Hu63] and [Mc73]. The expressions are simplest at $B=0$. In this limit, the energy is given by

$$E=-N_s J(\coth 2J)\left[1+\frac{2}{\pi}\kappa' K_1(\kappa)\right],\qquad(8.22\text{a})$$

and the specific heat is

$$C_B=N_s\frac{2}{\pi}(J\coth 2J)^2\left[2K_1(\kappa)-2E_1(\kappa)-(1-\kappa')\left[\frac{\pi}{2}+\kappa' K_1(\kappa)\right]\right],\qquad(8.22\text{b})$$

while the magnetization is given by

$$M=\pm N_s\frac{(1+z^2)^{1/4}(1-6z^2+z^4)^{1/8}}{(1-z^2)^{1/2}}\qquad(8.22\text{c})$$

for $J>J_c$ and vanishes for $J<J_c$. In these expressions,

$$\kappa=2\frac{\sinh 2J}{\cosh^2 2J}\leq 1,\quad \kappa'=2\tanh^2 2J-1,$$

the complete elliptic integrals of the first and second kinds are

$$K_1(\kappa)\equiv\int_0^{\pi/2}\frac{d\varphi}{(1-\kappa^2\sin^2\varphi)^{1/2}},\quad E_1(\kappa)\equiv\int_0^{\pi/2}d\varphi(1-\kappa^2\sin^2\varphi)^{1/2},$$

$z=e^{-2J}$, and $J_c=0.4406868$ is the critical value of J for which $\kappa=1$, where K_1 has a logarithmic singularity. Thus, all thermodynamic functions are singular at this coupling, strongly suggesting a phase transition. This is confirmed by the behavior of the magnetization, which vanishes below the critical coupling (or above the critical temperature), and can take on one of two equal and opposite values above this coupling.

A numerical solution of the Ising model is useful both as an illustration of the techniques we have been discussing and because it can be generalized readily to more complicated Hamiltonians [Fo63]. Because of the large number of terms involved, a direct evaluation of the sums in Eqs. (8.21) is out of the question. (For even a modest 16×16 lattice, there are $2^{256} \approx 10^{77}$ different configurations.) Hence, it is most efficient to generate spin configurations $\mathbf{S}$ with probability $w(\mathbf{S})$ using the Metropolis algorithm and then to average the required observables over these configurations. To implement the Metropolis algorithm, we could make our trial step from $\mathbf{S}$ to $\mathbf{S}_t$ by changing all of the spins randomly. This would, however, bring us to a configuration very different from $\mathbf{S}$, and so there would be a high probability of rejection. It is therefore better to take smaller steps, and so we consider trial configurations that differ from the previous one only by the flipping of one spin. This is done by sweeping systematically through the lattice and considering whether or not to flip each spin, one at a time. Hence, we consider two configurations, $\mathbf{S}$ and $\mathbf{S}_t$, differing only by the flipping of one spin, $S_\alpha \equiv S_{ij}$. Acceptance of this trial step depends upon the ratio of the weight functions,

$$r = \frac{w(\mathbf{S}_t)}{w(\mathbf{S})} = e^{-H(\mathbf{S}_t) + H(\mathbf{S})}.$$

Specifically, if $r > 1$ or if $r < 1$ but larger than a uniformly distributed random number between 0 and 1, then the spin S_α is flipped; otherwise, it is not. From (8.18), it is clear that only terms involving S_{ij} will contribute to r, so that after some algebra, we have

$$r = e^{-2S_\alpha(Jf + B)}; \quad f = S_{i+1j} + S_{i-1j} + S_{ij+1} + S_{ij-1}.$$

Here, f is the sum of the four spins neighboring the one being flipped. Because f can take on only 5 different values, $0, \pm 2, \pm 4$, only 10 different values of r can ever arise (there are two possible values of S_α); these can be conveniently calculated and stored in a table before the calculation begins so that exponentials need not be calculated repeatedly. Note that if we had used trial configurations that involved flipping several spins, the calculation of r would have been much more complicated.

   The program for Example 8, whose source listing is given in Appendix B and in the file EXAM8.BAS on the *Computational Physics* diskette, performs a Monte Carlo simulation of the Ising model using the algorithm just described. An initially random configuration of spins is used to start the Metropolis random walk and the lattice is shown after each successive sweep. Thermalization sweeps (no calculation of the observables) are allowed for; these permit the random walk to "settle down" before observables are accumulated. Values of the energy, magnetization, susceptibility, and specific heat per spin are displayed as the calculation proceeds, as is the fraction of the trial steps accepted.

One feature of this program requires further explanation. This is a simple technique used to monitor the sweep-to-sweep correlations in the observables inherent in the Metropolis algorithm. The basic observables (energy and magnetization) are computed every FREQ% sweeps. These values are then binned into "groups" with SIZE% members. For each group, the means and standard deviations of the energy and magnetization are calculated. As more groups are generated, their means are combined into a grand average. One way to compute the uncertainty in this grand average is to treat the group means as independent measurements and to use Eq. (8.3). Alternatively, the uncertainty can be obtained by averaging the standard deviations of the groups in quadrature. If the sampling frequency is sufficiently large, these two estimates will agree. However, if the sampling frequency is too small and there are significant correlations in the successive measurements, the values within each group will be too narrowly distributed, and the second estimate of the uncertainty will be considerably smaller than the first. These two estimates of the uncertainty for the grand average of the energy and magnetization per spin are therefore displayed. Note that this technique is not so easily implemented for the the specific heat and susceptibility, as they are themselves fluctuations in the energy and magnetization (Eqs. (8.21b,d)), and so only the uncertainties in their grand averages computed by the first method are displayed.

The following exercises will be useful in better understanding this example:

□ □ □ **Exercise 8.6** When $J=0$, the Hamiltonian (8.18) reduces to that for independent spins in an external magnetic field, a problem soluble by elementary means. For this case, obtain analytical expressions for the thermodynamic observables and verify that these are reproduced by the program.

□ □ □ **Exercise 8.7** Use Eqs. (8.22) to calculate and graph the exact energy, specific heat, and magnetization per spin for an infinite lattice at $B=0$ and for J running from 0 to 0.8.

□ □ □ **Exercise 8.8** Modify the code to compute the sweep-to-sweep correlation functions for the energy and the magnetization using (8.17). Using runs for a 16×16 lattice for $B=0$ and for several values of J between 0.1 and 0.6, estimate the proper sampling frequency at each coupling strength. Show that the two estimates of the uncertainties in the energy and magnetization agree when a proper sampling frequency is used and that they disagree when the samples are taken too often (a reasonable group size is 10). Also show that the sweep-to-sweep correlations become stronger when the system is close to the phase transition (critical slowing down).

Exercise 8.9 Run the code to obtain results for 8×8, 16×16, and 32×32 lattices at $B=0$ for a sequence of ferromagnetic couplings from 0.1 to 0.6; pay particular attention to the region near the expected phase transition. Compare your results with the exact behavior of the infinite lattice and show that the finite size smooths out the singularities in the thermodynamic observables. Notice that the size of the magnetic domains becomes very large near the critical coupling.   □ □ □

Exercise 8.10 Use the code to explore the thermodynamics of the model for finite B and for anti-ferromagnetic couplings ($J<0$). Also consider simulations of a model in which a given spin S_{ij} interacts with its neighbors ferromagnetically and with its diagonal neighbors $S_{i-1j-1}, S_{i+1j-1}, S_{i-1j+1}, S_{i+1j+1}$ anti-ferromagnetically.   □ □ □

Exercise 8.11 The "heat bath" algorithm is an alternative to the Metropolis algorithm for sampling the canonical ensemble. In this method, the particular spin being considered is set to +1 with probability $1/(1+g)$ and to -1 with probability $g/(1+g)$, where $g=\exp[2(Jf+B)]$. This can be interpreted as placing the spin in equilibrium with a heat bath at the specified temperature. Verify that this algorithm corresponds to putting $A=1$ and taking   □ □ □

$$T(\mathbf{S} \rightarrow \mathbf{S'}) = \frac{w(\mathbf{S'})}{w(\mathbf{S'})+w(\mathbf{S})}$$

in Eq. (8.16) and so leads to a correct sampling of spin configurations. Modify the code to use the heat-bath algorithm and compare its efficiency with that of the conventional Metropolis algorithm.

Project VIII: Quantum Monte Carlo for the H_2 molecule

In this project, we will consider Monte Carlo methods that can be used to calculate the *exact* properties of the ground states of quantum many-body systems. These methods are based on the formal similarity between the Schroedinger equation in imaginary time and a multi-dimensional diffusion equation (recall Section 7.4). Since the latter can be handled by Monte Carlo methods (ordinary diffusion arises as the result of many random microscopic collisions of the diffusing particles), these same techniques can also be applied to the Schroedinger equation. However, the Monte Carlo method is no panacea: it can make exact statements about only the ground state energy, it requires an already well-chosen variational wavefunction for the ground state, and it is generally intractable for fermion systems. However, there have been successful applications of these techniques to liquid ^{4}He [Ka81], the electron gas [Ce80], small molecules [Re82], and lattice gauge theories [Ch84]. More detailed discussions can be found in these references, as well as in [Ce79], which contains a good general overview.

VIII.1 Statement of the problem

The specific problem we will treat is the structure of the H_2 molecule: two protons bound by two electrons. This will be done within the context of the accurate Born-Oppenheimer approximation, which is based on the notion that the heavy protons move slowly compared to the much lighter electrons. The potential governing the protons' motion at a separation S, $U(S)$, is then the sum of the inter-proton electrostatic repulsion and the eigenvalue, $E_0(S)$, of the two-electron Schroedinger equation:

$$U(S) = \frac{e^2}{S} + E_0(S). \qquad \text{(VIII.1)}$$

The electronic eigenvalue is determined by the Schroedinger equation

$$H(S)\Psi_0(\mathbf{r}_1,\mathbf{r}_2;S) \equiv [K + V(S)]\Psi_0 = E_0(S)\Psi_0(\mathbf{r}_1,\mathbf{r}_2;S). \qquad \text{(VIII.2)}$$

Here, the electronic wavefunction, Ψ_0, is a function of the space coordinates of the two electrons, $\mathbf{r}_{1,2}$ and depends parametrically upon the inter-proton separation. If we are interested in the electronic ground state of the molecule and are willing to neglect small interactions involving the electrons' spin, then we can assume that the electrons are in an antisymmetric spin-singlet state; the Pauli principle then requires that Ψ_0 be symmetric under the interchange of $\mathbf{r}_1$ and $\mathbf{r}_2$. Thus, even though the electrons are two fermions, the spatial equation satisfied by their wavefunction is analogous to that for two bosons; the ground state wavefunction Ψ_0 will therefore have no nodes, and can be chosen to be positive everywhere.

The electron kinetic energy appearing in (VIII.2) is

$$K = -\frac{\hbar^2}{2m}(\nabla_1^2 + \nabla_2^2), \qquad \text{(VIII.3a)}$$

where m is the electron mass, while the potential V involves the attraction of the electrons by each nucleus and the inter-electron

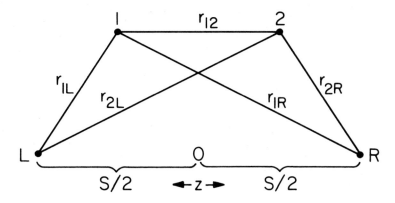

Figure VIII.1 Coordinates used in describing the H_2 molecule

repulsion:

$$V=-e^2\left[\frac{1}{r_{1L}}+\frac{1}{r_{2L}}+\frac{1}{r_{1R}}+\frac{1}{r_{2R}}\right]+\frac{e^2}{r_{12}}. \qquad \text{(VIII.3b)}$$

If we place the protons at locations $\pm\frac{1}{2}S$ on the z-axis, then the distance between electron 1 and the left or right proton is

$$r_{1L,R}=|\mathbf{r}_1\pm\tfrac{1}{2}S\hat{\mathbf{z}}|,$$

and the distances between electron 2 and the protons, $r_{2L,R}$ are given similarly. The interelectron distance is $r_{12}=|\mathbf{r}_1-\mathbf{r}_2|$. (See Figure VIII.1.)

Our goal in this project is therefore to solve the six-dimensional partial differential eigenvalue equation (VIII.2) for the lowest eigenvalue E_0 at each S, and so trace out the potential $U(S)$ via Eq. (VIII.1). This potential should look much like the 6-12 or Morse potentials studied in Chapter 1; the depth and location of the minimum and the curvature of U about this minimum are related to observable properties of the H_2 spectrum. We will also be able to calculate the exact ground state energies of two-electron atoms, $E_0(S=0)$. (These same systems were treated in the Hartree-Fock approximation in Project III.)

VIII.2 Variational Monte Carlo and the trial wavefunction

We begin by discussing a variational Monte Carlo solution to the eigenvalue problem. If $\Phi(\mathbf{r})$ is any trial wavefunction (which we can choose to be real) not orthogonal to the exact ground state Ψ_0, then an upper bound to the electronic eigenvalue is the variational energy

$$E_v = \frac{<\Phi|H|\Phi>}{<\Phi|\Phi>} \tag{VIII.4a}$$

$$= \frac{\int d\mathbf{r}\,\Phi^2(\mathbf{r})\left[\frac{1}{\Phi}H\Phi(\mathbf{r})\right]}{\int d\mathbf{r}\,\Phi^2(\mathbf{r})}\,, \tag{VIII.4b}$$

where we have used $\mathbf{r}$ as a short-hand notation for the six coordinates of the problem, $\mathbf{r}_{1,2}$, and have written (VIII.4b) in a somewhat unconventional way. Note that this last equation can be interpreted as giving the variational energy as the average of a "local energy",

$$\varepsilon(\mathbf{r}) \equiv \Phi^{-1}H\Phi(\mathbf{r}) = \Phi^{-1}K\Phi(\mathbf{r}) + V(\mathbf{r})$$

$$= -\frac{\hbar^2}{2m}\Phi^{-1}\sum_{i=1,2}\nabla_i^2\Phi + V(\mathbf{r})\,, \tag{VIII.5}$$

over all of the six-dimensional $\mathbf{r}$–space with weighting $w(\mathbf{r}) \sim \Phi^2(\mathbf{r})$. Hence, to evaluate the variational energy by a Monte Carlo quadrature, all we need do is generate configurations (values of $\mathbf{r}_{1,2}$) distributed according to Φ^2 (by the Metropolis algorithm, for example), and then average ε over these configurations. It should be clear that this method can be generalized readily to systems involving a larger number of coordinates.

The choice of Φ is constrained by the requirements that it be simple enough to allow a convenient evaluation of Φ^2 and ε, yet also be a good approximation to the true ground state, Ψ_0. Indeed, if we are fortunate enough to choose a Φ that is the exact solution, then ε is independent of $\mathbf{r}$, so that the Monte Carlo quadrature gives the exact energy with zero variance. Thus, not only will the variational bound become better as Φ better approximates Ψ, but the variance in the Monte Carlo quadrature will become smaller. The trial wavefunction should also be symmetric under the interchange of the two electrons' spatial coordinates, but it need not be normalized properly, as the normalization cancels in evaluating (VIII.4).

One plausible choice for the trial wavefunction is a correlated product of molecular orbitals:

$$\Phi(\mathbf{r}_1,\mathbf{r}_2) = \varphi(\mathbf{r}_1)\varphi(\mathbf{r}_2)f(r_{12}). \tag{VIII.6a}$$

Here, the first two factors are an independent-particle wavefunction placing each electron in a molecular orbital in which it is shared equally between the two protons. A simple choice for the molecular orbital is the symmetric linear combination of atomic orbitals centered about each proton,

$$\varphi(\mathbf{r}_i)=e^{-r_{iL}/a}+e^{-r_{iR}/a}, \qquad\qquad \text{(VIII.6b)}$$

with the variational parameter a to be determined below. The final factor in the trial wavefunction, f, expresses the correlation between the two electrons due to their Coulomb repulsion. That is, we expect f to be small when r_{12} is small and to approach a large constant value as the electrons become well separated. A convenient and reasonable choice is

$$f(r)=\exp\left[\frac{r}{\alpha(1+\beta r)}\right], \qquad\qquad \text{(VIII.6c)}$$

where α and β are additional positive variational parameters. Note that β controls the distance over which the trial wavefunction "heals" to its uncorrelated value as the two electrons separate.

The singularity of the Coulomb potential at short distances places additional constraints on the trial wavefunction. If one of the electrons (say 1) approaches one of the nuclei (say the left one) while the other electron remains fixed, then the potential term in ε becomes large and negative, since r_{1L} becomes small. This must be cancelled by a corresponding positive divergence in the kinetic energy term if we are to keep ε smooth and have a small variance in the Monte Carlo quadrature. Thus, the trial wavefunction should have a "cusp" at $\mathbf{r}_{1L}=0$, which means that the molecular orbital should satisfy

$$\lim_{r_{1L}\to 0}\left[-\frac{\hbar^2}{2m}\frac{1}{\varphi(\mathbf{r}_{1L})}\nabla_1^2\varphi(\mathbf{r}_{1L})-\frac{e^2}{r_{1L}}\right]=\text{finite terms.} \quad \text{(VIII.7)}$$

Similar conditions must also be satisfied whenever any one of the distances r_{1R}, $r_{2R,L}$, or r_{12} vanishes. Using (VIII.6) and a bit of algebra, it is easy to see that these constraints imply that a satisfies the transcendental equation

$$a=\frac{a_0}{(1+e^{-S/a})} \qquad\qquad \text{(VIII.8)}$$

and that $\alpha=2a_0$, where $a_0=\hbar^2/me^2$ is the Bohr radius. Thus, β is the only variational parameter at our disposal. Note that these Coulomb cusp conditions would not have been as important if we were to use some other quadrature method to evaluate the variational energy, as then each of the divergences in ε would be cancelled by a geometrical r^2 weighting in the associated distance.

With the trial wavefunction specified by Eqs. (VIII.6), explicit expressions can be worked out for $\varepsilon(\mathbf{r})$ in terms of the values and derivatives of φ and f. Note that this form of the trial function is not appropriate for very large proton separations, as it contains a finite amplitude to find the two electrons close to the same proton.

VIII.3 Monte Carlo evaluation of the exact energy

We now turn to a Monte Carlo method for evaluating the *exact* electronic eigenvalue, E_0. It is based upon evolution of the imaginary-time Schroedinger equation to refine the trial wave function to the exact ground state. In particular, the latter can be obtained by applying the operator $\exp(-Ht/\hbar)$ to the trial state Φ and considering the long-time limit $t \to \infty$. (See the discussion concerning eigenvalues of elliptic operators given in Section 7.4 for further details.) Thus, we define

$$\Psi(\mathbf{r},t) = \exp\left[\int_0^t E_n(t')dt' / \hbar \right] e^{-Ht/\hbar} \Phi(\mathbf{r}) , \qquad \text{(VIII.9)}$$

where $E_n(t')$ is an as-yet-undetermined c-number function. Note that as long as $<\Psi_0|\Phi> \neq 0$, $\Psi(t)$ will approach the (un-normalized) exact ground state Ψ_0 as t becomes large. Our method (Path Integral Monte Carlo) is equivalent to evaluating the path integral representation of (VIII.9) numerically. An alternative method (Green's Function Monte Carlo, [Ce79]) would be to filter with the inverse of the shifted Hamiltonian, instead of the exponential in (VIII.9).

To compute the exact ground state energy E_0, we consider a slight generalization of the variational energy (VIII.4) using the hermiticity of H:

$$E(t) = \frac{<\Phi|H|\Psi(t)>}{<\Phi|\Psi(t)>} = \frac{\int d\mathbf{r}\, \Phi(\mathbf{r})\Psi(\mathbf{r},t)\varepsilon(\mathbf{r})}{\int d\mathbf{r}\, \Phi(\mathbf{r})\Psi(\mathbf{r},t)} . \qquad \text{(VIII.10)}$$

It should be clear that $E(t=0)=E_v$, the variational energy associated with Φ, that $E(t \to \infty) = E_0$, and that $E(t)$ is independent of the function $E_n(t')$.

Equation (VIII.10) expresses the exact energy in a form suitable for Monte Carlo evaluation. To see this, we define $G(\mathbf{r},t) = \Phi(\mathbf{r})\Psi(\mathbf{r},t)$, so that (VIII.10) becomes

$$E(t) = \frac{\int d\mathbf{r}\, G(\mathbf{r},t)\varepsilon(\mathbf{r})}{\int d\mathbf{r}\, G(\mathbf{r},t)} . \qquad \text{(VIII.11)}$$

The exact energy $E(t)$ is thus the average of the local energy ε over the distribution $G(\mathbf{r},t)$. (Note that for the problem we are considering, G is positive definite since neither Φ nor Ψ has any nodes.) A Monte Carlo evaluation of E therefore requires an "ensemble" of N configurations $\{\mathbf{r}_1, \cdots ,\mathbf{r}_N\}$ distributed according to $G(\mathbf{r},t)$, through which $E(t)$ can be estimated by

$$E(t) \approx \frac{1}{N} \sum_{i=1}^{N} \varepsilon(\mathbf{r}_i), \qquad \text{(VIII.12)}$$

with an expression for the variance in analogy with (8.3). Note that if $\Phi(\mathbf{r})$ is in fact the **exact** ground state Ψ_0, then $\varepsilon(\mathbf{r}) = E_0$, independent of $\mathbf{r}$, and so $E(t) = E_0$ with zero variance.

Of course, the expressions above are of no practical use without a way to generate the ensemble of configurations. At $t=0$, $G(\mathbf{r},t) = |\Phi(\mathbf{r})|^2$, so that a method like that of Metropolis *et al.* can be used to generate the initial ensemble, typically having $N \approx 30$ members. To evolve the ensemble in time, note that since

$$\hbar \frac{\partial \Psi}{\partial t} = (E_n - H)\Psi,$$

G satisfies the evolution equation

$$\frac{\partial G}{\partial t} = \hbar^{-1} \left[E_n(t) - \Phi(\mathbf{r})H \frac{1}{\Phi(\mathbf{r})} \right] G(\mathbf{r},t) \qquad \text{(VIII.13a)}$$

$$= \frac{\hbar}{2m} \frac{\partial^2 G(\mathbf{r},t)}{\partial \mathbf{r}^2} - \frac{\partial}{\partial \mathbf{r}} \cdot [\mathbf{D}(\mathbf{r})G(\mathbf{r},t)]$$

$$-\hbar^{-1} [\varepsilon(\mathbf{r}) - E_n(t)]G(\mathbf{r},t) \ . \qquad \text{(VIII.13b)}$$

We have used here an obvious notation for the spatial derivatives.

Equation (VIII.13b) can be interpreted as a diffusion equation for G, with a drift function

$$\mathbf{D}(\mathbf{r}) = \frac{\hbar}{m} \frac{1}{\Phi(\mathbf{r})} \frac{\partial \Phi(\mathbf{r})}{\partial \mathbf{r}} = \frac{\hbar}{m} \frac{\partial \log \Phi}{\partial \mathbf{r}}. \qquad \text{(VIII.14)}$$

It shows that the kinetic energy acts to diffuse G, that D tends to keep G confined to regions where Φ is large, and that the "source" increases G where $\varepsilon(\mathbf{r})$ is smallest.

The evolution of G over a short time from t to $t + \Delta t$ can be represented through order Δt by the integral kernel

$$G(\mathbf{r},t + \Delta t) = \int d\mathbf{r}' \, P(\mathbf{r},\mathbf{r}';\Delta t) \, G(\mathbf{r}',t) , \qquad \text{(VIII.15)}$$

where

$$P(\mathbf{r},\mathbf{r}';\Delta t) = \exp \left\{ -[\varepsilon(\mathbf{r}) - E_n(t)]\Delta t / \hbar \right\} \times$$

$$\left[\frac{m}{2\pi\hbar\Delta t} \right]^3 \exp \left\{ \frac{-[\mathbf{r} - \mathbf{r}' - \mathbf{D}(\mathbf{r}')\Delta t]^2}{2\hbar\Delta t / m} \right\} . \qquad \text{(VIII.16)}$$

This kernel (which is positive definite) can be interpreted as the conditional probability for a configuration to evolve from $\mathbf{r}'$ at time t to $\mathbf{r}$ at time $t + \Delta t$. This probability contains a factor associated with the kinetic energy, which acts to diffuse the system about $\mathbf{r}'$ through a normalized Gaussian probability distribution with variance $\hbar\Delta t / m$ and mean $\mathbf{D}\Delta t$. The other factor in P is associated with the local energy, which acts to keep the system in regions of space where ε is most negative by enhancing the probability for jumps to occur to such locations. The quantum mechanical structure of the exact ground state is thus determined by the balance between these two competing tendencies.

The algorithm for evolving the ensemble should now be evident. A configuration at time t at the point $\mathbf{r}'$ generates a contribution to $G(\mathbf{r}, t + \Delta t)$ equal to $P(\mathbf{r}, \mathbf{r}'; \Delta t)$. This is realized by placing in the new ensemble a configuration $\mathbf{r}$ chosen according to the distribution

$$\exp\left\{ \frac{-[\mathbf{r} - \mathbf{r}' - \mathbf{D}(\mathbf{r}')\Delta t]^2}{2\hbar\Delta t / m} \right\} ,$$

and then weighting the importance of this configuration by

$$\exp\left\{ -[\varepsilon(\mathbf{r}) - E_n(t)]\Delta t / \hbar \right\} .$$

One way of effecting this weighting in practice is to replicate or delete the configuration in the new ensemble with probabilities given by this latter function. In this case, N fluctuates from time step to time step but can be held roughly constant by continuous adjustment of E_n. Indeed, to keep $\int d\mathbf{r}\, G(\mathbf{r}, t)$ (and hence N) constant, $E_n(t)$ should be equal to $E(t)$, so that E_n also furnishes an estimate of $E(t)$. An alternative way of effecting the weighting is to assign an "importance" (i.e., a weight W_i) to each configuration in the ensemble when averaging ε to compute the energy. That is, the summand in (VIII.12) is modified to $W_i \varepsilon(\mathbf{r}_i)$. At each time step, W_i for each configuration is multiplied by the first factor in (VIII.16), with E_n readjusted to keep average weight of each configuration in the ensemble equal to 1. This latter method requires less bookkeeping than does that of replicating and deleting, but can be inefficient, as a configuration that happens to acquire a very small weight is still followed during the evolution.

In summary, the method is as follows. The system is described by an ensemble of configurations, each having a relative weight in describing the properties of the system and initially distributed in $\mathbf{r}$ according to the trial function $|\Phi(\mathbf{r})|^2$. The overall efficiency of the method is closely related to the accuracy of this trial function. To evolve the ensemble in time, each member is moved in $\mathbf{r}$ with a shifted gaussian probability function [the second factor in Eq. (VIII.16)] and its weight is multiplied by the first factor in Eq.

(VIII.16). The quantity $E_n(t)$, which is adjusted after each time step to keep the average weight of the ensemble equal to 1, provides an estimate of the energy, as does the weighted average of $\varepsilon(\mathbf{r})$ over the ensemble at any time. Furthermore, once the total evolution time is sufficiently large to filter the trial wavefunction, continued evolution generates independent ensembles distributed according to the exact ground state wavefunction, which allows the statistics to be improved to any required accuracy.

Note that since the ensemble moves through configuration space at a rate determined by Δt, which must be sufficiently small so that Eqs. (VIII.15,16) are accurate, the estimates of the energy at successive time steps will not be statistically independent. Therefore, in forming averages and computing variances, care must be taken to use ensembles only at intervals of t sufficiently large that the values are uncorrelated. Such intervals are conveniently determined by examining the autocorrelation functions of the estimate. Alternatively, one can use the method of binning the values into groups, as discussed in connection with the Ising calculation in Section 8.4. It should also be noted that the finite time step requires that calculations be done for several different values of Δt and the results extrapolated to the $\Delta t = 0$ limit. A useful way to set a scale for Δt is to note that the average step size for the Gaussian distribution in (VIII.16) is roughly $(\hbar\Delta t/m)^{\frac{1}{2}}$; this must be small compared to the length scales in the wavefunction.

Expectation values of observables other than the energy in the exact ground state are not simply given by this method. To see this, consider the ground state expectation value of some observable A that does not commute with the Hamiltonian. Then

$$<\Psi_0|A|\Psi_0>\equiv\lim_{t\to\infty}\frac{<\Psi(t)|A|\Psi(t)>}{<\Psi(t)|\Psi(t)>}$$

$$=\frac{\int d\mathbf{r}\,\Psi^2(\mathbf{r},t)\Psi^{-1}A\Psi(\mathbf{r},t)}{\int d\mathbf{r}\,\Psi^2(\mathbf{r},t)} \qquad (\text{VIII.}17)$$

Evaluation of this integral requires an ensemble of configurations distributed as $|\Psi|^2$. However, the diffusion process described above generates an ensemble distributed as $G=\Phi\Psi$, which is not what is required. Although this ensemble can be used, through a rather complicated algorithm, to calculate exact ground state expectation values, a good estimate can be had by using it directly to calculate the first term in

$$<\Psi_0|A|\Psi_0>\approx\lim_{t\to\infty}2\,\frac{<\Phi|A|\Psi(t)>}{<\Phi|\Psi(t)>}-\frac{<\Phi|A|\Phi>}{<\Phi|\Phi>}, \qquad (\text{VIII.}18)$$

the second term being evaluated easily with an ensemble distributed according to Φ^2. Thus, this expression provides a way of perturbatively correcting the expectation value in the trial state; some algebra shows that it is accurate to second order in the error in the trial function, $\Psi_0 - \Phi$.

VIII.4 Solving the problem

The Monte Carlo methods described above can be applied to solve for the properties of the H_2 molecule through the following sequence of steps.

□ □ □   **Step 1** Verify that Eq. (VIII.8) and the choice $\alpha = 2a_0$ imply that the wavefunction (VIII.6) satisfies the Coulomb cusp condition. Derive explicit analytical expressions for $\varepsilon(\mathbf{r})$ (Eq. (VIII.5)) and $\mathbf{D}(\mathbf{r})$ (Eq. (VIII.14)) for the wavefunction (VIII.6).

□ □ □   **Step 2** Write a code that uses Monte Carlo quadrature to calculate the variational energy estimate associated with the trial wavefunction (VIII.6). You will need subroutines to evaluate Φ^2 and ε for a given configuration. Sample Φ^2 using the Metropolis algorithm. Study the auto-correlation function of ε to determine the minimum acceptable sampling frequency along the random walk.

□ □ □   **Step 3** For various inter-proton separations, S, find the parameter β that minimizes the electron eigenvalue and so determine the variational potential for the H_2 molecule. Verify that the uncertainties in your results behave with β as expected. Your value at $S = 0$ can be compared with the variational results obtained in Project III using scaled hydrogenic and full Hartree-Fock wavefunctions.

□ □ □   **Step 4** Verify the validity of Eqs. (VIII.15,16) by making a Taylor expansion of $G(\mathbf{r}', t)$ about $\mathbf{r}$ and then doing the resulting Gaussian $\mathbf{r}'$ integrals. (You will also have to make the approximation $\mathbf{D}(\mathbf{r}') \approx \mathbf{D}(\mathbf{r})$, which is accurate to $O(\Delta t)$.)

□ □ □   **Step 5** Test the Path Integral Monte Carlo scheme for finding the exact ground state energy of a Hamiltonian on the simple situation of a particle in a one-dimensional harmonic-oscillator potential. Write a program that assumes a gaussian trial function with an incorrect width and verify that the time evolution refines the energy in the expected way. An ensemble of 20-30 members should be sufficient. Investigate how the quality of your results varies with the error in the trial wavefunction and with the time step you use.

Step 6 Combine the programs you wrote in Steps 2 and 5 into one
that determines the exact eigenvalue of the ground state of the
two-electron problem and so determine the exact H_2 molecular
potential at various separations. In particular, determine the
location of the minimum in the potential and compare with the
empirical values given in Section 1.4. Use the trial functions determined in Step 2 and verify that your exact energies are always
lower than the variational values. Also verify that your exact
results are independent of the precise choice of β and that they
extrapolate smoothly as a function of Δt. Determine the binding
energy of the He atom by considering $S=0$ and compare with the
exact value given in the discussion of Project III.

□ □ □

Appendix A

Synopsis of the BASIC Language

This appendix contains a brief description of the language (IBM or Microsoft GW BASIC) used to write the programs in this book. It is by no means complete, giving neither all of the statements in the language nor all of the details of each statement. Rather, it is meant to provide the reader familiar with some other high-level language (e.g., FORTRAN, PASCAL, C) with enough background to read the codes. It is *not* meant for the programming novice, who should consult a text in computer programming for proper instruction. Further information about the BASIC language can be found in any of the user's manuals accompanying microcomputer systems or in the numerous texts available.

In the following, the BASIC statements are in uppercase, the variables to be supplied are in italic, and optional parameters are enclosed in brackets.

Line format

All lines of BASIC code begin with a sequential line number ranging from 0 to 65529. Several statements, separated by colons (:), can occupy the same line. Blank spaces are generally ignored, as are any characters following an apostrophe (') in a line.

Variables

A variable name can be any length, but only the first 40 characters are significant. The first character must be a letter (only uppercase is permitted). Such letters, numbers, and the decimal point are the only allowed characters, apart from the final character specifying the variable type. The four allowed variable types and their associated final characters are:

% integer variable (2 bytes, ranging from -32768 to +32767)

! single precision floating-point variable (4 bytes, 7-digit precision)

    #   double precision floating-point variable (8 bytes, 17-digit precision)

    $ string variable

Thus, *var!*, *var%*, *var#*, and *var$* are all distinct variables. The default specification (no special final character) is single precision floating-point.

Arrays

The statement

$$\text{DIM } variable\,(subscripts)\,[,variable\,(subscripts)]...$$

declares that *variable* is an array with dimensions specified by the subscripts. The maximum number of dimensions is 255, and the maximum size of each dimension is 32767; the subscript for each dimension begins from 0. Execution of a DIM statement also zeros each element of the array.

Arithmetic statements

The elementary arithmetic operations are, in order of increasing precedence,

$x + y$	the sum of x and y
$x - y$	the difference of x and y
x / y	the quotient of x and y
$x \char`^ y$	the y'th power of x
x MOD y	modulo arithmetic

Here, x and y are any two arithmetic statements or variables. In addition, BASIC also provides a number of built-in arithmetic functions:

ABS(x)	returns the absolute value of x
ASC($x\$$)	returns the ASCII code for the first character in $x\$$
ATN(x)	returns the arctangent of x
CINT(x)	converts x to an integer by rounding
COS(x)	returns the cosine of x
CSNG($x\%$)	converts $x\%$ to single precision

EXP(x)	returns the value of e (=2.71828...) raised to the x power
FIX(x)	converts x to an integer by truncation
LEN($x\$$)	returns the number of characters in $x\$$
LOG(x)	returns the natural logarithm (base e) of x
RANDOMIZE	seeds the random number generator accessed by RND by requesting a seed from the keyboard
RND	returns a random floating-point variable distributed uniformly between 0 and 1
SGN(x)	returns the sign (+1 or -1) of x
SIN(x)	returns the sine of x
SQR(x)	returns the square root of x
VAL($x\$$)	returns the numerical value of the string $x\$$

Logical statements and comparisons

A variable or expression is logically "true" if it is non-zero and is "false" if it is zero. However, -1 has been used exclusively for "true" in this book since NOT x is "false" only when x is -1. BASIC supports the following logical expressions:

NOT x	false if x is true and true if x is false
x AND y	true only if both x and y are true; false otherwise
x OR y	true if either x or y is true; false otherwise

Here, x and y are any two arithmetic statements or expressions. In addition, the following statements allow a comparison between variables or expressions:

$x=y$	true if x is equal to y; false otherwise
$x<>y$	true if x and y are unequal; false otherwise. Also can be written as $x><y$.
$x>y$	true if x is greater than y; false otherwise
$x<y$	true if x is less than y; false otherwise
$x>=y$	true if x is greater than or equal to y; false otherwise
$x<=y$	true if x is less than or equal to y; false otherwise.

Note that two strings can also be compared; they are equal if all of their characters are identical and unequal otherwise.

String functions

The following are BASIC functions that return string variables:

CHR$($n$)	returns the character whose ASCII code is n
INKEY$	returns the next character in the keyboard buffer
LEFT$($x\$,n$)	returns the leftmost n characters of $x\$$
RIGHT$($x\$,n$)	returns the rightmost n characters of $x\$$
SPACE$($n$)	returns a string of n spaces
STRING$($n,m$)	returns a string of n characters whose ASCII code is m
TIME$	returns an 8-character string of the form *hh*:*mm*:*ss* containing the current reading of the clock in hours (*hh*, from 00 to 23), minutes (*mm*, from 00 to 59), and seconds (*ss*, from 00 to 59)

Here, $x\$$ is any string variable and m and n are integers.

User-defined functions

BASIC allows for user-defined functions through the statement

$$\text{DEF FN}name\,[(arg\,[,arg\,]...)]=expression\,.$$

This defines a function of the specified arguments whose name is FN*name*; the value returned when FN*name* is used in a program statement is given by *expression* using the current values of the arguments and any other variables appearing in *expression*. The *args* used in a DEF FN statement are dummy arguments and do not affect program variables.

Control statements

The following statements control the flow of program execution:

STOP or END

> Terminates program execution.

GOTO *linenumber*

> Transfers control to the specified line.

IF *expression* THEN *statement*

> If *expression* is logically TRUE, then *statement* is executed, as are any statements following on the same line. Otherwise control transfers to the next line. If *statement* is GOTO *linenumber*, then either the verb GOTO or the word THEN can be omitted, but not both.

IF *expression* THEN *statement* 1 ELSE *statement* 2

> If *expression* is logically TRUE then *statement* 1 is executed and control transfers to the next line; if *expression* is logically FALSE, then *statement* 2 is executed, as are any statements following on the same line.

FOR $i=m$ TO n [STEP p]
statements
NEXT i

> This is the basic loop structure. The *statements* after FOR are executed until NEXT is reached, whereupon the value of i is incremented by p. Control is then transferred back to FOR where the value of x is checked before execution of the *statements* again. The lines between the FOR and NEXT statements are executed as long as $i \leq n$ (if p is positive) or $i \geq n$ (if p is negative). The default value of p is 1. FOR-NEXT loops can be nested.

WHILE *expression*
statements
WEND

> If *expression* is logically TRUE, the *statements* are executed and control is transferred back to the WHILE statement upon encountering WEND. If *expression* is logically FALSE, then control transfers to the line after WEND. WHILE-WEND loops can be nested.

GOSUB *linenumber*

> This is the form of a subroutine call. GOSUB transfers control to the subroutine beginning at *linenumber*. Execution then continues, with control being transferred back to the statement immediately following the GOSUB statement upon encountering a RETURN statement.

RETURN [*linenumber*]

> Resolves a subroutine call by transferring control to the line *linenumber*. If *linenumber* is omitted, control is transferred to the statement immediately following the calling GOSUB statement.

Input and output statements

The BASIC text display has 25 rows and 80 columns. The programs in this book use the graphics display only in the high-resolution mode (640 pixels horizontally × 200 pixels vertically). The following statements affect how information is presented on these displays, how data is transferred between the program and memory or disk files, and how data is input from the keyboard.

SCREEN *mode*

> Specifies the attributes of the screen. *mode* is either 0 (for text, 25 rows by 80 columns), 1 (for medium resolution graphics, 320×200 pixels, which is not used in this book) or 2 (for high resolution graphics, 640×200 pixels). Only *mode* =0 is permitted on a monochrome display.

KEY OFF

> Turns off the default display of BASIC function keys on the 25'th row.

CLS

Clears the screen.

COLOR [*m*] [,*n*]

The color of the characters and background are specified by *m* and *n*. For the monochrome display, 0 is black, 2-7 is green, and 10-15 is high intensity green; on the graphics display in text mode, *m* and *n* specify 16 different colors, as described in the BASIC manual.

LOCATE [*row*] [,*col*] [,*cursor*]

Places the cursor at *row*,*col* on the text screen. The allowed ranges are 1≤*row*≤25 and 1≤*col*≤80. If *cursor*=1 the cursor is visible, if *cursor*=0 it is not. If *row* is omitted, then the current row location of the cursor is assumed.

PRINT *expression* 1 [, *expression* 2,...] [;]

Displays the *expressions* on the screen. These can include string or numerical variables or mathematical expressions (e.g. X*5+2). The final semi-colon suppresses the carriage return after printing.

PRINT USING "*string*";*expression* 1 [, *expression* 2,...] [;]

Prints *expressions* in a format specified in *string*:

     #   represents each digit for a numerical variable; the decimal point must be indicated explicitly

     +   in front of the first digit (#) specifies that the sign (+ or -) will be printed

     ^^^^   specifies exponential notation

string can also include text to be displayed with the variables.

PSET (*x*,*y*)

Displays the pixel at location (*x*,*y*) on the graphics screen. In high resolution graphics, 0≤*x*≤639 and 0≤*y*≤199.

LINE [(*x* 1,*y* 1)]-(*x* 2,*y* 2) [,[*color*] [,B [,F]]]

Draws a line from the points (*x* 1,*y* 1) to the point (*x* 2,*y* 2) in the graphics mode. Including 'B' will draw a box for which the specified line is a diagonal, while including 'B,F' will fill the box with *color*. If (*x* 1,*y* 1) is omitted, it takes the default value of the last point referenced.

CIRCLE $(x,y),r$

Draws a circle of radius r with center (x,y) in graphics mode.

BEEP

Sounds a tone on the speaker for about 0.25 second.

INPUT[;] [*"prompt"*;] *variable* 1 [,*variable* 2, ...]

Requests the variables specified by displaying *prompt* and '?' on the screen and then waiting for a carriage return. If the number and types of variables input from the keyboard (separated by commas) do not match the number and types specified, a 'Redo from start' message appears and the system will wait again for a carriage return. The ';' immediately following INPUT suppresses a carriage return to the display after the variables have been entered.

OPEN *filename* FOR *mode* AS #*filenumber*

Opens the file specified by *filename* for the action specified by *mode*. This latter can be either INPUT or OUTPUT. The integer *filenumber* is 1, 2, or 3 and specifies the file in subsequent INPUT #, WRITE #, or CLOSE # statements.

INPUT #*filenumber*, *variable* 1 [,*variable* 2,...]

Reads the variables specified from the file specified by *filenumber*.

WRITE #*filenumber*, *expression* 1 [,*expression* 2, ...]

Writes the list of expressions specified on the file labeled by *filenumber*.

CLOSE #*filenumber*

Concludes input or output to the file specified by *filenumber*.

READ *variable* 1 [,*variable* 2, ...]

Assigns the variables specified to the values contained in the next DATA statement. READs access DATA statements in the order they appear, beginning with the DATA statement specified by the RESTORE command.

DATA *constant* 1 [,*constant* 2, ...]

   Specifies constants to be assigned to variables through a READ statement.

RESTORE *linenumber*

   Specifies that READ statements are to begin accessing the DATA statements at the line labeled by *linenumber*.

PEEK (*m*)

   Returns the byte (0-255) read from memory location *m* (0-65535) offest from the address specified by the last DEF SEG statement executed.

POKE *m*,*n*

   Deposits the byte *n* (0-255) at the memory location *m* (0-65535) offset from the address specified by the last DEF SEG statement executed.

DEF SEG [=*address*]

   Specifies the memory location from which POKE statements are offset.

Appendix B

Programs for the Examples

This appendix contains the source code listings for the programs described in each of the examples in the text. These are written in the BASIC language (Microsoft GW BASIC) standard on the IBM PC/XT/AT series of computers. The programs will run on these computers, as well as on other hardware compatible with these machines under the MS-DOS operating system.

In writing the codes, an attempt has been made to conform to general "good" programming practices, such as those discussed in [Ke78]. "Elegance" or machine efficiency has often been sacrificed in favor of intelligibility. However, to keep the codes of reasonable length, the "defensive" programming has not been airtight and it *is* possible to "crash" many of the programs by entering inappropriate parameters. Sample input parameters are given before each of the programs in this and the following appendix and, in any event, an examination of the source listing can quickly show what's gone wrong.

The programs are organized into subroutines, each performing limited and well-defined tasks. All subroutines begin at line numbers which are multiples of 1000. A header describes the purpose of the subroutine and the variables it uses. These are of four types: INPUT and OUTPUT variables (taken from other parts of the program or produced by the subroutine), GLOBAL variables (used elsewhere as well, similar to those contained in a FORTRAN COMMON block), and LOCAL variables (defined and used only within the subroutine). The distinction between the latter two types is useful since, although all BASIC variables are actually global, it is often convenient to use the same variable name in different subroutines, for example as loop indices.

The code for each program has been formatted to make it easier to read. To the extent possible, variable names have been chosen mnemonically and loops have been indented, as have sections of code controlled by IF statements. Running comments on the right-hand side of many statement lines indicate what computation is being done, and major sections are set off by blank comment lines. As much of the code tends to be input, output, and mundane "book keeping", the listings of the important sections are printed here in boldface.

The graphic display of results greatly enhances many of these programs. Although all of the programs will produce results on a system with only a text display (e.g., driven by an IBM mono-chrome display adapter), many are more effective when run on a system with a graphics capability and some are even best when separate graphics and text displays are available. All, however, use the graphics display in either high-resolution (200×640) graphics or text mode; color is used only in the latter. Each program begins with a title screen and enquires about the display configuration of the system; results will then be displayed in the most effective way possible for the responses given. If you are unsure about what display configuration your system has, it is best to answer NO in response to the prompt "Does your computer have a graphics capability?" the first time you run a program.

The switching of the default display in some of the programs is accomplished by two standard subroutines (e.g., beginning at lines 7000 and 8000 in the program for Example 1). These are appropriate for the IBM PC/XT/AT hardware, but can be modified simply to accomodate other machines. Another hardware dependence involves the display buffers in memory. Some programs use the technique of POKEing characters directly into these buffers to save time relative to BASIC's PRINT statement. When used, this technique is announced at the beginning of execution and the option of using the slower PRINT method is offered. If you are using a graphics display or BASIC dialect incompatible with the IBM standard, it is wisest to accept this choice, at least the first time you run a program.

B.1 Example 1

This program finds the semiclassical approximations to the bound state energies of the Lennard-Jones potential for the value of $\gamma=(2ma^2V_0/\hbar^2)^{\frac{1}{2}}$ input. The basic problem is to find, for each integer n, the value of ε_n for which Eq. (1.22) is satisfied. After the number of bound levels is estimated, (lines 1190-1200), the energy for each level is found (loop 230-300) using a secant search (loop 2070-2140) to locate the zero of $f=s-(n+\frac{1}{2})\pi$. Subroutine 3000 calculates the action, s, for a given energy by using simple searches to locate the inner (loop 3070-3120) and outer (loop (3140-3190) turning points and then using Simpson's rule to calculate the integral, with a special treatment for the square-root behavior near the turning points. After all of the energies have been found, the phase-space trajectories, $k(x)$, are graphed for each level by subroutine 5000.

An input value of

Gamma=50

will result in a representative output from this program.

```
10  '*************************************************************
20  'Example 1: Bohr-Sommerfeld quantization for bound states of
30  '              the Lennard-Jones potential
31  'COMPUTATIONAL PHYSICS by Steven E. Koonin
32  'Copyright 1985, Benjamin/Cummings Publishing Company
40  '*************************************************************
50  GOSUB 6000                              'display header screen
60  '
70  PI=3.14159                              'define constants, functions
80  MAX%=100                                'max number of bound states
90  DIM E(100)                              'array  for bound state energies
100 DIM XIN(100),XOUT(100)                  'arrays for turning points
110      'redimension E, XIN, and XOUT in you change MAX%
120 TOLX=.0005: TOLE=.0005                  'space and energy tolerances
130 NPTS%=40                                'number of integration points
140 NGRAPH%=100                             'number of points for graphing
150 XMIN=2^(1/6)                            'equilibrium position for 6-12
160 DEF FNV(X)=4*(X^(-12) -X^(-6))          'the Lennard-Jones potential
170      'If you change the potential, normalize to a minimum of -1
180      'and change XMIN to the new equilibrium value.
190 '
200 GOSUB 1000                              'input GAMMA, calculate NMAX%
210 '
220 E1=-1:  F1=0-PI/2                       'guesses for lowest state
230 FOR N%=0 TO NMAX%                       'find the NMAX% bound states
240     IF INKEY$="e" THEN GOTO 200         'typing e will end the program
250     E2=E1+ABS(E1)/4                      'convenient values to get
260     DE=2*TOLE                           ' secant search started
270     GOSUB 2000                          'search for bound state energy
280     E1=E2                               'initial guesses for next level
290     F1=F2-PI:                           ' extra PI from incrementing N%
300 NEXT N%
```

```
310 '
320 IF GRAPHICS% THEN GOSUB 5000                'graph the levels
330 GOTO 200                                    'get new GAMMA and begin again
340 '
1000 '***********************************************************************
1010 'subroutine to input GAMMA and calculate NMAX%
1020 'input   variables: none
1030 'output variables: GAMMA,NLEV%,NMAX%
1040 'global variables: GRAPHICS%,MONO%,MAX%,PI,TOLE
1050 'local   variables: E
1060 '***********************************************************************
1070 LOCATE 23,20,1: BEEP
1080 PRINT "Enter Gamma=sqr(2ma"+CHR$(253)+"V/hbar"+CHR$(253)+") or 0 to end";
1090 INPUT GAMMA
1100 IF GAMMA=>0 GOTO 1150
1110    LOCATE 24,24: BEEP
1120    PRINT "Gamma must be greater than zero";
1130    LOCATE 23,20: PRINT SPACE$(59);
1140    GOTO 1070                               'prompt for GAMMA again
1150 IF GAMMA=0 THEN END
1160 IF GRAPHICS% AND MONO% THEN GOSUB 8000     'switch to mono
1170 LOCATE ,,0
1180 '
1190 E=-TOLE: GOSUB 3000                        'find S for a very small E
1200 NMAX%=S/PI-.5: NLEV%=NMAX%+1               ' to get number of bound states
1210 IF NLEV%<=MAX% GOTO 1280                   'check if arrays large enough
1220    LOCATE 23,16,1:  BEEP
1230    PRINT USING  "The number of bound levels must be less than ###";MAX%
1240    LOCATE 24,25
1250    PRINT "Try a smaller value for gamma."
1260    LOCATE 23,10:  PRINT SPACE$(69)
1270    GOTO 1070
1280 CLS:  LOCATE 2,21,0
1290 PRINT USING "Gamma=####.##    Number of levels=###";GAMMA,NLEV%
1300 PRINT ""
1310 RETURN
1320 '
2000 '***********************************************************************
2010 'subroutine to search for the bound state energies
2020 'input   variables:  DE, E1, E2, F1, N%, S, XIN, XOUT
2030 'output variables:  E, E2, E(I%), F2, XIN(I%), XOUT(I%)
2040 'global variables:  TOLE
2050 'local   variables:  none
2060 '***********************************************************************
2070 WHILE ABS(DE)>=TOLE                        'secant search for energy
2080    E=E2: GOSUB 3000                        'calculate S at new energy
2090    F2=S-(N%+.5)*PI                         'calculate F at new energy
2100    IF F2=F1 THEN GOTO 2160                 'exit if F doesn't change
2110    DE=-F2*(E2-E1)/(F2-F1)                  'change in the energy
2120    E1=E2: F1=F2: E2=E1+DE                  'update energy and F
2130    IF E2>0 THEN E2=-TOLE                   'keep energy negative
2140 WEND
2150 '
2160 IF ((N%+1) MOD 18)=0 THEN GOSUB 4000    'clear screen if full
2170 PRINT USING "    N=##       Energy=+#.######";N%,E;
2180 PRINT USING "      Xin=##.######      Xout=##.######";XIN,XOUT
2190 '
2200 E(N%)=E:  XIN(N%)=XIN:  XOUT(N%)=XOUT  'save values for graphing
```

```
2210 '
2220 RETURN
2230 '
3000 '************************************************************
3010 'subroutine to calculate S (the action) in units of 2*hbar
3020 'input    variables: E
3030 'output variables: S,XIN,XOUT (the turning points)
3040 'global  variables: FNV(X),GAMMA,NPTS%,TOLX,XMIN
3050 'local    variables: DX,FAC,H,SUM
3060 '************************************************************
3070 XIN=XMIN: DX=.1                      'find inner t.p. by inward search
3080 WHILE DX>TOLX
3090     XIN=XIN-DX
3100     IF FNV(XIN)<E THEN GOTO 3120
3110        XIN=XIN+DX: DX=DX/2
3120 WEND
3130 '
3140 XOUT=XMIN: DX=.1                     'find outer t.p. by outward search
3150 WHILE DX>TOLX
3160     XOUT=XOUT+DX
3170     IF FNV(XOUT)<E THEN GOTO 3190
3180        XOUT=XOUT-DX: DX=DX/2
3190 WEND
3200 '
3210 H=(XOUT-XIN)/NPTS%: SUM=0           'Simpson's rule from XIN-H to XIN+H
3220 SUM=SUM+SQR(E-FNV(XIN+H))
3230 FAC=2
3240 FOR I%=2 TO NPTS%-2
3250     X=XIN+I%*H
3260     IF FAC=2 THEN FAC=4 ELSE FAC=2
3270        SUM=SUM+FAC*SQR(E-FNV(X))
3280 NEXT I%
3290 SUM=SUM+SQR(E-FNV(XOUT-H))
3300 SUM=SUM*H/3
3310 '
3320 SUM=SUM+(SQR(E-FNV(XIN +H)))*2*H/3   'special treatment for sqrt behavior
3330 SUM=SUM+(SQR(E-FNV(XOUT-H)))*2*H/3   ' in first and last intervals
3340 '
3350 S=GAMMA*SUM
3360 RETURN
3370 '
4000 '************************************************************
4010 'subroutine to clear the screen when full of output
4020 'input   variables: none
4030 'output variables: none
4040 'global  variables: GAMMA,NLEV%
4050 'local   variables: none
4060 '************************************************************
4070 LOCATE 24,31,1:   BEEP
4080 PRINT "Type c to continue";
4090 IF INKEY$<>"c" THEN GOTO 4090
4100 CLS:   LOCATE 2,21,0
4110 PRINT USING "Gamma=####.##    Number of levels=###";GAMMA,NLEV%
4120 PRINT ""
4130 RETURN
4140 '
5000 '************************************************************
5010 'subroutine to graph phase-space trajectories (wave number K vs. X)
```

```
5020 'input   variables: E,XIN,XOUT
5030 'output variables: none
5040 'global variables: GAMMA,MONO%,NGRAPH%,NMAX%
5050 'local   variables: H,I%,K,,KMIN%,KSCALE,K1%,K1OLD%,K2%,K2OLD%,MARK%,X,X%,
5060 '                   XMIN%,XOLD%,XSCALE,Y%
5070 '********************************************************************
5080 LOCATE 24,28,1:  BEEP                          'prompt and wait for command
5090 PRINT "Type g to begin graphing.";
5100 IF INKEY$<>"g" GOTO 5100
5110 LOCATE ,,0
5120 '
5130 IF MONO% THEN GOSUB 7000                       'switch to graphics screen
5140 SCREEN 2,0,0,0                                 'switch to hi res graphics
5150 CLS:  KEY OFF
5160 '
5170 LINE (60,180)-(640,180)                        'X axis
5180 LINE (60,1)-(60,180)                           'K axis
5190 '
5200 FOR MARK%=1 TO 5                               'X ticks
5210    X%=70+MARK%*568/5
5220    LINE (X%,178)-(X%,182)
5230 NEXT MARK%
5240 LINE (70,178)-(70,182)                         'tick for XIN(NMAX%)
5250 FOR I%=0 TO 5                                  'X labels
5260    IF I%<>5 THEN LOCATE 24,(9+I%*14) ELSE LOCATE 24,(6+I%*14)
5270    PRINT USING "#.##";XIN(NMAX%)+(XOUT(NMAX%)-XIN(NMAX%))*I%/5;
5280 NEXT I%
5290 LOCATE 22,78:  PRINT "X";                      'X legend
5300 '
5310 FOR MARK%=0 TO 4                               'K ticks
5320    Y%=MARK%*180/4
5330    LINE (58,Y%)-(62,Y%)
5340 NEXT MARK%
5350 FOR I%=2 TO -2 STEP -1                         'K labels
5360    LOCATE (12-5.5*I%),2
5370    PRINT USING "###.##";I%*GAMMA/2;
5380 NEXT I%
5390 LOCATE 4,3:  PRINT "K";                        'K legend
5400 '
5410 LOCATE 1,55                                    'print plot title
5420 PRINT "Phase space trajectories"
5430 LOCATE 2,61
5440 PRINT USING "Gamma=###.##";GAMMA;
5450 '
5460 XSCALE=569/(XOUT(NMAX%)-XIN(NMAX%))            'scales for the axes
5470 KSCALE=180/(2*GAMMA)
5480 '
5490 XMIN%=70+(XMIN-XIN(NMAX%))*XSCALE              'put a + at (XMIN,K=0)
5500 KMIN%=180-GAMMA*KSCALE
5510 LINE (XMIN%+2,KMIN%)-(XMIN%-2,KMIN%)
5520 LINE (XMIN%,KMIN%+2)-(XMIN%,KMIN%-2)
5530 '
5540 FOR N%=0 TO NMAX%                              'graph for each bound state
5550    H=(XOUT(N%)-XIN(N%))/NGRAPH%                'X step for graphing the state
5560    FOR I%=0 TO NGRAPH%                         'graph all points
5570       X=XIN(N%)+H*I%                           'X value of this point
5580       X%=70+(X-XIN(NMAX%))*XSCALE
5590       K=GAMMA*SQR(E(N%)-FNV(X))                'wavenumber at this point
```

```
5600       K1%=180-(K+GAMMA)*KSCALE
5610       K2%=180+(K-GAMMA)*KSCALE                  'K can be positive or negative
5620       IF I%=0 GOTO 5650                         'connect to previous point
5630          LINE (XOLD%,K1OLD%)-(X%,K1%)
5640          LINE (XOLD%,K2OLD%)-(X%,K2%)
5650       IF I%=0 OR I%=NGRAPH% THEN LINE (X%,K1%)-(X%,K2%)'close contour
5660       K1OLD%=K1%: K2OLD%=K2%                    'update previous values
5670       XOLD%=X%
5680     NEXT I%
5690 NEXT N%
5700 '
5710 LOCATE 19,61:  BEEP                             'prompt and wait for command
5720 PRINT "Type c to continue";
5730 IF INKEY$<>"c" THEN GOTO 5730
5740 IF MONO% THEN GOSUB 8000                        'switch back to mono screen
5750 SCREEN 0:  WIDTH 80:  LOCATE ,,1,12,13          'switch back to text mode
5760 RETURN
5770 '
6000 '**************************************************************************
6010 'subroutine to display the header screen
6020 'input   variables: none
6030 'output  variables: GRAPHICS%,MONO%
6040 'global  variables: none
6050 'local   variables: G$,M$,ROW%
6060 '**************************************************************************
6070 SCREEN 0: CLS: KEY OFF                          'program begins in text mode
6080 '
6090 LOCATE 1,30: COLOR 15                           'display book title
6100 PRINT "COMPUTATIONAL PHYSICS";
6110 LOCATE 2,40: COLOR 7
6120 PRINT "by";
6130 LOCATE 3,33: PRINT "Steven E. Koonin";
6140 LOCATE 4,14
6150 PRINT "Copyright 1985, Benjamin/Cummings Publishing Company";
6160 '
6170 LOCATE 5,10                                     'draw the box
6180 PRINT CHR$(201)+STRING$(59,205)+CHR$(187);
6190 FOR ROW%=6 TO 19
6200     LOCATE ROW%,10: PRINT CHR$(186);
6210     LOCATE ROW%,70: PRINT CHR$(186);
6220 NEXT ROW%
6230 LOCATE 20,10
6240 PRINT CHR$(200)+STRING$(59,205)+CHR$(188);
6250 '
6260 COLOR 15                                        'print title, etc.
6270 LOCATE  7,36:  PRINT "EXAMPLE 1";
6280 COLOR 7
6290 LOCATE  9,19:  PRINT "Bohr-Sommerfeld quantization for bound state"
6300 LOCATE 10,26:  PRINT "energies of the 6-12 potential"
6310 LOCATE 13,36:  PRINT "**********"
6320 LOCATE 15,26:  PRINT "Type 'e' to end while running."
6330 LOCATE 16,23:  PRINT "Type ctrl-break to stop at a prompt."
6340 '
6350 LOCATE 19,16: BEEP                              'get screen configuration
6360 INPUT "Does your computer have graphics capability (y/n)";G$
6370 IF LEFT$(G$,1)="y" OR LEFT$(G$,1)="n" GOTO 6400
6380     LOCATE 19,12:  PRINT SPACE$(58):  BEEP
6390     LOCATE 18,35:  PRINT "Try again,":  GOTO 6350
```

```
6400 IF LEFT$(G$,1)="y" GOTO 6430
6410    GRAPHICS%=0:  MONO%=-1
6420    RETURN
6430 GRAPHICS%=-1
6440 LOCATE 18,15:  PRINT SPACE$(55)
6450 LOCATE 19,15:  PRINT SPACE$(55)
6460 LOCATE 19,15: BEEP
6470 INPUT "Do you also have a separate screen for text (y/n)";M$
6480 IF LEFT$(M$,1)="y" OR LEFT$(M$,1)="n" GOTO 6500   'error trapping
6490    LOCATE 18,35:  PRINT "Try again,":  GOTO 6450
6500 IF LEFT$(M$,1)="y" THEN MONO%=-1 ELSE MONO%=0
6510 RETURN
6520 '
7000 '******************************************************************
7010 'subroutine to switch from mono to graphics screen
7020 'input   variables: none
7030 'output variables: none
7040 'global variables: none
7050 'local   variables: none
7060 '******************************************************************
7070 DEF SEG=0
7080 POKE &H410,  (PEEK(&H410) AND &HCF) OR &H10
7090 SCREEN 0:  WIDTH 40: LOCATE ,,1,6,7
7100 RETURN
7110 '
8000 '******************************************************************
8010 'subroutine to switch from graphics to mono screen
8020 'input   variables: none
8030 'output variables: none
8040 'global variables: none
8050 'local   variables: none
8060 '******************************************************************
8070 DEF SEG=0
8080 POKE &H410,  (PEEK(&H410) OR &H30)
8090 SCREEN 0:  WIDTH 80:  LOCATE ,,1,12,13
8100 RETURN
8110 '
```

B.2 Example 2

This program integrates trajectories in the Hénon-Heiles potential (2.35) and finds the corresponding surfaces of section. For the energy input (lines 1080-1190), the borders of the (y, p_y) surface of section are found (subroutine 2000) and initial conditions for a trajectory are then accepted within this border (subroutine 3000 or 5000; note that both x and p_x are initially 0). The time integration then proceeds (loop 330-520), with the 4'th-order Runge-Kutta step taken by subroutine 11000; derivatives of the coordinates and momenta are calculated by subroutine 12000. The (x, y) trajectory is displayed at each time step (subroutine 13000 or 14000) and whenever the trajectory crosses the y-axis (i.e., whenever $x=0$), a point on the surface of section is located and plotted (subroutine 15000) by temporarily switching to x as the independent variable, as described in the text. Integration continues until a command is received from the keyboard (processed by subroutine 9000). Note that a maximum of 2000 surface of section points can be accumulated.

Entering an energy of

$$E=0.1$$

and initial conditions of

$$Y=0.095, \quad Py=0.096$$

will result in a representative calculation.

```
10  '***************************************************************
20  'Example 2:  Trajectories in the Henon-Heiles potential
21  'COMPUTATIONAL PHYSICS by Steven E. Koonin
22  'Copyright 1985, Benjamin/Cummings Publishing Company
30  '***************************************************************
40  GOSUB 16000                          'display header screen
50  '
60  FIRST%=-1                            'define constants, functions
70  TSTEP=.12                            'time step
80  EINIT=0                              'first initial energy
90  TOLY=.0005                           'tolerance for Ymin, Ymax searches
100 '        constants for graphics
110 XSCALE=319/SQR(3)                    'scales for X-Y trajectory
120 YSCALE=130/(3/2)
130 NPTS%=50: DIM YBORD%(50)             'number and arrays of points to graph
140 DIM PYBORD%(50),PYNEGBORD%(50)       ' limit of the surface of section plot
150 DIM SSY%(2000),SSPY%(2000)           'arrays for surface of section points
160 REPLOT%=0                            'flag to replot surfaces of section
170 NCROSS%=0                            'count of surface of section points
180 CROSS%=0                             'flag to plot surface of section point
190 '        constants for Runge-Kutta integration
200 DIM K1(4),K2(4),K3(4),K4(4)          'arrays used in the R-K method
210 DIM VAR(4), OLDVAR(4) ,VARINIT(4)    ' the 4 variables are labeled
220 DIM F(4)                             ' 1) X; 2) Y; 3) Px; 4) Py
230 DEF FNV(X,Y)=(X*X+Y*Y)/2+X*X*Y-Y^3/3 'Henon-Heiles potential
```

```
240 DEF FNXDERIV(X,Y)=-(X+2*X*Y)              'X derivative of Henon-Heiles
250 DEF FNYDERIV(X,Y)=-(Y+X*X-Y*Y)           'Y derivative of Henon-Heiles
260 '
270 GOSUB 1000                               'input initial conditions
280 IF FIRST% THEN GOSUB 7000                'display commands on first time only
290 GOSUB 8000                               'prepare for output
300 '
310 N%=0:  H=TSTEP                           'start at time t=0
320 '                                        'beginning of integration loop
330 K$=INKEY$                                'main loop
340    IF K$<>"" THEN GOSUB 9000             'check for command
350    N%=N%+1                               'increment time-step counter
360    GOSUB 11000                           'take a Runge-Kutta step
370 '
380    T=N%*TSTEP                            'current time
390    EPOT=FNV(VAR(1),VAR(2))               'current energies
400    EKIN=.5*(VAR(3)^2+VAR(4)^2)
410    E=EKIN+EPOT
420    IF GRAPHICS% THEN GOSUB 13000 ELSE GOSUB 14000    'output
430 '
440    CHECK=VAR(1)*OLDVAR(1)                'quantity to signal X crossing 0
450 '
460    FOR J%=1 TO 4                         'replace old variables by current
470       OLDVAR(J%)=VAR(J%)
480    NEXT J%
490 '
500    IF CHECK<=0 THEN GOSUB 15000          'find ss point if X crossed 0
510 '
520 GOTO 330                                 'begin next integration step
530 '
1000 '*****************************************************************************
1010 'subroutine to input energy, and initial conditions
1020 'input  variables:  PY%, VAR(I%), Y%
1030 'output variables:  E, EINIT, OLDVAR(I%), PYINIT%, VAR(I%), VARINIT(I%),
1040 '                   YINIT%
1050 'global variables:  GRAPHICS%
1060 'local  variables:  I%
1070 '*****************************************************************************
1080 CLS
1090 LOCATE 3,23:  PRINT "For a bound trajectory 0 < E <= 1/6";
1100 IF FIRST% GOTO 1120
1110    LOCATE 4,26:  PRINT USING "Your last Energy was #.####";EINIT;
1120 LOCATE 5,36:  BEEP: INPUT; "E=";E
1130 IF E>0 AND E<=(1/6) GOTO 1210                   'error trapping for E value
1140    BEEP
1150    LOCATE 3,20:  PRINT SPACE$(40);
1160    LOCATE 4,20:  PRINT SPACE$(40);
1170    LOCATE 5,30:  PRINT SPACE$(20);
1180    LOCATE 2,35:  PRINT "Try again.";
1190    GOTO 1090
1200 '
1210 IF E<>EINIT THEN GOSUB 2000                     'find E-dependent quantities
1220 '
1230 IF GRAPHICS% THEN GOSUB 3000 ELSE GOSUB 5000    'input Y and Py
1240 '
1250 YINIT%=Y%:  PYINIT%=PY%:  EINIT=E               'save initial values
1260 FOR I%=1 TO 4
1270    VARINIT(I%)=VAR(I%)
```

```
1280     OLDVAR(I%)=VAR(I%)
1290 NEXT I%
1300 '
1310 RETURN
1320 '
2000 '*******************************************************************
2010 'subroutine to find energy dependent quantities
2020 'input   variables:   E
2030 'output variables:    DELTAY, PYBORD%(I%), PYNEGBORD%(I%), PYZERO%
2040 '                     SSPYSCALE, SSYSCALE, YBORD%(I%), YMAX, YMIN, YZERO%
2050 'global variables:    FNV(X,Y), GRAPHICS%, NPTS%, TOLY
2060 'local  variables:    DY, I%, PY, Y
2070 '*******************************************************************
2080 YMAX=0:  DY=.1                      'search outward from 0 to find ymax
2090 WHILE DY>TOLY
2100     YMAX=YMAX+DY
2110     IF FNV(0,YMAX)<E THEN GOTO 2130
2120        YMAX=YMAX-DY:  DY=DY/2
2130 WEND
2140 '
2150 YMIN=0:  DY=.1                      'search inward from 0 to find ymin
2160 WHILE DY>TOLY
2170     YMIN=YMIN-DY
2180     IF FNV(0,YMIN)<E THEN GOTO 2200
2190        YMIN=YMIN+DY:  DY=DY/2
2200 WEND
2210 '
2220 IF NOT GRAPHICS% THEN RETURN
2230 '
2240 DELTAY=YMAX-YMIN
2250 SSYSCALE=280/DELTAY                 'scales for surface of section plots
2260 SSPYSCALE=130/SQR(8*E)
2270 '
2280 YZERO%=35-YMIN*SSYSCALE             'scale (Y=0,Py=0) for plotting
2290 PYZERO%=150-SQR(2*E)*SSPYSCALE
2300 '
2310 FOR I%=0 TO NCROSS%                 'zero surface of section arrays
2320     SSY%(I%)=0                      ' since the energy has changed
2330     SSPY%(I%)=0
2340 NEXT I%
2350 NCROSS%=0                           'zero the crossing counter
2360 '
2370 FOR I%=0 TO NPTS%                   'find the border on the ss plot where
2380     Y=YMIN+DELTAY*I%/NPTS%          ' Px=0 from energy conservation
2390     PY=SQR(2*(E-FNV(0,Y)))
2400     YBORD%(I%)=35+(DELTAY*I%/NPTS%)*SSYSCALE
2410     PYBORD%(I%)=150-(PY+SQR(2*E))*SSPYSCALE
2420     PYNEGBORD%(I%)=150-(-PY+SQR(2*E))*SSPYSCALE
2430 NEXT I%
2440 '
2450 RETURN
2460 '
3000 '*******************************************************************
3010 'subroutine to input initial Y and Py (graphics only)
3020 'input   variables:   E, EINIT, PYBORD%(I%), PYINIT%, PYNEGBORD%(I%),
3030 '                     VARINIT(I%), YBORD%(I%), YINIT%
3040 'output variables:    OFFSET%, PICOFFSET%, PY%, REPLOT%, VAR(I%), XOLD%, Y%
3050 '                     YOLD%
```

```
3060 'global variables:  FNV(X,Y), NCROSS%, PYZERO%, SSPYSCALE, SSPY%(I%),
3070 '                   SSYSCALE, SSY%(I%), YMIN, YZERO%
3080 'local  variables:  CURSOR$, EPOT, I%, K$, OLDPYBORD%, OLDPYNEG%,
3090 '                   OLDYBORD%, REPLY$
3100 '**************************************************************
3110 CLS:  SCREEN 2,0,0,0
3120 LOCATE 1,16:  PRINT USING "E=#.###";E
3130 '
3140 OFFSET%=0: PICOFFSET%=0                        'draw axes for left-hand plot
3150 GOSUB 6000
3160 '
3170 OLDYBORD%=YBORD%(0)                            'draw surface of sect border
3180 OLDPYBORD%=PYBORD%(0):  OLDPYNEG%=PYNEGBORD%(0)
3190 FOR I%=1 TO NPTS%
3200    LINE (OLDYBORD%,OLDPYBORD%)-(YBORD%(I%),PYBORD%(I%))
3210    LINE (OLDYBORD%,OLDPYNEG%)-(YBORD%(I%),PYNEGBORD%(I%))
3220    OLDYBORD%=YBORD%(I%)
3230    OLDPYBORD%=PYBORD%(I%):  OLDPYNEG%=PYNEGBORD%(I%)
3240 NEXT I%                                        'close the border
3250 LINE (OLDYBORD%,OLDPYBORD%)-(OLDYBORD%,OLDPYNEG%)
3260 LINE (YBORD%(0),PYBORD%(0))-(YBORD%(0),PYNEGBORD%(0))
3270 '
3280 IF E<>EINIT GOTO 3350                          'draw axes for right-hand plot
3290    OFFSET%=40:  PICOFFSET%=320
3300    GOSUB 6000
3310    FOR I%=0 TO NCROSS%                         'plot previous ss points
3320       PSET (SSY%(I%),SSPY%(I%))
3330    NEXT I%
3340 '
3350 IF E=EINIT GOTO 3430                           'when the energy changes
3360    VARINIT(1)=0                                 ' zero all trajectory variables
3370    VARINIT(2)=0                                 ' except Px
3380    VARINIT(3)=SQR(2*E)
3390    VARINIT(4)=0
3400    YINIT%=YZERO%
3410    PYINIT%=PYZERO%
3420 '
3430 FOR I%=1 TO 4                                  'variables are assigned their
3440    VAR(I%)=VARINIT(I%)                          'initial values before the
3450 NEXT I%                                         'cursor begins to move
3460 Y%=YINIT%:  PY%=PYINIT%
3470 '
3480 LOCATE 22,4:   PRINT USING "X=#.###";VARINIT(1);
3490 LOCATE 22,19:  PRINT USING "Y=#.###";VARINIT(2);
3500 LOCATE 22,33:  PRINT USING "Px=#.###";VARINIT(3);
3510 LOCATE 22,49:  PRINT USING "Py=#.###";VARINIT(4);
3520 LOCATE 22,64:  PRINT USING "Pot E=#.###";FNV(0,VARINIT(2));
3530 IF E<>EINIT GOTO 3550
3540    LOCATE 23,17:  PRINT"The cursor begins at your last initial position.";
3550 LOCATE 24,2: BEEP
3560 PRINT  CHR$(24)+CHR$(25)+CHR$(26)+CHR$(27)+" keys move the ";
3570 PRINT  "cursor to the starting position. Type return when finished.";
3580 '
3590 PSET(Y%,PY%)
3600 '
3610 '
3620 K$=INKEY$                                      'loop to move around Y-Py plot
3630    IF K$=CHR$(13) GOTO 3910                     'typing return exits from loop
```

```
3640     IF LEN(K$)<2 GOTO 3620                    'cursors are special two
3650     CURSOR$=RIGHT$(K$,1)                      ' character ASCII codes
3660     PYOLD%=PY:   YOLD%=Y%
3670     IF CURSOR$=CHR$(77) THEN Y%=Y%+2          'cursor right
3680     IF CURSOR$=CHR$(75) THEN Y%=Y%-2          'cursor left
3690     IF CURSOR$=CHR$(72) THEN PY%=PY%-2        'cursor up
3700     IF CURSOR$=CHR$(80) THEN PY%=PY%+2        'cursor down
3710 '
3720     VAR(2)=YMIN+(Y%-35)/SSYSCALE              'Y position
3730     VAR(4)=-SQR(2*E)-(PY%-150)/SSPYSCALE      'Py position
3740     EPOT=FNV(0,VAR(2))                        'potential energy
3750 '
3760     IF EPOT+VAR(4)*VAR(4)/2<=E GOTO 3820      'keep cursor inside border
3770       Y%=YOLD%:   PY%=PYOLD%                  'reverse step
3780       VAR(2)=YMIN+(Y%-35)/SSYSCALE
3790       VAR(4)=-SQR(2*E)-(PY%-150)/SSPYSCALE
3800       GOTO 3620
3810 '
3820     PSET (YOLD%,PYOLD%),0:   PSET(Y%,PY%)     'erase old cursor and plot new
3830     VAR(3)=SQR(2*(E-EPOT)-VAR(4)*VAR(4))      'value of Px from the energy
3840     LOCATE 22,19:   PRINT USING "Y=#.###";VAR(2);
3850     LOCATE 22,33:   PRINT USING "Px=#.###";VAR(3);
3860     LOCATE 22,49:   PRINT USING "Py=#.###";VAR(4);
3870     LOCATE 22,64:   PRINT USING "Pot E=#.###";EPOT;
3880 GOTO 3620                                     'get next command
3890 '
3900 '
3910 SSY%(0)=355+(VAR(2)-YMIN)*SSYSCALE           'initial position is on
3920 SSPY%(0)=150-(VAR(4)+SQR(2*E))*SSPYSCALE     ' the surface of section
3930 '
3940 LINE (Y%+2,PY%)-(Y%-2,PY%)                    'mark initial position
3950 LINE (Y%,PY%+2)-(Y%,PY%-2)
3960 '
3970 XOLD%=1+(SQR(3)/2)*XSCALE                     'scale first point for
3980 YOLD%=150-(VAR(2)+.5)*YSCALE                  ' for the X-Y trajectory
3990 REPLOT%=0                                     'reset REPLOT% flag
4000 '
4010 IF E<>EINIT THEN GOTO 4090                    'ask if ss is to be plotted
4020   LOCATE 23,1:   PRINT SPACE$(79);
4030   LOCATE 24,1:   PRINT SPACE$(79);
4040   LOCATE 24,1
4050   PRINT "Would you like to plot all the surfaces of section for this";
4060   INPUT; " energy(y/n)";REPLY$
4070   IF REPLY$="y" THEN REPLOT%=-1 ELSE REPLOT%=0
4080 '
4090 RETURN
4100 '
5000 '*********************************************************************
5010 'subroutine to input initial conditions (mono only)
5020 'input   variables:   E, VARINIT(I%)
5030 'output variables:    VAR(I%)
5040 'global variables:    FIRST%, FNV(X,Y), YMAX, YMIN
5050 'local   variables:   EPOT, PX, PY, PYMAX, Y
5060 '*********************************************************************
5070 LOCATE 7,37:   PRINT "X=0.00";
5080 LOCATE 8,12
5090 PRINT "(Since all paths cross the Y-axis, this is not a restriction.)";
5100 '
```

```
5110 LOCATE 11,30:  PRINT USING "#.#### <= Y <= #.####";YMIN,YMAX;
5120 IF FIRST% GOTO 5140
5130   LOCATE 12,29:  PRINT USING "Your last Y was #.####";VARINIT(2);
5140 LOCATE 13,36:  INPUT; "Y=";Y
5150 IF Y>YMIN AND Y<YMAX GOTO 5230              'error trapping for Y value
5160   BEEP
5170   LOCATE 11,25:  PRINT SPACE$(30);
5180   LOCATE 12,25:  PRINT SPACE$(40);
5190   LOCATE 13,30:  PRINT SPACE$(30);
5200   LOCATE 10,35:  PRINT "Try again,";
5210   GOTO 5110
5220 '
5230 EPOT=FNV(0,Y)
5240 PYMAX=SQR(2*(E-EPOT))
5250 LOCATE 16,30:  PRINT USING "0.00 <= Py <= #.####";PYMAX;
5260 IF FIRST% GOTO 5280
5270   LOCATE 17,28:  PRINT USING "Your last Py was #.####";VARINIT(4);
5280 LOCATE 18,36:  INPUT; "Py=";PY
5290 IF PY>=0 AND PY<=PYMAX GOTO 5370            'error trapping for Py value
5300   BEEP
5310   LOCATE 16,25:  PRINT SPACE$(30);
5320   LOCATE 17,25:  PRINT SPACE$(30);
5330   LOCATE 18,35:  PRINT SPACE$(20);
5340   LOCATE 15,35:  PRINT "Try again,";
5350   GOTO 5250
5360 '
5370 PX=SQR(2*(E-EPOT)-PY*PY)                    'value of Px from the energy
5380 LOCATE 20,36:  PRINT USING "Px=#.###";PX;
5390 '
5400 LOCATE 24,31:  PRINT "Type c to continue.";
5410 IF INKEY$<>"c" GOTO 5410
5420 '
5430 VAR(1)=0:  VAR(2)=Y:  VAR(3)=PX:  VAR(4)=PY  'initial conditions
5440 '
5450 CLS
5460 '
5470 RETURN
5480 '
6000 '****************************************************************************
6010 'subroutine to draw and label axes for surface of section plot; horizontal
6020 '         location determined by OFFSET% and PICOFFSET%
6030 'input  variables:  E, OFFSET%, PICOFFSET%
6040 'output variables:  none
6050 'global variables:  DELTAY, PYZERO%, YMIN, YZERO%
6060 'local  variables:  I%, MARK%, PY%, Y%
6070 '****************************************************************************
6080 IF OFFSET%<>40 GOTO 6110                    'title on right side only
6090   LOCATE 1,51:  PRINT "Surface of Section"
6100 '
6110 LINE (PICOFFSET%+35,20)-(PICOFFSET%+35,150)   'Py axis
6120 LINE (PICOFFSET%+35,150)-(PICOFFSET%+315,150) 'Y axis
6130 '
6140 FOR MARK%=0 TO 3                            'Y ticks
6150   Y%=PICOFFSET%+35+MARK%*280/3
6160   LINE (Y%,148)-(Y%,152)
6170 NEXT MARK%
6180 FOR I%=-1 TO 2                              'Y label
6190   LOCATE 20,(OFFSET%+15+11*I%)
```

```
6200      PRINT USING "#.##";(YMIN+DELTAY*(I%+1)/3);
6210 NEXT I%
6220 LOCATE 20,(OFFSET%+33):   PRINT "Y";              'Y legend
6230 '
6240 FOR MARK%=0 TO 4                                  'Py ticks
6250     PY%=20+MARK%*130/4
6260     LINE (PICOFFSET%+33,PY%)-(PICOFFSET%+37,PY%)
6270 NEXT MARK%
6280 FOR I%=2 TO -2 STEP -1                            'Py labels
6290     LOCATE (11-I%*4),(OFFSET%+1)
6300     PRINT USING "#.##";I%*SQR(E/2)
6310 NEXT I%
6320 LOCATE 5,(OFFSET%+1):   PRINT "Py";               'Py legend
6330 '                                                 'mark center
6340 LINE (YZERO%+2+PICOFFSET%,PYZERO%)-(YZERO%-2+PICOFFSET%,PYZERO%)
6350 LINE (YZERO%+PICOFFSET%,PYZERO%+2)-(YZERO%+PICOFFSET%,PYZERO%-2)
6360 '
6370 RETURN
6380 '
7000 '*************************************************************
7010 'subroutine to display description of commands
7020 'input    variables:   none
7030 'output   variables:   none
7040 'global   variables:   FIRST%
7050 'local    variables:   none
7060 '*************************************************************
7070 FIRST%=0                            'this information is only printed once
7080 CLS
7090 LOCATE 2,23
7100 PRINT "Press p to pause during the integration;";
7110 LOCATE 3,20
7120 PRINT "pressing any key will then resume the program";
7130 LOCATE 7,18
7140 PRINT "Press i to begin again with new initial conditions";
7150 LOCATE 11,17
7160 PRINT "Press c to clear the X-Y trajectory (graphics only)";
7170 LOCATE 15,29
7180 PRINT "Press e to end the program";
7190 LOCATE 24,32:   BEEP:   PRINT "Press c to continue";
7200 IF INKEY$<>"c" THEN GOTO 7200
7210 '
7220 CLS
7230 '
7240 RETURN
7250 '
8000 '*************************************************************
8010 'subroutine to prepare screens for output
8020 'input    variables:   REPLOT%, SSPY(I%), SSY(I%)
8030 'output   variables:   OFFSET%, PICOFFSET%
8040 'global   variables:   GRAPHICS%
8050 'local    variables:   none
8060 '*************************************************************
8070 IF GRAPHICS% GOTO 8120
8080     LOCATE 23,25                    'heading for text only output
8090     PRINT "Surface of Section Coordinates"
8100     RETURN
8110 '
8120 IF REPLOT% GOTO 8200               'save the surface of section if it's
```

```
8130    CLS                                    ' being replotted, else start fresh
8140    GOSUB 10000                            'clear X-Y trajectory plot
8150    OFFSET%=40: PICOFFSET%=320             'redraw surface of section axes
8160    GOSUB 6000
8170    PSET(SSY%(0),SSPY%(0))                 '1'st point is on surface of section
8180    RETURN
8190    '
8200 LOCATE 1,1:    PRINT SPACE$(39);          'clear top line of the screen
8210 GOSUB 10000                               'clear X-Y trajectory plot
8220 LINE (0,163)-(640,199),0,BF               'clear lower lines of screen
8230 PSET(SSY%(0),SSPY%(0))                    '1'st point is on surface of sect
8240 RETURN
8250 '
9000 '************************************************************************
9010 'subroutine to check for command input
9020 'input  variables:   K$
9030 'output variables:   none
9040 'global variables:   GRAPHICS%
9050 'local  variables:   ANSWER$
9060 '************************************************************************
9070 IF K$<>"e" GOTO 9140                      'typing e will end the program
9080    IF GRAPHICS% THEN LOCATE 22,24 ELSE LOCATE 24,24
9090    BEEP: INPUT; "Do you really want to end (y/n)";ANSWER$
9100    IF ANSWER$="y" THEN END
9110    IF GRAPHICS% THEN LOCATE 22,24 ELSE LOCATE 24,24
9120    PRINT SPACE$(35);
9130 '
9140 IF K$="i" THEN RETURN 270                 'begin again with new init. cond.
9150 '
9160 IF K$="c" THEN GOSUB 10000                'clear trajectory
9170 '
9180 IF K$<>"p" THEN RETURN                    'pause during integration
9190    IF GRAPHICS% THEN LOCATE 22,27 ELSE LOCATE 24,27
9200    BEEP: PRINT "Press any key to continue";
9210    IF INKEY$="" GOTO 9210                 'wait for command
9220    IF GRAPHICS% THEN LOCATE 22,20 ELSE LOCATE 24,20
9230    PRINT SPACE$(40);                      'clear prompt
9240    RETURN
9250 '
10000 '************************************************************************
10010 'subroutine to clear the X-Y trajectory plot
10020 'input  variables:   none
10030 'output variables:   none
10040 'global variables:   none
10050 'local  variables:   none
10060 '************************************************************************
10070 LINE (0,0)-(315,160),0,BF                'clear left side of screen
10080 LINE (310,120)-(320,170),0,BF
10090 '
10100 LOCATE 1,14:    PRINT "X-Y Trajectory";  'draw and label triangle
10110 LINE (160,20)-(1,150):    LINE (1,150)-(320,150)
10120 LINE (320,150)-(160,20)
10130 '
10140 LOCATE 2,3:    PRINT "COMMANDS"          'display commands in a box
10150 LOCATE 3,2:    PRINT "p-pause";
10160 LOCATE 4,2:    PRINT "i-initial";
10170 LOCATE 5,4:    PRINT "cond."
10180 LOCATE 6,2:    PRINT "c-clear X-Y";
```

```
10190 LOCATE 7,2:  PRINT "e-end";
10200 LINE (1,6)-(99,56),1,B
10210 '
10220 RETURN
10230 '
11000 '********************************************************************
11010 'subroutine to perform the 4th order Runge Kutta integration
11020 'input  variables:  F(J%), H, OLDVAR(J%), VAR(J%)
11030 'output variables:  VAR(J%)
11040 'global variables:  none
11050 'local  variables:  J%, K1(J%), K2(J%), K3(J%), K4(J%)
11060 '********************************************************************
11070 GOSUB 12000                                'calculate K1
11080 FOR J%=1 TO 4                              'loop over 4 variables
11090     K1(J%)=H*F(J%)                         ' 1) X; 2)Y; 3) Px; 4)Py
11100     VAR(J%)=OLDVAR(J%)+K1(J%)/2
11110 NEXT J%
11120 '
11130 GOSUB 12000                                'calculate K2
11140 FOR J%=1 TO 4
11150     K2(J%)=H*F(J%)
11160     VAR(J%)=OLDVAR(J%)+K2(J%)/2
11170 NEXT J%
11180 '
11190 GOSUB 12000                                'calculate K3
11200 FOR J%=1 TO 4
11210     K3(J%)=H*F(J%)
11220     VAR(J%)=OLDVAR(J%)+K3(J%)
11230 NEXT J%
11240 '
11250 GOSUB 12000                                'calculate K4
11260 FOR J%=1 TO 4
11270     K4(J%)=H*F(J%)
11280     VAR(J%)=OLDVAR(J%)+(K1(J%)+2*K2(J%)+2*K3(J%)+K4(J%))/6  'new values
11290 NEXT J%
11300 RETURN
11310 '
12000 '********************************************************************
12010 'subroutine to calculate the derivatives of x,y,px,py
12020 'input  variables:  CROSS%, FNXDERIV(X,Y), FNYDERIV(X,Y), VAR(I%)
12030 'output variables:  F(I%)
12040 'global variables:  none
12050 'local  variables:  none
12060 '********************************************************************
12070 IF CROSS% THEN GOTO 12150                  'find surface of section
12080 '
12090   F(1)=VAR(3)                              'dX/dt=Px
12100   F(2)=VAR(4)                              'dY/dt=Py
12110   F(3)=FNXDERIV(VAR(1),VAR(2))             'dPx/dt=-dV/dx
12120   F(4)=FNYDERIV(VAR(1),VAR(2))             'dPy/dt=-dV/dy
12130   RETURN
12140 '
12150 F(1)=1                                     'for surface of section all
12160 F(2)=VAR(4)/VAR(3)                         ' derivatives are divided by
12170 F(3)=FNXDERIV(VAR(1),VAR(2))/VAR(3)        ' Px
12180 F(4)=FNYDERIV(VAR(1),VAR(2))/VAR(3)
12190 RETURN
12200 '
```

```
13000 '*******************************************************************
13010 'subroutine to plot the X-Y trajectory (graphics only)
13020 'input  variables:  E, EINIT, EKIN, EPOT, T, VAR(I%)
13030 'output variables:  none
13040 'global variables:  XSCALE, YSCALE
13050 'local  variables:  X%, XOLD%, Y%, YOLD%
13060 '*******************************************************************
13070 X%=1+(VAR(1)+SQR(3)/2)*XSCALE                'extend the X-Y trajectory
13080 Y%=150-(VAR(2)+.5)*YSCALE                    ' to current point
13090 LINE (X%,Y%)-(XOLD%,YOLD%)
13100 XOLD%=X%:   YOLD%=Y%                         'this point is now old one
13110 '
13120 LOCATE 23,3:    PRINT USING "time=####.##";T;    'display current values
13130 LOCATE 23,23:   PRINT USING "X=#.###";VAR(1);
13140 LOCATE 23,38:   PRINT USING "Y=#.###";VAR(2);
13150 LOCATE 23,54:   PRINT USING "Px=#.###";VAR(3);
13160 LOCATE 23,70:   PRINT USING "Py=#.###";VAR(4);
13170 LOCATE 24,2:    PRINT USING "Energy=#.#####";E;
13180 LOCATE 24,19:   PRINT USING "Initial Energy=#.####";EINIT;
13190 LOCATE 24,43:   PRINT USING "Kinetic=#.#####";EKIN;
13200 LOCATE 24,61:   PRINT USING "Potential=#.#####";EPOT;
13210 '
13220 RETURN
13230 '
14000 '*******************************************************************
14010 'subroutine to output parameters to a monochrome screen
14020 'input  variables:  E, EINIT, EKIN, EPOT, T
14030 'output variables:  none
14040 'global variables:  none
14050 'local  variables:  none
14060 '*******************************************************************
14070 LOCATE 25,4,0:   PRINT USING "Time=####.##";T;
14080 LOCATE 25,20:    PRINT USING "Energy=#.#####";E;
14090 LOCATE 25,38:    PRINT USING "Initial=#.####";EINIT;
14100 LOCATE 25,55:    PRINT USING "Kin=#.###";EKIN;
14110 LOCATE 25,68:    PRINT USING "Pot=#.###";EPOT;
14120 '
14130 RETURN
14140 '
15000 '*******************************************************************
15010 'subroutine to find the point on the surface of section
15020 'input  variables:  OLDVAR(I%), VAR(I%)
15030 'output variables:  CROSS%, H, NCROSS%, SSY(I%), SSPY(I%) VAR(I%)
15040 'global variables:  E, GRAPHICS%, SSPYSCALE, SSYSCALE, YMIN
15050 'local  variables:  I%
15060 '*******************************************************************
15070 H=-VAR(1)                                    'distance in X to integrate
15080 NCROSS%=NCROSS%+1:   CROSS%=-1               'increment counter and set flag
15090 '
15100 IF NCROSS%<>1950 GOTO 15160                  'warn if array is getting full
15110   LOCATE 22,1:  BEEP
15120   PRINT USING "At time t=####.##, you have";T;
15130   PRINT " 1950 out of 2000 possible surface of";
15140   PRINT " section points.";
15150 '
15160 GOSUB 11000                                  'take the Runge-Kutta step in X
15170 '
15180 IF GRAPHICS% GOTO 15250                      'output ss point if mono only
```

```
15190    LOCATE 24,30
15200    PRINT USING "Y=#.###      Py=#.###";VAR(2),VAR(4)
15210    LOCATE 1,1:  PRINT "COMMANDS              p:pause";
15220    PRINT "             i:change initial conditions            e:end";
15230    GOTO 15290
15240    '
15250    SSY%(NCROSS%)=355+(VAR(2)-YMIN)*SSYSCALE'location of this ss point
15260    SSPY%(NCROSS%)=150-(VAR(4)+SQR(2*E))*SSPYSCALE
15270    PSET (SSY%(NCROSS%),SSPY%(NCROSS%))        'plot point on graphics
15280    '
15290    H=TSTEP:  CROSS%=0                      'restore step and reset flag
15300    FOR I%=1 TO 4                           'restore trajectory variables
15310       VAR(I%)=OLDVAR(I%)
15320    NEXT I%
15330    '
15340    RETURN
15350    '
16000    '*******************************************************************
16010    'subroutine to display header screen
16020    'input  variables:  none
16030    'output variables:  GRAPHICS%, MONO%
16040    'global variables:  none
16050    'local  variables:  G$, M$, ROW%
16060    '*******************************************************************
16070    SCREEN 0                                'program begins in text mode
16080    CLS: KEY OFF
16090    '
16100    LOCATE 1,30: COLOR 15                   'display book title
16110    PRINT "COMPUTATIONAL PHYSICS";
16120    LOCATE 2,40: COLOR 7
16130    PRINT "by";
16140    LOCATE 3,33: PRINT "Steven E. Koonin";
16150    LOCATE 4,14
16160    PRINT "Copyright 1985, Benjamin/Cummings Publishing Company";
16170    '
16180    LOCATE 5,10                             'draw box
16190    PRINT CHR$(201)+STRING$(59,205)+CHR$(187);
16200    FOR ROW%=6 TO 19
16210       LOCATE ROW%,10: PRINT CHR$(186);
16220       LOCATE ROW%,70: PRINT CHR$(186);
16230    NEXT ROW%
16240    LOCATE 20,10: PRINT CHR$(200)+STRING$(59,205)+CHR$(188);
16250    '
16260    COLOR 15                                'print title, etc
16270    LOCATE 7,36:  PRINT "EXAMPLE 2";
16280    COLOR 7
16290    LOCATE 10,21:  PRINT "Trajectories in the Henon-Heiles potential"
16300    LOCATE 13,36:  PRINT "*********"
16310    LOCATE 15,25:  PRINT "Type 'e' to end while running."
16320    LOCATE 16,22:  PRINT "Type ctrl-break to stop at a prompt."
16330    '
16340    LOCATE 19,16:  BEEP                     'get screen configuration
16350    INPUT "Does your computer have graphics capability (y/n)";G$
16360    IF LEFT$(G$,1)="y" OR LEFT$(G$,1)="n" GOTO 16390
16370       LOCATE 19,12:  PRINT SPACE$(58):  BEEP
16380       LOCATE 18,35:  PRINT "Try again,":  GOTO 16340
16390    IF LEFT$(G$,1)="y" GOTO 16420
16400       GRAPHICS%=0:  MONO%=-1
```

```
16410    GOTO 16520
16420 GRAPHICS%=-1
16430 LOCATE 18,15:  PRINT SPACE$(55)
16440 LOCATE 19,15:  PRINT SPACE$(55)
16450 LOCATE 19,15:  BEEP
16460 INPUT "Do you also have a separate screen for text (y/n)";M$
16470 IF LEFT$(M$,1)="y" OR LEFT$(M$,1)="n" GOTO 16500
16480   LOCATE 19,12:  PRINT SPACE$(58):  BEEP
16490   LOCATE 18,35:  PRINT "Try again,":  GOTO 16450
16500 IF LEFT$(M$,1)="y" THEN MONO%=-1 ELSE MONO%=0
16510 '
16520 LOCATE 21,9                                'preface to the code
16530 PRINT "A bound trajectory in the Henon-Heiles potential occurs if the"
16540 LOCATE 22,11
16550 PRINT "energy<1/6 and the initial position is inside the triangle"
16560 LOCATE 23,10
16570 PRINT "whose vertices are (1,0), (+sqr(3)/2,.5), and (-sqr(3)/2,.5)."
16580 LOCATE 24,31: BEEP:  PRINT "Press c to continue";
16590 IF INKEY$<>"c" THEN GOTO 16590
16600 '
16610 IF GRAPHICS% AND MONO% THEN GOSUB 17000        'switch to graphics screen
16620 '
16630 RETURN
16640 '
17000 '**************************************************************
17010 'subroutine to switch from monochrome to graphics screen
17020 'input  variables:   none
17030 'output variables:   none
17040 'global variables:   none
17050 'local  variables:   none
17060 '**************************************************************
17070 DEF SEG=0
17080 POKE &H410, (PEEK(&H410) AND &HCF) OR &H10
17090 SCREEN 2,0,0,0:  KEY OFF:  CLS
17100 RETURN
17110 '
18000 '**************************************************************
18010 'subroutine to switch from graphics to monochrome screen
18020 'input  variables:   none
18030 'output variables:   none
18040 'global variables:   none
18050 'local  variables:   none
18060 '**************************************************************
18070 DEF SEG=0
18080 POKE &H410, (PEEK(&H410) OR &H30)
18090 SCREEN 0:  WIDTH 80:  LOCATE ,,1,12,13
18100 RETURN
```

B.3 Example 3

This program finds the stationary states of the one-dimensional Schroedinger equation for a particle in a potential normalized so that its maximum and minimum values are +1 and -1, respectively. The mass of the particle is specified by the parameter γ, as discussed in the text. Three analytical forms of the potential are available (lines 80-160); the potential also can be altered pointwise using the cursor (subroutine 3000). For each level of the 50 levels sought above the starting energy (loop 280-560), a search is made to find the energy for which Eq. (3.23) is satisfied. A simple search is used until f changes sign, whereupon the secant method is employed (lines 390-440). To find f for a given energy, the Schroedinger equation is integrated leftward and rightward (subroutine 9000) using the Numerov algorithm and the solutions are matched at the leftmost turning point (line 9200). As the search proceeds, the wavefunction is displayed (subroutine 11000) and the number of nodes is counted. When the eigenenergy is found, the level is displayed on a plot of the potential (subroutine 13000).

Using the analytical form of a parabolic well with $\gamma=30$, an initial energy of -0.99, and an energy increment of DE=0.02 results in a typical calculation.

```
10 '************************************************************************
20 'Example 3:  Bound states in a one-dimensional potential
21 'COMPUTATIONAL PHYSICS by Steven E. Koonin
22 'Copyright 1985, Benjamin/Cummings Publishing Company
30 '************************************************************************
40 GOSUB 15000                                    'display header screen
50 '
60 FIRST%=-1:  VSAVE%=0:  ENDING%=0               'set flags,constants,functions
70 NPTS%=160                                      'number of lattice points
80 'square-well potential
90  '      DEF FNV1(X)=-1
100        XMIN1=-2: XMAX1=2: DX1=(XMAX1-XMIN1)/NPTS%
110 'parabolic-well potential
120        DEF FNV2(X)=-(1-.5*X*X)
130        XMIN2=-2: XMAX2=2: DX2=(XMAX2-XMIN2)/NPTS%
140 'Lennard-Jones potential
150        DEF FNV3(X)=4*(X^(-12)-X^(-6))
160        XMIN3=.9: XMAX3=1.9: DX3=(XMAX3-XMIN3)/NPTS%
170 '
180 DIM V(160):    DIM V%(160)                    'If you change NPTS%,
190 DIM PSI(160):  DIM PSI%(160)                  ' redimension these arrays
200 TOLF=.00005                                   'matching tolerance
210 '
220 GOSUB 1000                                    'input potential
230 IF FIRST% THEN GOSUB 5000                     'display commands
240 IF GRAPHICS% THEN GOSUB 6000                  'graph potential
250 GOSUB 7000                                    'input GAMMA, E, DE
260 '
270 EIGENE=-1                                     'minimum value for E
```

```
280 FOR N%=1 TO 50                                    'loop over many levels
290    SECANT%=0:  START%=-1:  F=TOLF                 'set values to begin E search
300    WHILE ABS(F)=>TOLF
310       K$=INKEY$                                   'check for command input
320       IF K$<>"" THEN GOSUB 8000
330       IF ENDING% GOTO 580                         'stop iterations
340       GOSUB 9000                                  'integrate Schrodinger eq.
350 '
360       IF GRAPHICS% THEN GOSUB 11000 ELSE GOSUB 12000      'output
370 '
380       IF START% THEN FSTART=F                     'Save first F value
390       IF F*FSTART<0 THEN SECANT%=-1               'then when F switches sign
400       IF F=FOLD GOTO 430                          ' use the secant search
410          IF SECANT% THEN DE=-F*(E-EOLD)/(F-FOLD)
420 '
430       EOLD=E:  FOLD=F                             'save old values
440       E=E+DE                                      'increment Energy
450       START%=0                                    'no longer first time
460       IF E<-1 THEN GOSUB 10000                    'keep E>-1
470    WEND
480 '
490    GOSUB 13000                                    'output
500 '
510    DE=ABS((EOLD-EIGENE)/3)                        'guess for next DE, but use
520    IF ABS(DESAVE)<DE THEN DE=ABS(DESAVE)          ' input DE if it is smaller
530 '
540    EIGENE=EOLD                                    'save eigenvalue
550    E=EOLD+DE                                      'increment E to begin
560 NEXT N%                                           ' next search
570 '
580 ENDING%=0:  GOSUB 14000                           'options for continuing
590 '
1000 '***************************************************************************
1010 'subroutine to input the 1-dimensional potential
1020 'input   variables:   DX1, DX2, DX3, FNV1, FNV2, FNV3, OPT$, SAVE%, XMIN1,
1030 '                     XMIN2, XMIN3
1040 'output variables:    COLUMN%, DX, POT$, V(I%), V%(I%), VSCALE, XMIN
1050 'global variables:    GRAPHICS%, NPTS%
1060 'local   variables:   D$, E$, F$, I%
1070 '***************************************************************************
1080 IF VSAVE% THEN GOTO 1730                         'if VSAVE%, then begin
1090 '                                                'with current V(x)
1100 GOSUB 2000                                       'input analytic form
1110 '
1120 IF OPT$<>"1" THEN GOTO 1200                      'define DE, XMIN, V(I%)
1130    XMIN=XMIN1:  DX=DX1                           'depending on choice
1140    FOR I%=0 TO NPTS%                             'of analytic potential
1150       V(I%)=FNV1(XMIN+DX*I%)
1160       IF V(I%)>1 THEN V(I%)=1                    'keep -1<=V<=1
1170    NEXT I%
1180    COLUMN%=35:  POT$="Square Well"               'text output variables
1190 '
1200 IF OPT$<>"2" THEN GOTO 1280                      'Parabolic Well
1210    XMIN=XMIN2:  DX=DX2
1220    FOR I%=0 TO NPTS%
1230       V(I%)=FNV2(XMIN+DX*I%)
1240       IF V(I%)>1 THEN V(I%)=1
1250    NEXT I%
```

```
1260     COLUMN%=33:   POT$="Parabolic Well"
1270 '
1280 IF OPT$<>"3" THEN GOTO 1360                        'Lennard Jones Potl.
1290    XMIN=XMIN3:   DX=DX3
1300    FOR I%=0 TO NPTS%
1310       V(I%)=FNV3(XMIN+DX*I%)
1320       IF V(I%)>1 THEN V(I%)=1
1330    NEXT I%
1340    COLUMN%=29:   POT$="Lennard-Jones Potential"
1350 '
1360 IF NOT GRAPHICS% THEN RETURN
1370 '
1380 VSCALE=170/2                                       'rescale V for graphing
1390 FOR I%=0 TO NPTS%
1400    V%(I%)=170-(V(I%)+1)*VSCALE
1410 NEXT I%
1420 '
1430 CLS                                                'directions for altering V
1440 LOCATE 2,2,0
1450 PRINT "You may now use the keyboard to alter V(x) at any of the 160";
1460 LOCATE 2,62
1470 PRINT " lattice points:";
1480 LOCATE 6,17
1490 PRINT CHR$(24)+CHR$(25)+" keys increase and decrease V(x) at fixed x.";
1500 LOCATE 9,7
1510 PRINT "PgUp and Home keys increase and decrease x without altering V(x).";
1520 LOCATE 12,12
1530 PRINT CHR$(26)+CHR$(27)+" keys increase and decrease x and change";
1540 LOCATE 12,54
1550 PRINT " V(x+dx) to V(x).";
1560 LOCATE 16,17
1570 PRINT "When you are finished, the code will normalize ";
1580 LOCATE 17,24
1590 PRINT "the new potential so that Vmin=-1";
1600 LOCATE 19,13
1610 PRINT "You will then be able to alter the potential again, so";
1620 LOCATE 20,25
1630 PRINT " don't be afraid to experiment.";
1640 '
1650 LOCATE 24,8:   BEEP
1660 INPUT;"Do you wish to 1)alter the potential or 2)use the analytic form";D$
1670 IF D$="2" THEN RETURN
1680 IF D$="1" THEN GOTO 1730
1690    LOCATE 24,1:    PRINT SPACE$(79);:   BEEP
1700    LOCATE 23,35:   PRINT "Try again";
1710    GOTO 1650
1720 '
1730 GOSUB 3000                                         'alter V from keyboard
1740 VSAVE%=0                                           'reset flag
1750 '
1760 LOCATE 24,4:   BEEP
1770 PRINT "Are you 1)happy with the potential or 2)do you wish to alter";
1780 LOCATE 24,64
1790 INPUT; " it further";E$
1800 IF E$="1" THEN RETURN
1810 IF E$="2" GOTO 1850
1820    LOCATE 23,35:   PRINT "Try again,";
1830    GOTO 1760
```

```
1840 '
1850 LOCATE 23,1:  PRINT SPACE$(79);:  LOCATE 24,1:  PRINT SPACE$(79);
1860 LOCATE 24,13:  BEEP
1870 PRINT "Would you like to 1)begin again with the analytic form";
1880 LOCATE 25,18
1890 INPUT; "or 2)alter the potential now on the screen";F$
1900 IF F$="1" THEN GOTO 1100
1910 IF F$="2" THEN GOTO 1950
1920    LOCATE 23,25                                    'error trapping
1930    PRINT "Try again,";
1940    GOTO 1860
1950 LOCATE 23,1:  PRINT SPACE$(79);
1960 LOCATE 24,1:  PRINT SPACE$(79);
1970 LOCATE 25,1:  PRINT SPACE$(79);
1980 GOTO 1730
1990 '
2000 '*******************************************************************
2010 'subroutine to choose the analytic form of the potential
2020 'input  variables:   none
2030 'output variables:   OPT$
2040 'global variables:   none
2050 'local  variables:   none
2060 '*******************************************************************
2070 CLS
2080 LOCATE 1,8,0                                    'display options for V(x)
2090 PRINT "In this program you have a choice of one-dimensional potentials:";
2100 LOCATE 5,20
2110 PRINT "1)Square well:  V(x)=-1  (-2 <= x <= 2)";
2120 LOCATE 8,15
2130 PRINT "2)Parabolic well:  V(x)=-(1-.5*x*x)  (-2 <= x <= 2)";
2140 LOCATE 11,5
2150 PRINT "3)Lennard Jones Potential:  V(x)=4*(x^(-12)-x^(-6))";
2160 LOCATE 11,56
2170 PRINT "  (.9 <= x <= 1.9)";
2180 LOCATE 15,6
2190 PRINT "All of these potentials have walls (V(x)=infinity";
2200 LOCATE 15,55
2210 PRINT ") at xmax and xmin,";
2220 LOCATE 16,6
2230 PRINT "giving the boundary conditions that the wave ";
2240 LOCATE 16,51
2250 PRINT "function is zero there.";
2260 LOCATE 17,10
2270 PRINT "We also impose the condition that -1 <= V(x) <= 1 for all x.";
2280 LOCATE 20,3
2290 PRINT "You may also exit the program and insert your own potential ";
2300 LOCATE 20,63
2310 PRINT "into the code";
2320 LOCATE 21,20
2330 PRINT "(see subroutine beginning at line 1000).";
2340 '
2350 LOCATE 24,12:  BEEP
2360 INPUT; "Enter the number of the desired potential, or e to exit";OPT$
2370 IF (OPT$="1" OR OPT$="2" OR OPT$="3") THEN CLS:  LOCATE ,,0:  RETURN
2380 IF OPT$="e" THEN END
2390    LOCATE 24,1:  PRINT SPACE$(79);:  LOCATE 23,20
2400    PRINT "Try again, the responses are 1,2,3, or e:";
2410    GOTO 2350
```

```
2420 '
3000 '************************************************************************
3010 'subroutine to alter the analytic potential from the graphics screen
3020 'input   variables:  DX, V(I%), V%(I%), VSCALE, XMIN
3030 'output variables:  LENGTH%, V(I%), V%(I%), XROW%
3040 'global variables:  NPTS%
3050 'local   variables:  B, CURSOR$, I%, IOLD%, K$, NORM, V%, VMAX, VMIN, VOLD%
3060 '************************************************************************
3070 LENGTH%=170:  XROW%=10                      'set values for subroutine
3080 GOSUB 4000                                  'which draws the axes
3090 '
3100 FOR I%=0 TO NPTS%                           'graph the potential
3110    LINE (I%*4,V%(I%))-((I%*4)+2,V%(I%))
3120 NEXT I%
3130 '
3140 LOCATE 24,8: BEEP                           'display instructions
3150 PRINT  CHR$(24)+CHR$(25)+CHR$(26)+CHR$(27)+" Home and PgUp keys move the";
3160 LOCATE 24,40:  PRINT " cursor to alter the potential.";
3170 LOCATE 25,27:  PRINT "Type return when finished.";
3180 '
3190 V%=V%(80):  I%=80                           'begin at the bottom of the well
3200 LINE (I%*4+1,V%+1)-(I%*4-1,V%-1)
3210 LINE (I%*4-1,V%+1)-(I%*4+1,V%-1)
3220 LOCATE 23,27:  PRINT USING "X=##.###     V(x)=##.###";(XMIN+DX*I%),V(I%);
3230 '
3240 K$=INKEY$
3250    IF K$=CHR$(13) GOTO 3690                 'typing return exits from loop
3260    IF LEN(K$)<2 GOTO 3240                   'cursors are special two
3270    CURSOR$=RIGHT$(K$,1)                     'character ASCII codes
3280    VOLD%=V%:  IOLD%=I%
3290    IF (CURSOR$=CHR$(72) OR CURSOR$=CHR$(80)) THEN GOTO 3410
3300    IF (CURSOR$=CHR$(75) OR CURSOR$=CHR$(77)) THEN GOTO 3480
3310    '
3320    IF CURSOR$=CHR$(73) THEN I%=I%+1         'cursor moves along
3330    IF CURSOR$=CHR$(71) THEN I%=I%-1         'the potential curve
3340    IF I%<0 THEN I%=NPTS%                    'cursor wraps around x axis
3350    IF I%>NPTS% THEN I%=0
3360    V%=(170-(V(I%)+1)*VSCALE)                'evaluate new variables
3370    LINE (IOLD%*4,VOLD%)-(IOLD%*4+2,VOLD%)   'redraw line at old cursor pos.
3380    LINE (I%*4,V%)-(I%*4+2,V%),0             'erase line at new cursor pos.
3390    GOTO 3580
3400    '
3410    IF CURSOR$=CHR$(72) THEN V%=V%-1         'cursor up
3420    IF CURSOR$=CHR$(80) THEN V%=V%+1         'cursor down
3430    IF V%>170 THEN V%=170                    'keep potential -1<V<1
3440    IF V%<0 THEN V%=0
3450    V(I%)=-1-(V%-170)/VSCALE
3460    GOTO 3580
3470    '
3480    IF CURSOR$=CHR$(77) THEN I%=I%+1         'cursor right
3490    IF CURSOR$=CHR$(75) THEN I%=I%-1         'cursor left
3500    IF I%<0 THEN I%=NPTS%                    'cursor wraps around x axis
3510    IF I%>NPTS% THEN I%=0
3520    V%=(170-(V(I%)+1)*VSCALE)
3530    LINE (IOLD%*4,VOLD%)-(IOLD%*4+2,VOLD%)
3540    LINE (I%*4,V%)-(I%*4+2,V%),0
3550    V%=VOLD%                                 'V at the new position
3560    V(I%)=V(IOLD%)                           'is the same as V at last pos.
```

```
3570        '
3580        LINE (IOLD%*4+1,VOLD%+1)-(IOLD%*4-1,VOLD%-1),0      'erase old cursor
3590        LINE (IOLD%*4-1,VOLD%+1)-(IOLD%*4+1,VOLD%-1),0
3600        LINE (I%*4+1,V%+1)-(I%*4-1,V%-1)                    'draw new cursor
3610        LINE (I%*4-1,V%+1)-(I%*4+1,V%-1)
3620        LINE (0,0)-(0,170)                                  'redraw axes
3630        LINE (0,85)-(640,85)
3640        '
3650        LOCATE 23,27
3660        PRINT USING "X=##.###      V(x)=##.###";(XMIN+DX*I%),V(I%);
3670 GOTO 3240
3680 '
3690 VMIN=V(0)                                                 'search for min V
3700 FOR I%=1 TO NPTS%
3710     IF V(I%)<VMIN THEN VMIN=V(I%)
3720 NEXT I%
3730 '
3740 LENGTH%=170:   XROW%=10
3750 GOSUB 4000
3760 '
3770 B=1+VMIN                                                  'shift V so VMIN=-1
3780 FOR I%=0 TO NPTS%
3790     V(I%)=V(I%)-B
3800     V%(I%)=170-(V(I%)+1)*VSCALE
3810     LINE (I%*4,V%(I%))-(I%*4+2,V%(I%))
3820 NEXT I%
3830 '
3840 RETURN
3850 '
4000 '***************************************************************************
4010 'subroutine to draw and label axes for V(x) vs. x
4020 'input  variables:  DX, LENGTH%, XMIN, XROW%
4030 'output variables:  none
4040 'global variables:  none
4050 'local  variables:  COL%, I%, MARK%, ROW%
4060 '***************************************************************************
4070 CLS:   SCREEN 2,0,0,0
4080 '
4090 LINE (0,0)-(0,LENGTH%)                                    'draw and label axes
4100 LINE (0,LENGTH%/2)-(640,LENGTH%/2)
4110 LOCATE 3,2:   PRINT "V(x)";
4120 LOCATE (XROW%+2),78:   PRINT "x";
4130 '
4140 FOR I%=0 TO 4
4150     MARK%=I%*(LENGTH%/4)
4160     LINE (0,MARK%)-(3,MARK%)
4170 NEXT I%
4180 '
4190 IF LENGTH%=86 GOTO 4260
4200     FOR I%=0 TO 4
4210         ROW%=I%*5.5:   IF I%=0 THEN ROW%=1
4220         LOCATE ROW%,2:   PRINT USING "##.#";(1-I%*.5);
4230     NEXT I%
4240     GOTO 4300
4250 '
4260 LOCATE 1,2:   PRINT "1.0";
4270 LOCATE 6,2:   PRINT "0.0";
4280 LOCATE 11,2:   PRINT "-1.0";
```

```
4290 '
4300 FOR I%=1 TO 4
4310     MARK%=I%*(640/4)
4320     LINE (MARK%,(LENGTH%/2-2))-(MARK%,(LENGTH%/2+2))
4330     COL%=I%*20-2:  IF I%=4 THEN COL%=COL%-2
4340     LOCATE XROW%,COL%:  PRINT USING "##.#";(XMIN+DX*I%*40);
4350 NEXT I%
4360 '
4370 RETURN
4380 '
5000 '********************************************************************
5010 'subroutine to display commands
5020 'input  variables:  none
5030 'output variables:  FIRST%
5040 'global variables:  none
5050 'local  variables:  none
5060 '********************************************************************
5070 CLS
5080 LOCATE 1,12,0
5090 PRINT "The commands available while the program is running are:";
5100 '
5110 LOCATE 5,14
5120 PRINT "p to pause -- typing any key will resume the program";
5130 LOCATE 8,35
5140 PRINT "e to end";
5150 LOCATE 11,24
5160 PRINT "n to change the energy and DE";
5170 LOCATE 14,20
5180 PRINT "d to change only DE (energy increment)";
5190 '
5200 LOCATE 24,31,0:  BEEP
5210 PRINT "Type c to continue";
5220 IF INKEY$<>"c" THEN GOTO 5220
5230 '
5240 FIRST%=0                                    'This screen is printed once.
5250 '
5260 LOCATE ,,0: CLS                             'turn off cursor
5270 '
5280 RETURN
5290 '
6000 '********************************************************************
6010 'subroutine to plot the potential on the upper half of the graphics screen
6020 'input  variables:  V%(I%)
6030 'output variables:  LENGTH%, XROW%
6040 'global variables:  NPTS%
6050 'local  variables:  VHALF
6060 '********************************************************************
6070 LENGTH%=86:  XROW%=5
6080 GOSUB 4000
6090 '
6100 VHALF=86/170                               'rescale V to fit on upper
6110 FOR I%=1 TO NPTS%                          'half of the screen
6120     LINE ((I%-1)*4,VHALF*V%(I%-1))-(I%*4,VHALF*V%(I%))
6130 NEXT I%
6140 '
6150 RETURN
6160 '
7000 '********************************************************************
```

```
7010 'subroutine to input GAMMA, E, DE
7020 'input  variables:   COLUMN%, POT$
7030 'output variables:   DE, DESAVE, E, GAMMA
7040 'global variables:   GRAPHICS%, TOLE
7050 'local  variables:   LINE1, LINE2
7060 '*****************************************************************
7070 LOCATE 24,1,0:  PRINT SPACE$(79);                  'input GAMMA
7080 LOCATE 25,1:  PRINT SPACE$(79);
7090 LOCATE 24,26
7100 PRINT "Gamma=sqr[2m(a^2)V/hbar^2]=";
7110 LOCATE 24,54:  INPUT; GAMMA
7120 IF GAMMA>0 GOTO 7170
7130   LOCATE 24,1:  PRINT SPACE$(79);:  BEEP          'error trapping
7140   LOCATE 25,28:  PRINT "Gamma must be positive,";
7150   GOTO 7090
7160 '
7170 LOCATE 24,1,0:  PRINT SPACE$(79);                  'input E
7180 LOCATE 25,1:  PRINT SPACE$(79);
7190 LOCATE 24,36
7200 INPUT; "Energy=";E
7210 IF E>-1  GOTO 7260                                 'error trapping
7220   LOCATE 24,1:  PRINT SPACE$(79);:  BEEP
7230   LOCATE 25,20:  PRINT "Since the minimum value of V=-1, E > -1";
7240   GOTO 7190
7250 '
7260 LOCATE 24,1,0:  PRINT SPACE$(79);                  'input DE
7270 LOCATE 25,1:  PRINT SPACE$(79);
7280 LOCATE 24,22:
7290 INPUT; "DE=increment in eigenenergy search=";DE
7300 IF DE<1 AND DE>TOLE GOTO 7340                      'error trapping
7310   LOCATE 24,1:  PRINT SPACE$(79);:  BEEP
7320   LOCATE 25,33:  PRINT USING "#.### < DE < 1";TOLE;
7330   GOTO 7280
7340 DESAVE=DE
7350 '
7360 LOCATE 24,1:  PRINT SPACE$(79);:  LOCATE 25,1:  PRINT SPACE$(79);
7370 '
7380 IF NOT GRAPHICS% THEN LOCATE 21,COLUMN%:  PRINT POT$;  'print type of pot
7390 '
7400 IF GRAPHICS% THEN LINE1=1 ELSE LINE1=22            'display GAMMA
7410 LOCATE LINE1,34,0:  PRINT USING "Gamma=###.##";GAMMA
7420 '
7430 IF GRAPHICS% THEN LINE2=25 ELSE LINE2=1            'display commands
7440 LOCATE LINE2,10,0
7450 PRINT "  p:pause      e:end      n:change energy      d:change DE";
7460 '
7470 RETURN
7480 '
8000 '*****************************************************************
8010 'subroutine to check for commands
8020 'input  variables:   K$
8030 'output variables:   DE, DESAVE, E, ENDING%, SECANT%, START%
8040 'global variables:   GRAPHICS%
8050 'local  variables:   A$
8060 '*****************************************************************
8070 LOCATE 23,1:  PRINT SPACE$(79);
8080 '
8090 IF K$<>"e" GOTO 8160                               'end the program
```

```
8100    LOCATE 23,21:  BEEP
8110    INPUT; "Do you want to end iterations(yes/no)";A$
8120    IF A$="yes" THEN ENDING%=-1
8130    LOCATE 23,1,0:  PRINT SPACE$(79);
8140    RETURN
8150    '
8160  IF K$<>"p" GOTO 8230                        'pause
8170    LOCATE 23,28,1:  BEEP
8180    PRINT "Type any key to continue";
8190    IF INKEY$="" GOTO 8190
8200    LOCATE 23,1,0:  PRINT SPACE$(79);
8210    RETURN
8220    '
8230  IF K$<>"n" THEN GOTO 8320                   'change Energy
8240    BEEP
8250    LOCATE 23,36:  INPUT; "Energy=";E
8260    LOCATE 23,1:  PRINT SPACE$(79);
8270    LOCATE 23,38:  INPUT; "DE=";DE
8280    DESAVE=DE
8290    LOCATE 23,1,0:  PRINT SPACE$(79);
8300    START%=-1:  SECANT%=0:  RETURN
8310    '
8320  IF K$<>"d" THEN RETURN                      'change energy increment
8330    IF NOT SECANT% THEN GOTO 8370
8340      LOCATE 23,19
8350      PRINT "You cannot change DE during secant search.";
8360      RETURN
8370    E=E-DE
8380    LOCATE 23,38:  INPUT; "DE=";DE
8390    E=E+DE:  DESAVE=DE
8400    LOCATE 23,1,0:  PRINT SPACE$(79);
8410    '
8420  RETURN
8430    '
9000  '*******************************************************************
9010  'subroutine to integrate Schroedinger equation
9020  'input   variables:  DX, GAMMA
9030  'output variables:  F, ILEFT%, IMATCH%, IRIGHT%, NCROSS%, NODES%, PSI(I%)
9040  '                    PSIMAX
9050  'global variables:  NPTS%
9060  'local   variables:  C, I%, K, KI, KIM1, KIP1, KLAST, LEFT%, NORM
9070  '                    PSIMMTCH, PSIPMTCH, RIGHT%
9080  '*******************************************************************
9090  C=(DX*DX/12)*GAMMA*GAMMA                    'evaluate constant
9100  IMATCH%=0                                   'clear matching point
9110  '
9120  PSI(0)=0                                    'left hand side bound. cond.
9130  PSI(1)=9.999999E-10
9140  '
9150  KIM1=C*(E-V(0))                            'initial K*K values
9160  KI=C*(E-V(1))
9170  '
9180  FOR I%=1 TO NPTS%-1                        'Numerov algorithm, S=0
9190    KIP1=C*(E-V(I%+1))
9200    IF (KI*KIP1<=0 AND KI>0) THEN IMATCH%=I%: GOTO 9290    'matching point
9210    PSI(I%+1)=(PSI(I%)*(2-10*KI)-PSI(I%-1)*(1+KIM1))/(1+KIP1)
9220    IF ABS(PSI(I%+1))<(1E+10) GOTO 9260       'if PSI grows too large
9230      FOR K%=1 TO I%+1                        '  rescale all previous
```

```
9240        PSI(K%)=PSI(K%)*9.999999E-06            ' points
9250     NEXT K%
9260     KIM1=KI:  KI=KIP1                          'roll values of K*K
9270 NEXT I%
9280 '
9290 IF IMATCH%=0 THEN IMATCH%=NPTS%-10            'if there is no turning point
9300                                               ' match wave functions near rhs
9310 PSIMMTCH=PSI(IMATCH%)                         'save value for normalization
9320 '
9330 PSI(NPTS%)=0                                  'rhs boundary conditions
9340 PSI(NPTS%-1)=9.999999E-10
9350 '
9360 KIP1=C*(E-V(NPTS%))                           'initial K*K values
9370 KI=C*(E-V(NPTS%-1))
9380 '
9390 FOR I%=NPTS%-1 TO IMATCH%+1 STEP -1           'Numerov algorithm, S=0
9400     KIM1=C*(E-V(I%-1))
9410     PSI(I%-1)=(PSI(I%)*(2-10*KI)-PSI(I%+1)*(1+KIP1))/(1+KIM1)
9420     IF ABS(PSI(I%-1))<(1E+10) GOTO 9460       'if PSI grows too large
9430        FOR K%=NPTS%-1 TO I%-1 STEP -1         ' renormalize all
9440           PSI(K%)=PSI(K%)*9.999999E-06        ' previous points
9450        NEXT K%
9460     KIP1=KI:  KI=KIM1                         'roll values of K*K
9470 NEXT I%
9480 KIM1=C*(E-V(IMATCH%-1))                       'finds value need for log deriv
9490 PSIPMTCH=(PSI(IMATCH%)*(2-10*KI)-PSI(IMATCH%+1)*(1+KIP1))/(1+KIM1)
9500 '
9510 NORM=PSI(IMATCH%)/PSIMMTCH                    'norm PSI left to PSI right
9520 FOR I%=1 TO IMATCH%-1                         ' keeping sign the same
9530     PSI(I%)=PSI(I%)*ABS(NORM)
9540 NEXT I%
9550 '
9560 IF NORM>0 THEN GOTO 9620                      'flip sign of PSI right if
9570    FOR I%=IMATCH% TO NPTS%                    ' NORM is negative
9580        PSI(I%)=-PSI(I%)
9590    NEXT I%
9600    PSIPMTCH=-PSIPMTCH
9610 '
9620 PSIMAX=ABS(PSI(1))                            'find maximum PSI value
9630 FOR I%=1 TO NPTS%
9640     IF ABS(PSI(I%))>PSIMAX THEN PSIMAX=ABS(PSI(I%))
9650 NEXT I%
9660 '
9670 F=(PSI(IMATCH%-1)-PSIPMTCH)/PSIMAX           'evaluate matching condition
9680 '
9690 NODES%=0                                      'count nodes
9700 FOR I%=2 TO NPTS%-1
9710     IF SGN(PSI(I%))*SGN(PSI(I%-1))<0 THEN NODES%=NODES%+1
9720 NEXT I%
9730 '
9740 LEFT%=0:  RIGHT%=0:                           'set flags
9750 I%=0                                          'look for left turning point
9760 WHILE NOT LEFT%
9770     IF (E-V(I%))>=0! THEN ILEFT%=I%:  LEFT%=-1
9780     I%=I%+1
9790 WEND
9800 I%=NPTS%                                      'look for right turning point
9810 WHILE NOT RIGHT%
```

```
9820    IF (E-V(I%))>=0! THEN IRIGHT%=I%:   RIGHT%=-1
9830    I%=I%-1
9840 WEND
9850 '
9860 KLAST=E-V(ILEFT%):  NCROSS%=0
9870 FOR I%=ILEFT%+1 TO IRIGHT%-1                    'look for entry into classically
9880    K=E-V(I%)                                    ' forbidden regions
9890    IF K*KLAST<=0 AND K<0 THEN NCROSS%=NCROSS%+1
9900    KLAST=K
9910 NEXT I%
9920 '
9930 RETURN
9940 '
10000 '****************************************************************
10010 'subroutine to keep E>-1
10020 'input  variables:  none
10030 'output variables:  DE, DESAVE, E, SECANT%, START%
10040 'global variables:  none
10050 'local  variables:  none
10060 '****************************************************************
10070 LOCATE 23,1:  PRINT SPACE$(79);
10080 LOCATE 23,26
10090 INPUT; "E is now <-1, enter new E";E          'prompt for new E
10100 LOCATE 23,1:  PRINT SPACE$(79);
10110 LOCATE 23,37:  INPUT; "DE=";DE
10120 LOCATE 23,1:  PRINT SPACE$(79);
10130 '
10140 START%=-1:  DESAVE=DE:  SECANT%=0             'reset flags
10150 '
10160 RETURN
10170 '
11000 '****************************************************************
11010 'subroutine to graph the wavefunction
11020 'input  variables:  F, IMATCH%, NCROSS%, NODES%, PSI(I%), PSIMAX
11030 'output variables:  none
11040 'global variables:  NPTS%
11050 'local  variables:  PSI%(I%), PSISCALE
11060 '****************************************************************
11070 LINE (0,90)-(640,191),0,BF                    'erase lower half of screen
11080 LINE (0,90)-(0,174)
11090 LINE (0,132)-(640,132)                        'redraw x axis
11100 LOCATE 14,2:  PRINT "PSI(x)";
11110 LOCATE 18,78:  PRINT "x";
11120 FOR I%=1 TO 4
11130    MARK%=I%*(640/4)
11140    LINE (MARK%,130)-(MARK%,134)
11150 NEXT I%
11160 '
11170 PSISCALE=84/(2*PSIMAX)
11180 '
11190 PSI%(0)=174-(PSI(0)+PSIMAX)*PSISCALE          'graph PSI left
11200 FOR I%=1 TO IMATCH%
11210    PSI%(I%)=174-(PSI(I%)+PSIMAX)*PSISCALE
11220    LINE ((I%-1)*4,PSI%(I%-1))-(I%*4,PSI%(I%))
11230 NEXT I%
11240 '
11250 PSI%(NPTS%)=174-(PSI(NPTS%)+PSIMAX)*PSISCALE          'graph PSI right
11260 FOR I%=NPTS%-1 TO IMATCH% STEP -1
```

```
11270    PSI%(I%)=174-(PSI(I%)+PSIMAX)*PSISCALE
11280    LINE ((I%+1)*4,PSI%(I%+1))-(I%*4,PSI%(I%))
11290 NEXT I%
11300 '
11310 IF NCROSS% < 1 THEN GOTO 11350
11320    LOCATE 23,17,0
11330    PRINT "Caution:  the wave function may be inaccurate.";
11340 '
11350 LOCATE 24,10,0
11360 PRINT USING "Energy=###.#####      DE=##.##^^^^      ";E,DE;
11370 LOCATE 24,47
11380 PRINT USING "F=##.##^^^^      Nodes=###";F,NODES%;
11390 '
11400 RETURN
11410 '
12000 '*******************************************************************
12010 'subroutine to output wave function for mono screen
12020 'input  variables:  DE, E, F, GAMMA, NODES%
12030 'output variables:  none
12040 'global variables:  none
12050 'local  variables:  none
12060 '*******************************************************************
12070 LOCATE 25,2,0
12080 PRINT USING "Gamma=###.##      Energy=##.#####      ";GAMMA,E;
12090 LOCATE 25,38
12100 PRINT USING "DE=#.##^^^^      F=##.##^^^^      Nodes=###";DE,F,NODES%;
12110 '
12120 RETURN
12130 '
13000 '*******************************************************************
13010 'subroutine to output eigenenergy
13020 'input  variables:  DX, EOLD, F, ILEFT%, IRIGHT%, NODES%, XMIN
13030 'output variables:  none
13040 'global variables:  GRAPHICS%
13050 'local  variables:  E%, XIN, XOUT
13060 '*******************************************************************
13070 IF NOT GRAPHICS% THEN GOTO 13170              'graphics output
13080    E%=86-(EOLD+1)*86/2                         'graph eigenenergy
13090    LINE (4*ILEFT%,E%)-(4*IRIGHT%,E%)
13100    LOCATE 23,1:  PRINT SPACE$(79);
13110    LOCATE 23,21,1:  BEEP
13120    PRINT "Bound state found.  Type c to continue";
13130    IF INKEY$<>"c" GOTO 13130
13140    LOCATE 23,1,0:  PRINT SPACE$(79);
13150    RETURN
13160 '
13170 XIN=XMIN+DX*ILEFT%                            'text output
13180 XOUT=XMIN+DX*IRIGHT%                          'calculate the turning points
13190 '
13200 LOCATE 23,1,0                                 'output description of state
13210 PRINT USING "Eigen energy=##.#####      Nodes=###      ";EOLD,NODES%;
13220 LOCATE 23,39
13230 PRINT USING "Xin=##.###      Xout=##.###";XIN,XOUT;
13240 LOCATE 23,64
13250 PRINT USING "    DE=##.##^^^^";DE;
13260 '
13270 LOCATE 24,1,0:  PRINT " "                      'scroll the screen
13280 '
```

```
13290 LOCATE 1,10,0                              'display commands
13300 PRINT "   p:pause        e:end      n:change energy       d:change DE";
13310 '
13320 RETURN
13330 '
14000 '********************************************************************
14010 'subroutine to print options for continuing the program
14020 'input   variables:  none
14030 'output  variables:  VSAVE%
14040 'global  variables:  GRAPHICS%
14050 'local   variables:  A%
14060 '********************************************************************
14070 CLS
14080 LOCATE 1,21
14090 PRINT "Options for continuing the calculation";
14100 LOCATE 3,10:  PRINT "1)Change gamma";
14110 LOCATE 5,10:  PRINT "2)Begin again with a new potential";
14120 LOCATE 7,10:  PRINT "3)Alter the current potential (graphics only)";
14130 LOCATE 9,10:  PRINT "4)End the program";
14140 LOCATE 11,25:  PRINT SPACE$(50)
14150 LOCATE 11,25:  BEEP:  INPUT "Enter the number of you choice";A%
14160 IF A%>=1 AND A%<=4 THEN GOTO 14180 ELSE GOTO 14150
14170 '
14180 IF NOT GRAPHICS% AND A%=3 THEN A%=2
14190 CLS
14200 IF A%<>1 THEN GOTO 14220
14210    RETURN 240
14220 IF A%<>2 THEN GOTO 14240
14230    RETURN 220
14240 IF A%<>3 THEN END
14250    VSAVE%=-1
14260    RETURN 220
14270 '
15000 '********************************************************************
15010 'subroutine to display the header screen
15020 'input   variables: none
15030 'output  variables: GRAPHICS%,MONO%
15040 'global  variables: none
15050 'local   variables: G$,M$,ROW%
15060 '********************************************************************
15070 SCREEN 0: CLS: KEY OFF                   'program begins in text mode
15080 '
15090 LOCATE 1,30: COLOR 15                    'display book title
15100 PRINT "COMPUTATIONAL PHYSICS";
15110 LOCATE 2,40: COLOR 7
15120 PRINT "by";
15130 LOCATE 3,33: PRINT "Steven E. Koonin";
15140 LOCATE 4,14
15150 PRINT "Copyright 1985, Benjamin/Cummings Publishing Company";
15160 '
15170 LOCATE 5,10                              'draw the box
15180 PRINT CHR$(201)+STRING$(59,205)+CHR$(187);
15190 FOR ROW%=6 TO 19
15200    LOCATE ROW%,10: PRINT CHR$(186);
15210    LOCATE ROW%,70: PRINT CHR$(186);
15220 NEXT ROW%
15230 LOCATE 20,10
15240 PRINT CHR$(200)+STRING$(59,205)+CHR$(188);
```

```
15250 '
15260 COLOR 15                                        'print title, etc.
15270 LOCATE  7,36: PRINT "EXAMPLE 3";
15280 COLOR 7
15290 LOCATE 10,15: PRINT "Quantum bound states in a one-dimensional potential"
15300 LOCATE 13,36: PRINT "**********"
15310 LOCATE 15,25: PRINT "Type 'e' to end while running."
15320 LOCATE 16,22: PRINT "Type ctrl-break to stop at a prompt."
15330 '
15340 LOCATE 19,16: BEEP                              'get screen configuration
15350 INPUT "Does your computer have graphics capability (y/n)";G$
15360 IF LEFT$(G$,1)="y" OR LEFT$(G$,1)="n" GOTO 15390
15370    LOCATE 19,12:  PRINT SPACE$(58):  BEEP
15380    LOCATE 18,35:  PRINT "Try again,":  GOTO 15340
15390 IF LEFT$(G$,1)="y" GOTO 15420
15400    GRAPHICS%=0:  MONO%=1
15410    RETURN
15420 GRAPHICS%=-1
15430 LOCATE 18,15:  PRINT SPACE$(55)
15440 LOCATE 19,15:  PRINT SPACE$(55)
15450 LOCATE 19,15: BEEP
15460 INPUT "Do you also have a separate screen for text (y/n)";M$
15470 IF LEFT$(M$,1)="y" OR LEFT$(M$,1)="n" GOTO 15490  'error trapping
15480    LOCATE 18,35:  PRINT "Try again,":  GOTO 15440
15490 IF LEFT$(M$,1)="y" THEN MONO%=-1 ELSE MONO%=0
15500 '
15510 IF MONO% THEN GOSUB 16000
15520 '
15530 RETURN
15540 '
16000 '**********************************************************************
16010 'subroutine to switch from mono to graphics screen
16020 'input   variables: none
16030 'output variables: none
16040 'global variables: none
16050 'local  variables: none
16060 '**********************************************************************
16070 DEF SEG=0
16080 POKE &H410, (PEEK(&H410) AND &HCF) OR &H10
16090 SCREEN 2,0,0,0:  WIDTH 80: LOCATE ,,1,6,7
16100 RETURN
16110 '
17000 '**********************************************************************
17010 'subroutine to switch from graphics to mono screen
17020 'input   variables: none
17030 'output variables: none
17040 'global variables: none
17050 'local  variables: none
17060 '**********************************************************************
17070 DEF SEG=0
17080 POKE &H410, (PEEK(&H410) OR &H30)
17090 SCREEN 0:  WIDTH 80:  LOCATE ,,1,12,13
17100 RETURN
17110 '
```

B.4 Example 4

This program calculates the Born and eikonal scattering amplitudes and cross sections (Eqs. (4.21) and (4.29,30), respectively) for an electron incident on a square-well, gaussian well, or Lenz-Jensen potential, as described in the text. The total cross section in each case is calculated by integrating the differential cross section over the forward and backward angles (loops 2160-2200 and 2250-2290). The total cross section is also calculated from the eikonal scattering amplitude using the optical theorem (4.31) (line 2350). The differential cross section itself is calculated in the Born and eikonal approximations by subroutines 4000 and 6000, respectively, with the Bessel function J_0 calculated by subroutine 7000. All integrations are done by the same order Gauss-Legendre quadrature, the allowed numbers of points being listed in line 12030. After the numerical results are presented, a semi-log plot of the differential cross sections is displayed.

Choosing a square-well potential of depth

$$\text{Vzero}=20$$

and an incident electron energy of

$$E=20$$

with 20 quadrature points results in a typical output from this program.

```
10  '***********************************************************
20  'Example 4:  Born and eikonal approximations to quantum scattering
21  'COMPUTATIONAL PHYSICS by Steven E. Koonin
22  'Copyright 1985, Benjamin/Cummings Publishing Company
30  '***********************************************************
40  GOSUB 9000                           'display header screen
50  '
60  FIRST%=-1                            'define constants, functions
70  PI=3.14159
80  SQHALF=SQR(.5)
90  E2=14.409                            'electron charge squared
100 HBARM=7.6359                         '(h-bar)^2/(electron mass)
110 '
120 DIM XLEG(48),WLEG(48)                'quadrature abscissae and wts.
130 DIM DEGREE(96)                       'angles at which to calculate
140 DIM SIGE(96),SIGB(96)                'arrays for dsigma/domega
150 DIM CHI(48)                          'eikonal profile function
160 RMAX=2                               'V is zero beyond RMAX
170 '
180 GOSUB 1000                           'input V, Vzero, Z, E, NLEG%
190 '
200 GOSUB 2000                           'find SIGMA total
210 '
220 LOCATE 24,31:  PRINT "Type c to continue";
230 IF INKEY$<>"c" GOTO 230
240 '
```

```
250 IF GRAPHICS% THEN GOSUB 8000                         'sigma plot
260 '
270 GOTO 180                                             'begin again with new input
280 '
1000 '*******************************************************************************
1010 'subroutine to input Z, Energy, number of quadrature points
1020 'input   variables:   none
1030 'output variables:   E, K, NLEG%, OPT$, RCUT, V$, V%, VZERO, Z, Z6
1040 'global variables:   FIRST%
1050 'local   variables:   ESAVE, ILLEGAL%, NSAVE%, VZSAVE, ZSAVE
1060 '*******************************************************************************
1070 CLS
1080 '
1090 LOCATE 1,26                                         'input potential
1100 PRINT "The possible potentials are";
1110 LOCATE 2,11
1120 PRINT "1)Lenz-Jensen potential for an electron and a neutral atom";
1130 LOCATE 3,21
1140 PRINT "2)Square well        3)Gaussian well";
1150 LOCATE 4,11
1160 INPUT; "Enter the number corresponding to your choice or 0 to end";OPT$
1170 IF OPT$="0" OR OPT$="1" OR OPT$="2" OR OPT$="3" THEN GOTO 1220
1180    LOCATE 5,18:  BEEP
1190    PRINT "Try again, the responses are 0, 1, 2, or 3";
1200    LOCATE 4,1:   PRINT SPACE$(80);
1210    GOTO 1150
1220 IF OPT$="0" THEN END
1230 IF OPT$="1" GOTO 1360
1240 '
1250 IF FIRST% GOTO 1280                                 'input well depth
1260 LOCATE 8,26
1270 PRINT USING "Your last Vzero was=+###.##";VZSAVE
1280 LOCATE 9,27
1290 INPUT; "Vzero=depth of the well=";VZERO
1300 VZSAVE=VZERO
1310 IF OPT$="2" THEN V$="Square Well" ELSE V$="Gaussian Well"
1320 IF OPT$="2" THEN V%=11 ELSE V%=13
1330 RCUT=2
1340 GOTO 1510
1350 '
1360 IF FIRST% THEN GOTO 1390                            'input nuclear charge
1370    LOCATE 8,30
1380    PRINT USING "Your last Z was ###";ZSAVE;
1390 LOCATE 9,23
1400 INPUT; "Z=charge of the atomic nucleus=";Z
1410 IF Z>0 THEN GOTO 1460
1420    LOCATE 10,25:  BEEP
1430    PRINT "Try again, Z must be positive";
1440    LOCATE 9,1:   PRINT SPACE$(79);
1450    GOTO 1390
1460 ZSAVE=Z
1470 V$="Lenz-Jensen":  V%=11
1480 RCUT=1
1490 Z6=Z^.166667
1500 '
1510 IF FIRST% GOTO 1540                                 'input beam energy
1520    LOCATE 13,25
1530    PRINT USING "Your last Energy was #####.###";ESAVE;
```

```
1540 LOCATE 14,22
1550 INPUT; "E=energy of the projectile in eV=";E
1560 IF E>0 THEN GOTO 1610
1570    LOCATE 15,25:   BEEP
1580    PRINT "Try again, E must be positive";
1590    LOCATE 14,1:    PRINT SPACE$(79)'
1600    GOTO 1540
1610 ESAVE=E
1620 K=SQR(2*E/HBARM)                              'calculate wave number
1630 '
1640 IF FIRST% GOTO 1670                           'input number of points
1650    LOCATE 18,18
1660    PRINT USING "Your last number of quadrature points was ##";NLEG%;
1670 LOCATE 19,25
1680 INPUT;  "Number of quadrature points=";NLEG%
1690 ILLEGAL%=0                                    'set flag
1700 GOSUB 12000                                   'set up weights and
1710 IF NOT ILLEGAL% THEN GOTO 1760               ' abscissae
1720    LOCATE 20,12:   BEEP
1730    PRINT "The allowed values are 2,3,4,5,6,8,10,12,16,20,24,32,40,48";
1740    LOCATE 19,1:    PRINT SPACE$(79);
1750    GOTO 1670
1760 NSAVE%=NLEG%
1770 '
1780 CLS
1790 '
1800 IF OPT$="1" GOTO 1840
1810    LOCATE 1,12
1820    PRINT USING "Vzero=+###.##      Energy=#####.##     ";VZERO,E;
1830    GOTO 1860
1840 LOCATE 1,15
1850 PRINT USING "Z=###       Energy=#####.##      ";Z,E;
1860 PRINT USING "number of points=##";NLEG%
1870 '
1880 FIRST%=0                                      'no longer first time
1890 '
1900 RETURN
1910 '
2000 '****************************************************************
2010 'subroutine to calculate SIGMA total
2020 'input   variables:   IMFEO, SIGBT, SIGET
2030 'output variables:    COSTH, IMFEO, L%, SIGBT, SIGET, SIGOPT
2040 'global variables:    K, NLEG%, PI, SQHALF, XLEG(I%)
2050 'local  variables:    COSCUT, I%, SIGBTFOR, SIGBTBACK, SIGETFOR, SIGETBACK,
2060 '                     SINHALF, THETACUT
2070 '****************************************************************
2080 SINHALF=PI/(K*RCUT)                           'find the angle at which
2090 IF SINHALF>SQHALF THEN SINHALF=SQHALF        ' q*rcut=2*pi, to divide
2100 THETACUT=2*ATN(SINHALF/SQR(1-SINHALF*SINHALF)) ' integral into forward
2110 COSCUT=COS(THETACUT)                          ' and backward scattering
2120 '
2130 '                                             'sigma for small angles
2140 SIGBT=0:  SIGET=0:                            'zero total sigmas
2150 IMFEO=0                                       'zero imaginary forward amp.
2160 FOR I%=1 TO NLEG%                             'the forward scattering is
2170    COSTH=COSCUT+(XLEG(I%)+1)*(1-COSCUT)/2     ' from 0 to THETACUT
2180    L%=I%
2190    GOSUB 3000                                 'calculate sigma at this angle
```

```
2200 NEXT I%
2210 SIGBTFOR=PI*(1-COSCUT)*SIGBT              'factor of (1-COSCUT)/2 comes
2220 SIGETFOR=PI*(1-COSCUT)*SIGET              ' from change of variables
2230 '
2240 SIGBT=0:   SIGET=0                        'zero total sig for back scatt
2250 FOR I%=1 TO NLEG%                         'contributions from back angle
2260     COSTH=-1+(XLEG(I%)+1)*(COSCUT+1)/2    ' from THETACUT to 180
2270     L%=I%+NLEG%
2280     GOSUB 3000                            'calculate sigma at this angle
2290 NEXT I%
2300 SIGBTBACK=PI*(COSCUT+1)*SIGBT
2310 SIGETBACK=PI*(COSCUT+1)*SIGET
2320 '
2330 SIGBT=SIGBTFOR+SIGBTBACK                  'sigma total is the sum of
2340 SIGET=SIGETFOR+SIGETBACK                  ' forward and backward parts
2350 SIGOPT=4*PI/K*IMFE0                       'sigma total from optical thm
2360 '
2370 PRINT "":   PRINT ""                      'scroll text
2380 IF NLEG%<10 GOTO 2460                     'repeat E if it has scrolled
2390   IF OPT$="1" GOTO 2430                   ' off the screen
2400     LOCATE ,12
2410     PRINT USING "Vzero=+###.##      Energy=#####.##       ";VZERO,E;
2420     GOTO 2450
2430     LOCATE ,15
2440     PRINT USING "Z=###     Energy=#####.##     ";Z,E;
2450     PRINT USING "number of points=##";NLEG%
2460 LOCATE ,25
2470 PRINT USING "sigma total (Born)=+#.###^^^^";SIGBT
2480 LOCATE ,24
2490 PRINT USING "sigma total (eikonal)=+#.###^^^^";SIGET
2500 LOCATE ,13
2510 PRINT USING"sigma total (optical thm using eikonal amp)=+#.###^^^^";SIGOPT
2520 LOCATE ,32
2530 PRINT USING "Theta cut=###.##";(180*THETACUT/PI)
2540 '
2550 RETURN
2560 '
3000 '*********************************************************************
3010 'subroutine to calculate the differential cross section at a given angle
3020 '          and its contribution to the total cross section
3030 'input   variables:  COSTH, L%
3040 'output variables:   DEGREE(I%), Q, SIGB(I%), SIGE(I%), SIGBT, SIGET
3050 'global variables:   K, PI, WLEG(I%)
3060 'local   variables:  THETA
3070 '*********************************************************************
3080 IF INKEY$="e" THEN RETURN 180             'typing e ends the program
3090 '
3100 THETA=ATN(SQR(1-COSTH^2)/COSTH)          'the integration variable
3110 IF THETA<0 THEN THETA=THETA+PI           ' is cos theta
3120 DEGREE(L%)=THETA*180/PI
3130 Q=2*K*SIN(THETA/2)                        'momentum transfer
3140 '
3150 GOSUB 4000                                'calculate Born approx sigma
3160 SIGBT=SIGBT+SIGB(L%)*WLEG(I%)             'Born approx total sigma
3170 GOSUB 6000                                'calculate eikonal approx sigma
3180 SIGET=SIGET+SIGE(L%)*WLEG(I%)             'eikonal approx total sigma
3190 '
3200 PRINT USING "Theta=###.##     q=###.##      ";DEGREE(L%),Q;
```

```
3210 PRINT USING "sigma Born=+#.###^^^^     ";SIGB(L%);
3220 PRINT USING "sigma eikonal=+#.###^^^^";SIGE(L%)
3230 '
3240 RETURN
3250 '
4000 '*******************************************************
4010 'subroutine to calculate (dsigma/domega) using the Born approximation
4020 'input    variables:  L%, Q
4030 'output variables:  R, SIGB(I%)
4040 'global variables:  HBARM, NLEG%, RMAX, WLEG(I%), XLEG(I%)
4050 'local    variables:  FBORN, J%
4060 '*******************************************************
4070 FBORN=0                                   'zero the Born amp
4080 '
4090 FOR J%=1 TO NLEG%
4100     R=RMAX*(XLEG(J%)+1)/2                  'scale the variable
4110     GOSUB 5000                             'calculate the potential
4120     FBORN=FBORN+SIN(Q*R)*V*R*WLEG(J%)      'amplitude integral
4130 NEXT J%
4140 '
4150 FBORN=-RMAX/(Q*HBARM)*FBORN                'multiply by constants
4160 SIGB(L%)=FBORN*FBORN                       'sigma=amplitude^2
4170 '
4180 RETURN
4190 '
5000 '*******************************************************
5010 'subroutine to calculate V(r) using either a square or gaussian well
5020 '             or the Lenz-Jensen potential
5030 'input    variables:  R
5040 'output variables:  V
5050 'global variables:  E2, OPT$, VZERO, Z, Z6
5060 'local    variables:  RR,R1S,U
5070 '*******************************************************
5080 IF OPT$="2" THEN V=-VZERO:   RETURN         'square well
5090 '
5100 IF OPT$="3" THEN V=-VZERO*EXP(-2*R*R):   RETURN   'gaussian well
5110 '
5120 RR=R: R1S=.529/Z                            'Lenz-Jensen potential,
5130 IF R<R1S THEN RR=R1S                        ' cutoff at radius of
5140 U=4.5397*Z6*SQR(RR)                         ' the 1s shell
5150 V=-((Z*E2)/RR)*EXP(-U)*(1+U+U*U*(.3344+U*(.0485+.002647*U)))
5160 '
5170 RETURN
5180 '
6000 '*******************************************************
6010 'subroutine to calculate (dsigma/domega) using the eikonal approximation
6020 'input    variables:  BESSJO, L%, Q, V
6030 'output variables:  IMFEO, R, SIGE(I%), X
6040 'global variables:  HBARM, K, NLEG%, RMAX, WLEG(I%), XLEG(I%)
6050 'local    variables:  B, CHI(I%), IMFE, K%, M%, REFE, ZMAX, ZZ
6060 '*******************************************************
6070 REFE=0:  IMFE=0                             'zero real and imag amp
6080 '
6090 FOR K%=1 TO NLEG%                           'integrate over b for F
6100     B=RMAX*(XLEG(K%)+1)/2
6110     IF L%>1 THEN GOTO 6240                  'calculate CHI(b) only
6120         LOCATE 24,23: PRINT "Patience please, I'm working on it";   ' once
6130         CHI(K%)=0
```

```
6140        ZMAX=SQR(RMAX*RMAX-B*B)                              'r^2=z^2+b^2
6150        FOR M%=1 TO NLEG%
6160          ZZ=ZMAX*(XLEG(M%)+1)/2
6170          R=SQR(ZZ*ZZ+B*B)
6180          GOSUB 5000                                         'calculate V
6190          CHI(K%)=CHI(K%)+V*WLEG(M%)
6200        NEXT M%
6210        CHI(K%)=-(ZMAX/(2*K*HBARM))*CHI(K%)                  'contribution to CHI
6220        IMFE0=IMFE0+B*SIN(CHI(K%))^2*WLEG(K%)                'calculate Im Fe(0)
6230        '
6240      X=Q*B
6250      GOSUB 7000                                             'calculate bessel funct
6260        '
6270      REFE=REFE+B*BESSJ0*SIN(2*CHI(K%))*WLEG(K%)
6280      IMFE=IMFE+B*BESSJ0*SIN(CHI(K%))*SIN(CHI(K%))*WLEG(K%)
6290        '
6300 NEXT K%
6310 '
6320 IMFE=K*RMAX*IMFE                                            'multiply by constants
6330 REFE=K*RMAX*REFE/2
6340 SIGE(L%)=IMFE*IMFE+REFE*REFE                                'sigma=|f|^2
6350 '
6360 IF L%=1 THEN IMFE0=K*RMAX*IMFE0
6370 IF L%=1 THEN LOCATE 24,1: PRINT SPACE$(79);: LOCATE 3,1 'erase message
6380 '
6390 RETURN
6400 '
7000 '***************************************************************
7010 'subroutine to compute J0 by polynomial approximation
7020 'input   variables:  X
7030 'output variables:  BESSJ0
7040 'global variables:  none
7050 'local   variables:  TEMP, Y, Y2
7060 '***************************************************************
7070 Y=X/3: Y2=Y*Y
7080 '
7090 IF ABS(X)>3 THEN GOTO 7170
7100 '
7110      TEMP=-.0039444+.00021*Y2: TEMP=.0444479+Y2*TEMP
7120      TEMP=-.3163866+Y2*TEMP:    TEMP=1.2656208#+Y2*TEMP
7130      TEMP=-2.2499997#+Y2*TEMP
7140      BESSJ0=1+Y2*TEMP
7150      RETURN
7160 '
7170 Y=1/Y
7180 TEMP=-7.2805E-04+1.4476E-04*Y:  TEMP=1.37237E-03+TEMP*Y
7190 TEMP=-9.512E-05+TEMP*Y:         TEMP=-.0055274+TEMP*Y
7200 TEMP=-7.7E-07+TEMP*Y:           TEMP=.79788456#+TEMP*Y
7210 TEMP2=-2.9333E-04+1.3558E-04*Y: TEMP2=-5.4125E-04+TEMP2*Y
7220 TEMP2=2.62573E-03+TEMP2*Y:      TEMP2=-3.954E-05+TEMP2*Y
7230 TEMP2=-4.166397E-02+TEMP2*Y:    TEMP2=X-.78539816#+TEMP2*Y
7240 BESSJ0=TEMP*COS(TEMP2)/SQR(X)
7250 RETURN
7260 '
8000 '***************************************************************
8010 'subroutine to graph the partial cross section vs theta
8020 'input   variables: DEGREE(I%), SIGB(I%), SIGE(I%), SIGBT, SIGET, SIGOPT,
8030 '                   V$, V%
```

```
8040 'output variables:   none
8050 'global variables:   E, MONO%, NLEG%, OPT$, VZERO, Z
8060 'local   variables:  COL%, EXPMAX%, EXPMIN%, LSIGE, LSIGB, MARK%, NORD%,
8070 '                    ROW%, S%, SB%, SE%, SBOLD%, SEOLD%, SIGMAX, SIGMIN,
8080 '                    SIGSCALE, T%, TOLD%, TSCALE
8090 '***********************************************************************
8100 IF MONO% THEN GOSUB  10000                           'switch to graphics
8110 CLS:  SCREEN 2,0,0,0
8120 '
8130 SIGMAX=0                                             'find largest sigma
8140 FOR L%=1 TO 2*NLEG%
8150    IF SIGE(L%)>SIGMAX THEN SIGMAX=SIGE(L%)
8160    IF SIGB(L%)>SIGMAX THEN SIGMAX=SIGB(L%)
8170 NEXT L%
8180 '
8190 SIGMIN=SIGMAX                                        'find smallest sigma
8200 FOR L%=1 TO 2*NLEG%
8210    IF SIGE(L%)<SIGMIN THEN SIGMIN=SIGE(L%)
8220    IF SIGB(L%)<SIGMIN THEN SIGMIN=SIGB(L%)
8230 NEXT L%
8240 '
8250 EXPMAX%=INT(LOG(SIGMAX)/LOG(10)+1)                   'find max and min
8260 EXPMIN%=INT(LOG(SIGMIN)/LOG(10))                     ' exponents for sigma
8270 NORD%=EXPMAX%-EXPMIN%                                'orders of magnitude
8280 '
8290 TSCALE=584/180                                       'theta scale
8300 SIGSCALE=175/NORD%                                   'sigma scale
8310 '
8320 LINE (56,0)-(56,175)                                 'sigma axis
8330 LINE (56,175)-(640,175)                              'theta axis
8340 LOCATE 3,1:  PRINT "Sigma";                          'sigma legend
8350 LOCATE 23,68:  PRINT "Theta";                        'theta legend
8360 '
8370 LOCATE 1,(80-V%)                                     'display name of pot
8380 PRINT V$;
8390 '
8400 IF OPT$="1" THEN GOTO 8440                           'display E, Z, Vzero
8410 LOCATE 2,54
8420 PRINT USING "Vzero=+###.##    E=#####.##";VZERO,E;
8430 GOTO 8460
8440 LOCATE 2,62
8450 PRINT USING "Z=###    E=#####.##";Z,E;
8460 LOCATE 3,57
8470 PRINT "Key:  X=Born, o=eikonal";
8480 LOCATE 24,2
8490 PRINT USING "sigma total:  Born=+#.###^^^^      ";SIGBT;
8500 PRINT USING "eikonal=+#.###^^^^     optical=+#.###^^^^";SIGET,SIGOPT;
8510 '
8520 FOR MARK%=0 TO 4                                     'theta ticks
8530    T%=56+MARK%*584/4
8540    LINE (T%,173)-(T%,177)
8550    COL%=6+MARK%*18
8560    IF MARK%=4 THEN COL%=77                           'theta labels
8570    LOCATE 23,COL%
8580    PRINT USING "###";MARK%*45;
8590 NEXT MARK%
8600 '
8610 FOR MARK%=0 TO NORD%                                 'sigma ticks
```

```
8620    S%=MARK%*175/NORD%
8630    LINE (54,S%)-(58,S%)
8640    ROW%=MARK%*22/NORD%:   IF ROW%=0 THEN ROW%=1        'sigma labels
8650    LOCATE ROW%,1
8660    PRINT USING "1E+##";(EXPMAX%-MARK%)
8670 NEXT MARK%
8680 '
8690 FOR I%=1 TO 2*NLEG%                                    'graph sig eikonal
8700    T%=56+DEGREE(I%)*TSCALE                             ' on semilog plot
8710    LSIGE=LOG(SIGE(I%))/LOG(10)
8720    SE%=175-(LSIGE-EXPMIN%)*SIGSCALE
8730    CIRCLE (T%,SE%),3                                   'mark points with
8740    IF I%=1 THEN GOTO 8760                              ' a circle
8750      LINE (TOLD%,SEOLD%)-(T%,SE%)
8760    TOLD%=T%:   SEOLD%=SE%
8770 NEXT I%
8780 '
8790 FOR I%=1 TO 2*NLEG%                                    'graph sig Born
8800    T%=56+DEGREE(I%)*TSCALE                             ' on semilog plot
8810    LSIGB=LOG(SIGB(I%))/LOG(10)
8820    SB%=175-(LSIGB-EXPMIN%)*SIGSCALE
8830    LINE (T%+2,SB%+2)-(T%-2,SB%-2)                      'mark points with
8840    LINE (T%+2,SB%-2)-(T%-2,SB%+2)                      ' an X
8850    IF I%=1 GOTO 8870
8860      LINE (TOLD%,SBOLD%)-(T%,SB%)
8870    TOLD%=T%
8880    SBOLD%=SB%
8890 NEXT I%
8900 '
8910 LOCATE 25,31,1:  BEEP:  PRINT "Type c to continue";
8920 IF INKEY$<>"c" THEN GOTO 8920
8930 LOCATE ,,0
8940 '
8950 IF MONO% THEN GOSUB 11000                             'return to text screen
8960 CLS
8970 '
8980 RETURN
8990 '
9000 '*****************************************************************
9010 'subroutine to display the header screen
9020 'input   variables: none
9030 'output variables: GRAPHICS%,MONO%
9040 'global  variables: none
9050 'local   variables: G$,M$,ROW%
9060 '*****************************************************************
9070 SCREEN 0: CLS: KEY OFF                                'program begins in text mode
9080 '
9090 LOCATE 1,30: COLOR 15                                 'display book title
9100 PRINT "COMPUTATIONAL PHYSICS";
9110 LOCATE 2,40: COLOR 7
9120 PRINT "by";
9130 LOCATE 3,33: PRINT "Steven E. Koonin";
9140 LOCATE 4,14
9150 PRINT "Copyright 1985, Benjamin/Cummings Publishing Company";
9160 '
9170 LOCATE 5,10                                           'draw the box
9180 PRINT CHR$(201)+STRING$(59,205)+CHR$(187);
9190 FOR ROW%=6 TO 19
```

```
9200      LOCATE ROW%,10: PRINT CHR$(186);
9210      LOCATE ROW%,70: PRINT CHR$(186);
9220 NEXT ROW%
9230 LOCATE 20,10
9240 PRINT CHR$(200)+STRING$(59,205)+CHR$(188);
9250 '
9260 COLOR 15                                    'print title, etc.
9270 LOCATE  7,36:  PRINT "EXAMPLE 4";
9280 COLOR 7
9290 LOCATE 10,14:
9300 PRINT "Born and eikonal approximations to quantum scattering";
9310 LOCATE 13,36:  PRINT "**********";
9320 LOCATE 15,25:  PRINT "Type 'e' to end while running."
9330 LOCATE 16,22:  PRINT "Type ctrl-break to stop at a prompt."
9340 '
9350 LOCATE 19,16: BEEP                          'get screen configuration
9360 INPUT "Does your computer have graphics_capability (y/n)";G$
9370 IF LEFT$(G$,1)="y" OR LEFT$(G$,1)="n" GOTO 9400
9380    LOCATE 19,12:  PRINT SPACE$(58):  BEEP
9390    LOCATE 18,35:  PRINT "Try again,":  GOTO 9350
9400 IF LEFT$(G$,1)="y" GOTO 9430
9410    GRAPHICS%=0:  MONO%=-1
9420    GOTO 9520
9430 GRAPHICS%=-1
9440 LOCATE 18,15:  PRINT SPACE$(55)
9450 LOCATE 19,15:  PRINT SPACE$(55)
9460 LOCATE 19,15: BEEP
9470 INPUT "Do you also have a separate screen for text (y/n)";M$
9480 IF LEFT$(M$,1)="y" OR LEFT$(M$,1)="n" GOTO 9500  'error trapping
9490    LOCATE 18,35: PRINT "Try again,":  GOTO 9450
9500 IF LEFT$(M$,1)="y" THEN MONO%=-1 ELSE MONO%=0
9510 '
9520 LOCATE 21,10
9530 PRINT "This program calculates the total and partial cross sections";
9540 LOCATE 22,13
9550 PRINT "for scattering of electrons from a potential using both";
9560 LOCATE 23,23
9570 PRINT "the Born and eikonal approximations";
9580 LOCATE 24, 9
9590 PRINT "All energies are in eV, angles in degrees, lengths in Angstroms";
9600 LOCATE 25,32,1:  BEEP
9610 PRINT "Type c to continue";
9620 IF INKEY$<>"c" GOTO 9620
9630 LOCATE ,,0
9640 '
9650 IF MONO% AND GRAPHICS% THEN GOSUB 11000          'switch to text screen
9660 '
9670 RETURN
9680 '
10000 '***************************************************************
10010 'subroutine to switch from mono to graphics screen
10020 'input   variables: none
10030 'output variables: none
10040 'global variables: none
10050 'local   variables: none
10060 '***************************************************************
10070 DEF SEG=0
10080 POKE &H410,  (PEEK(&H410) AND &HCF) OR &H10
```

```
10090 SCREEN 0:   WIDTH 80: LOCATE ,,0
10100 RETURN
10110 '
11000 '******************************************************************
11010 'subroutine to switch from graphics to mono screen
11020 'input   variables: none
11030 'output variables: none
11040 'global  variables: none
11050 'local   variables: none
11060 '******************************************************************
11070 DEF SEG=0
11080 POKE &H410, (PEEK(&H410) OR &H30)
11090 SCREEN 0:   WIDTH 80:  LOCATE ,,0
11100 RETURN
11110 '
12000 '******************************************************************
12010 'subroutine to establish Gauss-Legendre weights and abscissae
12020 '              allowed numbers of quadrature points are
12030 '              2,3,4,5,6,8,10,12,16,20,24,32,40, and 48
12040 'input   variables:  NLEG%
12050 'output variables:  ILLEGAL%, WLEG(I%), XLEG(I%)
12060 'global  variables:  none
12070 'local   variables:  ILEG%
12080 '******************************************************************
12090 IF NLEG%= 2 THEN GOTO 12330          'branch to the table for the
12100 IF NLEG%= 3 THEN GOTO 12360          ' input value of NLEG%
12110 IF NLEG%= 4 THEN GOTO 12400
12120 IF NLEG%= 5 THEN GOTO 12440
12130 IF NLEG%= 6 THEN GOTO 12490
12140 IF NLEG%= 8 THEN GOTO 12540
12150 IF NLEG%=10 THEN GOTO 12600
12160 IF NLEG%=12 THEN GOTO 12670
12170 IF NLEG%=16 THEN GOTO 12750
12180 IF NLEG%=20 THEN GOTO 12850
12190 IF NLEG%=24 THEN GOTO 12970
12200 IF NLEG%=32 THEN GOTO 13110
12210 IF NLEG%=40 THEN GOTO 13290
12220 IF NLEG%=48 THEN GOTO 13510
12230 '
12240 ILLEGAL%=-1                          'if NLEG% is a disallowed value
12250 RETURN                               ' return and prompt again
12260 '
12270 FOR ILEG%=1 TO FIX(NLEG%/2)          'since the weights and abscissa
12280    XLEG(NLEG%-ILEG%+1)=-XLEG(ILEG%)  ' are even and odd functions
12290    WLEG(NLEG%-ILEG%+1)=WLEG(ILEG%)   ' functions respectively, the
12300 NEXT ILEG%                           ' other half of the array is
12310 RETURN                               ' trivial to fill out
12320 '
12330 XLEG(1)=.577350269189626#  : WLEG(1)=1#                'NLEG%=2
12340 GOTO 12270
12350 '
12360 XLEG(1)=.774596669241483#  : WLEG(1)=.555555555555556#  'NLEG%=3
12370 XLEG(2)=0                   : WLEG(2)=.888888888888889#
12380 GOTO 12270
12390 '
12400 XLEG(1)=.861136311594053#  : WLEG(1)=.347854845137454#  'NLEG%=4
12410 XLEG(2)=.339981043584856#  : WLEG(2)=.652145154862546#
12420 GOTO 12270
```

```
12430  '
12440  XLEG(1)=.906179845938664#    :  WLEG(1)=.236926885056189#    'NLEG%=5
12450  XLEG(2)=.538469310105683#    :  WLEG(2)=.478628670499366#
12460  XLEG(3)=0                    :  WLEG(3)=.568888888888889#
12470  GOTO 12270
12480  '
12490  XLEG(1)=.932469514203152#    :  WLEG(1)=.17132449237917#     'NLEG%=6
12500  XLEG(2)=.661209386466265#    :  WLEG(2)=.360761573048139#
12510  XLEG(3)=.238619186083197#    :  WLEG(3)=.467913934572691#
12520  GOTO 12270
12530  '
12540  XLEG(1)=.960289856497536#    :  WLEG(1)=.101228536290376#    'NLEG%=8
12550  XLEG(2)=.796666477413627#    :  WLEG(2)=.222381034453374#
12560  XLEG(3)=.525532409916329#    :  WLEG(3)=.313706645877887#
12570  XLEG(4)=.18343464249565#     :  WLEG(4)=.362683783378362#
12580  GOTO 12270
12590  '
12600  XLEG(1)=.973906528517172#    :  WLEG(1)=.066671344308688#    'NLEG%=10
12610  XLEG(2)=.865063366688985#    :  WLEG(2)=.149451349150581#
12620  XLEG(3)=.679409568299024#    :  WLEG(3)=.219086362515982#
12630  XLEG(4)=.433395394129247#    :  WLEG(4)=.269266719309996#
12640  XLEG(5)=.148874338981631#    :  WLEG(5)=.295524224714753#
12650  GOTO 12270
12660  '
12670  XLEG(1)=.981560634246719#    :  WLEG(1)=.047175336386512#    'NLEG%=12
12680  XLEG(2)=.904117256370475#    :  WLEG(2)=.106939325995318#
12690  XLEG(3)=.769902674194305#    :  WLEG(3)=.160078328543346#
12700  XLEG(4)=.587317954286617#    :  WLEG(4)=.203167426723066#
12710  XLEG(5)=.36783149899818#     :  WLEG(5)=.233492536538355#
12720  XLEG(6)=.125233408511469#    :  WLEG(6)=.249147045813403#
12730  GOTO 12270
12740  '
12750  XLEG(1)=.98940093499165#     :  WLEG(1)=.027152459411754#    'NLEG%=16
12760  XLEG(2)=.944575023073233#    :  WLEG(2)=.062253523938648#
12770  XLEG(3)=.865631202387832#    :  WLEG(3)=.095158511682493#
12780  XLEG(4)=.755404408355003#    :  WLEG(4)=.124628971255534#
12790  XLEG(5)=.617876244402644#    :  WLEG(5)=.149595988816577#
12800  XLEG(6)=.45801677657227#     :  WLEG(6)=.169156519395003#
12810  XLEG(7)=.281603550779259#    :  WLEG(7)=.182603415044924#
12820  XLEG(8)=.095012509837637#    :  WLEG(8)=.189450610455069#
12830  GOTO 12270
12840  '
12850  XLEG(1)=.993128599185094#    :  WLEG(1)=.017614007139152#    'NLEG%=20
12860  XLEG(2)=.96397192727913#     :  WLEG(2)=.040601429800386#
12870  XLEG(3)=.912234428251325#    :  WLEG(3)=.062672048334109#
12880  XLEG(4)=.839116971822218#    :  WLEG(4)=.083276741576704#
12890  XLEG(5)=.74633190646015#     :  WLEG(5)=.10193011981724#
12900  XLEG(6)=.636053680726515#    :  WLEG(6)=.118194531961518#
12910  XLEG(7)=.510867001950827#    :  WLEG(7)=.131688638449176#
12920  XLEG(8)=.373706088715419#    :  WLEG(8)=.142096109318382#
12930  XLEG(9)=.227785851141645#    :  WLEG(9)=.149172986472603#
12940  XLEG(10)=.076526521133497#   :  WLEG(10)=.152753387130725#
12950  GOTO 12270
12960  '
12970  XLEG(1)=.995187219997021#    :  WLEG(1)=.012341229799987#    'NLEG%=24
12980  XLEG(2)=.974728555971309#    :  WLEG(2)=.028531388628933#
12990  XLEG(3)=.938274552002732#    :  WLEG(3)=.044277438817419#
13000  XLEG(4)=.886415527004401#    :  WLEG(4)=.059298584915436#
```

```
13010 XLEG(5)=.820001985973902#  : WLEG(5)=.07334648141108#
13020 XLEG(6)=.740124191578554#  : WLEG(6)=.086190161531953#
13030 XLEG(7)=.648093651936975#  : WLEG(7)=.097618652104113#
13040 XLEG(8)=.545421471388839#  : WLEG(8)=.107444270115965#
13050 XLEG(9)=.433793507626045#  : WLEG(9)=.115505668053725#
13060 XLEG(10)=.315042679696163# : WLEG(10)=.121670472927803#
13070 XLEG(11)=.191118867473616# : WLEG(11)=.125837456346828#
13080 XLEG(12)=.064056892862605# : WLEG(12)=.127938195346752#
13090 GOTO 12270
13100 '
13110 XLEG(1)=.997263861849481#  : WLEG(1)=.00701861000947#     'NLEG%=32
13120 XLEG(2)=.985611511545268#  : WLEG(2)=.016274394730905#
13130 XLEG(3)=.964762255587506#  : WLEG(3)=.025392065309262#
13140 XLEG(4)=.934906075937739#  : WLEG(4)=.034273862913021#
13150 XLEG(5)=.896321155766052#  : WLEG(5)=.042835898022226#
13160 XLEG(6)=.849367613732569#  : WLEG(6)=.050998059262376#
13170 XLEG(7)=.794483795967942#  : WLEG(7)=.058684093478535#
13180 XLEG(8)=.732182118740289#  : WLEG(8)=.065822222776361#
13190 XLEG(9)=.663044266930215#  : WLEG(9)=.072345794108848#
13200 XLEG(10)=.587715757240762# : WLEG(10)=.07819389578707#
13210 XLEG(11)=.506899908932229# : WLEG(11)=.083311924226946#
13220 XLEG(12)=.421351276130635# : WLEG(12)=.087652093004403#
13230 XLEG(13)=.331868602282127# : WLEG(13)=.091173878695763#
13240 XLEG(14)=.239287362252137# : WLEG(14)=.093844399080804#
13250 XLEG(15)=.144471961582796# : WLEG(15)=.095638720079274#
13260 XLEG(16)=.048307665687738# : WLEG(16)=.096540088514727#
13270 GOTO 12270
13280 '
13290 XLEG(1)=.998237709710559#  : WLEG(1)=.004521277098533#    'NLEG%=40
13300 XLEG(2)=.990726238699457#  : WLEG(2)=.010498284531152#
13310 XLEG(3)=.977259949983774#  : WLEG(3)=.016421058381907#
13320 XLEG(4)=.957916819213791#  : WLEG(4)=.022245849194166#
13330 XLEG(5)=.932812808278676#  : WLEG(5)=.027937006980023#
13340 XLEG(6)=.902098806968874#  : WLEG(6)=.033460195282547#
13350 XLEG(7)=.865959503212259#  : WLEG(7)=.038782167974472#
13360 XLEG(8)=.824612230833311#  : WLEG(8)=.043870908185673#
13370 XLEG(9)=.778305651426519#  : WLEG(9)=.048695807635072#
13380 XLEG(10)=.727318255189927# : WLEG(10)=.053227846983936#
13390 XLEG(11)=.671956684614179# : WLEG(11)=.057439769099391#
13400 XLEG(12)=.61255388966798#  : WLEG(12)=.061306242492928#
13410 XLEG(13)=.549467125095128# : WLEG(13)=.064804013456601#
13420 XLEG(14)=.483075801686178# : WLEG(14)=.067912045815233#
13430 XLEG(15)=.413779204371605# : WLEG(15)=.070611647391286#
13440 XLEG(16)=.341994090825758# : WLEG(16)=.072886582395804#
13450 XLEG(17)=.268152185007253# : WLEG(17)=.074723169057968#
13460 XLEG(18)=.192697580701371# : WLEG(18)=.076110361900626#
13470 XLEG(19)=.116084070675255# : WLEG(19)=.077039818164247#
13480 XLEG(20)=.03877241750605#  : WLEG(20)=.077505947978424#
13490 GOTO 12270
13500 '
13510 XLEG(1)=.998771007252426#  : WLEG(1)=.003153346052305#    'NLEG%=48
13520 XLEG(2)=.99353017226635#   : WLEG(2)=.007327553901276#
13530 XLEG(3)=.984124583722826#  : WLEG(3)=.011477234579234#
13540 XLEG(4)=.970591592546247#  : WLEG(4)=.015579315722943#
13550 XLEG(5)=.95298770316043#   : WLEG(5)=.019616160457355#
13560 XLEG(6)=.931386690706554#  : WLEG(6)=.023570760839324#
13570 XLEG(7)=.905879136715569#  : WLEG(7)=.027426509708356#
13580 XLEG(8)=.876572020274247#  : WLEG(8)=.031167227832798#
```

```
13590 XLEG(9)=.843588261624393#  :  WLEG(9)=.03477722256477#
13600 XLEG(10)=.807066204029442#  :  WLEG(10)=.03824135106583#
13610 XLEG(11)=.76715903251574#   :  WLEG(11)=.041545082943464#
13620 XLEG(12)=.724034130923814#  :  WLEG(12)=.044674560856694#
13630 XLEG(13)=.677872379632663#  :  WLEG(13)=.04761665849249#
13640 XLEG(14)=.628867396776513#  :  WLEG(14)=.050359035553854#
13650 XLEG(15)=.577224726083972#  :  WLEG(15)=.052890189485193#
13660 XLEG(16)=.523160974722233#  :  WLEG(16)=.055199503699984#
13670 XLEG(17)=.466902904750958#  :  WLEG(17)=.057277292100403#
13680 XLEG(18)=.408686481990716#  :  WLEG(18)=.059114839698395#
13690 XLEG(19)=.34875588629216#   :  WLEG(19)=.060704439165893#
13700 XLEG(20)=.287362487355455#  :  WLEG(20)=.062039423159892#
13710 XLEG(21)=.224763790394689#  :  WLEG(21)=.063114192286254#
13720 XLEG(22)=.161222356068891#  :  WLEG(22)=.063924238584648#
13730 XLEG(23)=.097004699209462#  :  WLEG(23)=.06446616443595#
13740 XLEG(24)=.032380170962869#  :  WLEG(24)=.064737696812683#
13750 GOTO 12270
13760 '
```

B.5 Example 5

This program fits electron-nucleus elastic scattering cross sections to determine the nuclear charge density using the method described in the text. The target nucleus is selected and the experimental data are defined in subroutine 1000, while the initial density and Fourier expansion coefficients are assumed to be that of a Fermi function (subroutine 3000). Before the iterations begin, the inner profile functions defined by Eq. (5.66) are calculated and stored (lines 2290-2420), as are the Bessel functions $J_{0,1}$ at the quadrature points (lines 2450-2520 and subroutine 9000) and the outer contributions to the scattering amplitude at each of the data points (lines 2550-2650). All integrals are done by 20-point Gauss-Legendre quadrature. At each iteration of the fit (loop 390-470), the improved expansion coefficients are found (subroutine 8000) by constructing the matrix A and vector B defined by Eqs. (5.49) (lines 8080-8450), inverting the matrix (subroutine 11000), and the solving Eq. (5.51) (lines 8500-8560). Note that since Eq. (5.47b) is homogeneous, it can be suitably scaled to the same order of magnitude as the rest of the rows of A (line 8390). At each iteration, χ^2 is calculated (subroutine 4000), as is the density in coordinate space (loop 4360-4560). Plots of the fit to the cross section (subroutine 5000), the fractional error in the fit, and the density (subroutine 6000) are displayed, as is quantitative information about the fit (subroutine 7000).

A demonstration of this program's output can be had by choosing the nucleus ^{40}Ca, entering a boundary radius of 7 fm, using 8 sine functions, and also using 8 sine functions initially in the fitting procedure (see discussion immediately above Exercise 5.8). Iterations improving the fit will then continue until "e" is pressed.

```
10  '*********************************************************************
20  'Example 5: Determining nuclear charge densities
21  'COMPUTATIONAL PHYSICS by Steven E. Koonin
22  'Copyright 1985, Benjamin/Cummings Publishing Company
30  '*********************************************************************
40  DIM THETA(100),QEFF(100)              'exptl angles and effective Q
50  DIM SIGE(100),DSIGE(100)              'exptl sigma and errors
60  DIM REFOUT(100),IMFOUT(100)           'real and imag of outer f
70  DIM REFT(100),IMFT(100)               'real and imag of total f
80  DIM SIGT(100)                         'theoretical cross section
90  DIM CHIN(20,15)                       'chi(n) at the quadrature pts.
100 DIM JTABLE(20,100)                    'Bessel fnctn at quad. pts.
110 DIM CZERO(15)                         'sine coefficients of rho
120 DIM SCHI(20),CCHI(20),S2CHI(20)       'trig fnctns of chi at quad.pts
130 DIM W(15)                             'array for derivs of sigma
140 DIM A#(16,16),BVECTOR(16)             'matrix to be inverted and B
150 DIM R%(20),Q#(20),P#(20)              'temp. arrays for inversion
160 DIM XLEG(20),WLEG(20)                 'quad. abscissae and weights
```

```
170 DIM RHO(90),DRHO(90)                        'rho(r) and its error
180 '
190 DEF FNFF(R)=1/(1+EXP(4.4*(R-RTARGET)/THICK)) 'fermi-function shape
200 DEF FNG(Y)=LOG((1+SQR(1-Y^2))/Y)-SQR(1-Y^2) 'function phi in Eq. (5.65)
210 PI=3.141593: ALPHA=1/137.036: HBARC=197.329 'physical constants
220 NLEG%=20                                     'number of quadrature points
230 '
240 GOSUB 12000                                 'display header screen
250 GOSUB 1000                                  'pick target,read in the data
260 GOSUB 2000                                  'define chi(n),Bessel, outer f
270 '
280 LOCATE 22,10
290 BEEP: INPUT "Enter initial number of sine functions";NSTART%
300 IF NSTART%=NBASIS% THEN GOTO 330
310    LOCATE 23,10
320    BEEP: INPUT "Enter how often to increment number of sines";FREQ%
330 NSAVE%=NBASIS%: NBASIS%=NSTART%
340 IF FREQ%<=0 THEN FREQ%=1
350 '
360 GOSUB 3000                                  'initial C's from Fermi rho
370 GOSUB 4000                                  'calculate initial fit, display
380 '
390 FOR ITER%=1 TO 100
400    IF NBASIS%>=NSAVE% OR ITER% MOD FREQ% <>0 THEN GOTO 430
410       NBASIS%=NBASIS%+1
420       CZERO(NBASIS%)=0
430    GOSUB 8000                               'find new C's
440    IF INKEY$="e" THEN GOTO 480
450    GOSUB 4000                               'calculate and display fit
460    IF INKEY$="e" THEN GOTO 480
470 NEXT ITER%
480 LOCATE 25,30: PRINT "Press c to continue";: BEEP
490 IF INKEY$="c" THEN GOTO 250 ELSE GOTO 490
500 '
1000 '*****************************************************************
1010 'subroutine to read in the data
1020 'input   variables: none
1030 'output variables: ATARGET,DSIGE,EBEAM,KBEAM,NPOINTS%,QMAX,SIGE,THETA,
1040 '                   TARGET$,VC1,ZA,ZTARGET
1050 'global variables: ALPHA,HBARC,ITER%,PI
1060 'local   variables: CHOICE%,I%,QLAB,RECOIL
1070 '*****************************************************************
1080 CLS                                        'decide which target
1090 LOCATE 10,10
1100 PRINT "The nuclei for which this program has data are:"
1110 LOCATE 12,20: PRINT "1) 40Ca"
1120 LOCATE 14,20: PRINT "2) 58Ni"
1130 LOCATE 16,20: PRINT "3) 208Pb"
1140 LOCATE 18,10: BEEP
1150 INPUT "Enter the number of your choice or 0 to end";CHOICE%
1160 IF CHOICE%=0 THEN END
1170 IF CHOICE%=1 THEN RESTORE 15000
1180 IF CHOICE%=2 THEN RESTORE 16000
1190 IF CHOICE%=3 THEN RESTORE 17000
1200 '
1210 READ TARGET$,ZTARGET,ATARGET,EBEAM,NPOINTS%          'read the data
1220 ZA=ZTARGET*ALPHA: KBEAM=EBEAM/HBARC
1230 VC1=1+4/3*ZTARGET*ALPHA*HBARC/(EBEAM*1.07*ATARGET^(1/3))'Coul. correct.
```

```
1240 QMAX=0
1250 FOR I%=1 TO NPOINTS%
1260     READ THETA(I%),SIGE(I%),DSIGE(I%)
1270     IF DSIGE(I%)>=SIGE(I%) THEN DSIGE(I%)=.8*SIGE(I%)      'keep error small
1280     THETA(I%)=THETA(I%)*PI/180                             'angle in radians
1290     QLAB=2*KBEAM*SIN(THETA(I%)/2)
1300     QEFF(I%)=QLAB*VC1                                      'Coul. correction
1310     IF QEFF(I%)>QMAX THEN QMAX=QEFF(I%)                    'find max Q in data
1320     RECOIL=1+2*EBEAM/940/ATARGET*SIN(THETA(I%)/2)^2        'take out recoil
1330     SIGE(I%)=SIGE(I%)*RECOIL
1340     DSIGE(I%)=DSIGE(I%)*RECOIL
1350 NEXT I%
1360 ITER%=0                                                   'zero iteration #
1370 RETURN
1380 '
2000 '**********************************************************************
2010 'subroutine to define the chi(n), Bessel, and outer f~functions
2020 'input   variables: none
2030 'output variables: CHIN,IMFOUT,JTABLE,NBASIS%,REFOUT,RMAX
2040 'global variables: A#,ALPHA,ATARGET,BESSJ0,BESSJ1,EBEAM,KBEAM,NLEG%,
2050 '                   NPOINTS%,ORDER%,PI,QEFF,QMAX,TARGET$,X,XLEG,WLEG,ZA
2060 'local   variables: B,ILEG%,IMF,JLEG%,KB$,M%,N%,R,REF,SUM
2070 '**********************************************************************
2080 CLS                                                       'input radius and # of sines
2090 LOCATE 5,10
2100 PRINT "For the target "+TARGET$
2110 LOCATE 5,31
2120 PRINT USING "the data are at a beam energy of ###.## MeV";EBEAM
2130 LOCATE 6,15
2140 PRINT USING "The maximum momentum transfer covered is ##.### fm~-1";QMAX
2150 LOCATE 7,15
2160 PRINT USING "The nuclear radius is about #.### fm";1.07*ATARGET^.3333
2170 LOCATE 10,10: BEEP
2180 INPUT "Enter the boundary radius to use";RMAX
2190 LOCATE 12,10: BEEP
2200 INPUT "Enter the number of sine functions to use (<=15)";NBASIS%
2210 LOCATE 14,15
2220 PRINT USING "For these parameters, Qmax is ##.### fm~-1";NBASIS%*PI/RMAX
2230 LOCATE 16,15: BEEP
2240 INPUT "Would you like to change these parameters (y/n)";KB$
2250 IF KB$="y" THEN GOTO 2080
2260 '
2270 GOSUB 10000                                               'set up quadrature points
2280 LOCATE 18,25: PRINT "Calculating the chi(n)";
2290 FOR N%=1 TO NBASIS%
2300     FOR ILEG%=1 TO NLEG%                                  'find chi(n) at each b
2310         B=(1+XLEG(ILEG%))/2
2320         SUM=0
2330         FOR JLEG%=1 TO NLEG%                               'r integral for chi(n)
2340             R=B+(1-B)/2*(1+XLEG(JLEG%))
2350             SUM=SUM+WLEG(JLEG%)*R*SIN(N%*PI*R)*FNG(B/R)
2360         NEXT JLEG%
2370         CHIN(ILEG%,N%)=-4*PI*ALPHA*RMAX^2*SUM*(1-B)/2
2380     NEXT ILEG%
2390     FOR M%=1 TO NBASIS%                                   'zero the A# matrix
2400         A#(N%,M%)=0
2410     NEXT M%
2420 NEXT N%
```

```
2430 '
2440 LOCATE 19,25: PRINT "Calculating Bessel table";
2450 ORDER%=0
2460 FOR ILEG%=1 TO NLEG%                              'fill Bessel fntcn table
2470     B=(1+XLEG(ILEG%))/2
2480     FOR I%=1 TO NPOINTS%
2490         X=QEFF(I%)*RMAX*B: GOSUB 9000
2500         JTABLE(ILEG%,I%)=BESSJ0*B*WLEG(ILEG%)/2
2510     NEXT I%
2520 NEXT ILEG%
2530 '
2540 LOCATE 20,25: PRINT "Calculating outer f";
2550 FOR I%=1 TO NPOINTS%                              'loop over data points
2560     X=QEFF(I%)*RMAX                               'find and store Real outer f
2570     ORDER%=0: GOSUB 9000
2580     REF=(1+4*(ZA/X)^2-(2/X)-4*ZA^2*(5-ZA^2))*BESSJ0
2590     ORDER%=1: GOSUB 9000
2600     REF=REF+16*ZA^2/X^3*BESSJ1
2610     REFOUT(I%)=2*KBEAM/QEFF(I%)^2*ZA*REF
2620     IMF=BESSJ0*(1-8*(1-2*ZA^2)/X^2)              'find and store Imag outer f
2630     IMF=IMF-X*BESSJ1/2*(1-4*(1-ZA^2)/X^2)
2640     IMFOUT(I%)=-8*KBEAM*ZA^2/(QEFF(I%)*X)^2*IMF
2650 NEXT I%
2660 RETURN
2670 '
3000 '*************************************************************************
3010 'subroutine to calculate the initial expansion coefficients
3020 'input   variables: none
3030 'output  variables: CZERO
3040 'global  variables: ATARGET,NBASIS%,PI,RMAX,ZTARGET
3050 'local   variables: DR,N%,RAD,RRHO,RTARGET,SN,SUM,THICK
3060 '*************************************************************************
3070 RTARGET=1.07*ATARGET^.3333: THICK=2.4            'params of Fermi density
3080 FOR N%=1 TO NBASIS%                              'zero expansion coefficients
3090     CZERO(N%)=0
3100 NEXT N%
3110 DR=.1
3120 FOR RAD=DR TO RMAX STEP DR                       'r integrals for C's
3130     RRHO=RAD*FNFF(RAD)
3140     FOR N%=1 TO NBASIS%
3150         CZERO(N%)=CZERO(N%)+RRHO*SIN(N%*PI*RAD/RMAX)
3160     NEXT N%
3170 NEXT RAD
3180 '
3190 SUM=0                                            'normalize the C's to charge Z
3200 FOR N%=1 TO NBASIS%
3210     SN=-RMAX^2/N%/PI
3220     IF N% MOD 2 =1 THEN SN=-SN
3230     SUM=SUM+SN*CZERO(N%)
3240 NEXT N%
3250 SUM=ZTARGET/(4*PI*SUM)
3260 FOR N%=1 TO NBASIS%
3270     CZERO(N%)=CZERO(N%)*SUM
3280 NEXT N%
3290 RETURN
3300 '
4000 '*************************************************************************
4010 'subroutine to calculate chisquare and rho and the display the fit
```

```
4020 'input   variables: CZERO
4030 'output variables: CHISQ,IMFT,NDOF%,REFT,SIGT
4040 'global variables: CCHI,CHIN,DSIGE,GRAPHICS%,JTABLE,KBEAM,MONO%,NBASIS%,
4050 '                  NLEG%,NPOINTS%,PI,QMAX,RMAX,SCHI,S2CHI,SIGE,VC1,W,ZA
4060 'local  variables: CHI,DRRHO,I%,ILEG%,IMF,J%,N%,RAD,REF,RRHO,SUM,TENMORE%
4070 '*********************************************************************
4080 FOR ILEG%=1 TO NLEG%                          'calculate total chi at each b
4090    CHI=-ZA*LOG((1+XLEG(ILEG%))/2)
4100    FOR N%=1 TO NBASIS%
4110       CHI=CHI+CZERO(N%)*CHIN(ILEG%,N%)
4120    NEXT N%
4130    SCHI(ILEG%)=SIN(2*CHI)                     'store trig functions of chi
4140    CCHI(ILEG%)=COS(2*CHI)
4150    S2CHI(ILEG%)=SIN(CHI)^2
4160 NEXT ILEG%
4170 '
4180 QMAX=NBASIS%*PI/RMAX                          'largest Q described by basis
4190 CHISQ=0: NDOF%=0                              'calculate fit sigma,chisquare
4200 FOR I%=1 TO NPOINTS%
4210    REF=0: IMF=0:
4220    FOR ILEG%=1 TO NLEG%
4230       REF=REF+JTABLE(ILEG%,I%)*SCHI(ILEG%)    'real and imag of inner f
4240       IMF=IMF+JTABLE(ILEG%,I%)*S2CHI(ILEG%)
4250    NEXT ILEG%
4260    REFT(I%)=REFOUT(I%)+KBEAM*RMAX^2*REF        'store real and imag of total f
4270    IMFT(I%)=IMFOUT(I%)+2*KBEAM*RMAX^2*IMF
4280 '                                             'sigma in mb, with Coul corr.
4290    SIGT(I%)=10*COS(THETA(I%)/2)^2*(REFT(I%)^2+IMFT(I%)^2)*VC1^4
4300    IF QEFF(I%)>QMAX THEN GOTO 4330
4310       CHISQ=CHISQ+((SIGE(I%)-SIGT(I%))/DSIGE(I%))^2
4320       NDOF%=NDOF%+1
4330 NEXT I%
4340 NDOF%=NDOF%-NBASIS%                           'degrees of freedom
4350 '
4360 FOR J%=1 TO 90                                'calculate rho, its error
4370    RHO(J%)=0: DRHO(J%)=0
4380    RAD=J%*.1                                  'interval is 0.1 fm to 9 fm
4390    IF RAD>RMAX THEN GOTO 4560
4400    RRHO=0
4410    DRRHO=0
4420    FOR N%=1 TO NBASIS%
4430       W(N%)=SIN(N%*PI*RAD/RMAX)               'W is temp sine storage
4440       RRHO=RRHO+CZERO(N%)*W(N%)
4450       SUM=0
4460       IF NOT GRAPHICS% THEN GOTO 4530         'only graphics needs error
4470          FOR M%=1 TO N%
4480             FAC=2
4490             IF M%=N% THEN FAC=1
4500             SUM=SUM+FAC*A#(N%,M%)*W(M%)
4510          NEXT M%
4520          DRRHO=DRRHO+W(N%)*SUM
4530    NEXT N%
4540    RHO(J%)=RRHO/RAD
4550    DRHO(J%)=SQR(ABS(DRRHO))/RAD
4560 NEXT J%
4570 '                                             'display depends on screens
4580 IF NOT GRAPHICS% THEN GOTO 4700
4590    GOSUB 5000                                 'display log plot of sigma
```

```
4600  '                                              'wait 10 seconds
4610    TENMORE%= (VAL(RIGHT$(TIME$,2))+10) MOD 60
4620    IF VAL(RIGHT$(TIME$,2))=TENMORE% THEN GOTO 4640 ELSE GOTO 4620
4630  '
4640    GOSUB 6000                                   'display fractional error,rho
4650    IF MONO% GOTO 4700
4660  '                                              'wait 10 seconds
4670      TENMORE%= (VAL(RIGHT$(TIME$,2))+10) MOD 60
4680      IF VAL(RIGHT$(TIME$,2))=TENMORE% THEN GOTO 4710 ELSE GOTO 4680
4690  '
4700  GOSUB 14000                                    'switch to mono
4710  GOSUB 7000                                     'display numerical values
4720  RETURN
4730  '
5000  '****************************************************************
5010  'subroutine to display a log plot of the fit
5020  'input   variables: none
5030  'output variables: none
5040  'global variables: CHISQ,DSIGE,ITER%,NBASIS%,NDOF%,NPOINTS%,QEFF,SIGE,
5050  '                   SIGT,TARGET$
5060  'local   variables: I%,VSCALE,X%,Y1%,Y2%
5070  '****************************************************************
5080  GOSUB 13000                                    'switch to graphics
5090  LOCATE 2,43                                    'plot title
5100  PRINT "Log sigma (mb/sr) for "+TARGET$+"
5110  LOCATE 3,43
5120  PRINT USING "Iteration ## with ## sine functions";ITER%,NBASIS%
5130  LOCATE 4,43
5140  PRINT USING "Chisq = ##.###^^^^ for ### DoF";CHISQ,NDOF%
5150  VSCALE=190/LOG(10000!/1E-12)                   'range is 1e4 to 1e-12
5160  LINE (30,0)-(630,190),1,B                      'draw the box
5170  FOR I%=-11 TO 3                                'vertical ticks
5180      Y%=190-LOG((10^I%)/1E-12)*VSCALE
5190      LINE (30,Y%)-(35,Y%): LINE (630,Y%)-(625,Y%)
5200  NEXT I%
5210  LOCATE  1,1: PRINT " +4";: LOCATE 7,1: PRINT "  0";'vertical scale
5220  LOCATE 13,1: PRINT " -4";: LOCATE 19,1: PRINT " -8";
5230  LOCATE 24,1: PRINT "-12";
5240  FOR I%=1 TO 7                                  'horizontal ticks
5250      X%=30+600*I%/8
5260      LINE (X%,190)-(X%,187): LINE (X%,0)-(X%,3)
5270  NEXT I%
5280  LOCATE 25, 4: PRINT "0";: LOCATE 25,23: PRINT "1"; 'horizontal scale
5290  LOCATE 25,42: PRINT "2";: LOCATE 25,60: PRINT "3";
5300  LOCATE 25,79: PRINT "4";
5310  LOCATE 25,45: PRINT "Qeff (fm^-1)";            'horizontal legend
5320  LOCATE  6,50: PRINT "Data";: LINE (440,40)-(440,45)'symbol legend
5330  LOCATE  7,50: PRINT "Fit"; : CIRCLE (440,52),2
5340  FOR I%=1 TO NPOINTS%                           'plot the points
5350      X%=30+150*QEFF(I%)
5360      Y1%=190-LOG((SIGE(I%)+DSIGE(I%))/1E-12)*VSCALE
5370      Y2%=190-LOG((SIGE(I%)-DSIGE(I%))/1E-12)*VSCALE
5380      LINE (X%,Y1%)-(X%,Y2%)                     'data, with error bar
5390      Y1%=190-LOG(SIGT(I%)/1E-12)*VSCALE
5400      IF Y1%<0 THEN Y1%=0
5410      IF Y1%>190 THEN Y1%=190
5420      CIRCLE (X%,Y1%),2                          'fit
5430  NEXT I%
```

```
5440 RETURN
5450 '
6000 '*******************************************************************
6010 'subroutine to display plots of fractional error and rho
6020 'input   variables: none
6030 'output variables: none
6040 'global variables: CHISQ,DRHO,DSIGE,ITER%,NBASIS%,NDOF%,NPOINTS%,QEFF,
6050 '                   RHO,RMAX,SIGE,SIGT,TARGET$
6060 'local   variables: I%,J%,RAD%,X%,Y%,Y1%,Y2%
6070 '*******************************************************************
6080 CLS
6090 LINE (30,0)-(630,86),1,B                        'draw box
6100 LINE (30,43)-(630,43)                           'draw zero line
6110 FOR I%=1 TO 4                                    'vertical ticks
6120     Y%=43*(1-I%/5)
6130     LINE (30,Y%)-(35,Y%): LINE (630,Y%)-(625,Y%)
6140     Y%=43*(1+I%/5)
6150     LINE (30,Y%)-(35,Y%): LINE (630,Y%)-(625,Y%)
6160 NEXT I%
6170 LOCATE  1,2: PRINT "+1";: LOCATE 6,3: PRINT "0"; 'vertical scale
6180 LOCATE 11,2: PRINT "-1";
6190 FOR I%=1 TO 7                                    'horizontal ticks
6200     X%=30+600*I%/8
6210     LINE (X%,0)-(X%,3): LINE (X%,86)-(X%,83)
6220 NEXT I%
6230 LOCATE 12,23: PRINT "1";: LOCATE 12,42: PRINT "2";'horizontal scale
6240 LOCATE 12,61: PRINT "3";: LOCATE 12,79: PRINT "4";
6250 LOCATE 12,46: PRINT "Qeff (fmr-1)";             'horizontal legend
6260 LOCATE  2, 6: PRINT "Fractional error in fit";  'plot title
6270 LOCATE  9, 6: PRINT USING "Iteration ##";ITER%;
6280 LOCATE 10, 6: PRINT USING "Chisq=+#.###^^^^ for ## DoF";CHISQ,NDOF%
6290 FOR I%=1 TO NPOINTS%                            'plot the points
6300     X%=30+QEFF(I%)/4*600
6310     Y1%=43*(1-(SIGE(I%)+DSIGE(I%)-SIGT(I%))/SIGE(I%))
6320     IF Y1%<0 THEN Y1%=0
6330     IF Y1%>86 THEN Y1%=86
6340     Y2%=43*(1-(SIGE(I%)-DSIGE(I%)-SIGT(I%))/SIGE(I%))
6350     IF Y2%<0 THEN Y2%=0
6360     IF Y2%>86 THEN Y2%=86
6370     IF Y1%=86 AND Y2%=86 THEN Y1%=76
6380     IF Y1%=0 AND Y2%=0 THEN Y2%=10
6390     LINE (X%,Y1%)-(X%,Y2%)
6400 NEXT I%
6410 '                                               'now plot rho
6420 LINE (30,189)-(30,91)                           'vertical axis
6430 FOR I%=0 TO 8 STEP 2                            'vertical ticks,scale
6440     LINE (30,189-89*I%/8)-(34,189-89*I%/8)
6450     J%=25-3*I%/2
6460     IF I%=0 THEN J%=24
6470     LOCATE J%,1
6480     PRINT USING ".##";.01*I%;
6490 NEXT I%
6500 LOCATE 17,1: PRINT "rho";                       'vertical legend
6510 LINE (30,189)-(630,189)                         'horizontal axis
6520 FOR I%=0 TO 9                                   'horizontal ticks,scale
6530     X%=30+I%*600/9
6540     LINE (X%,189)-(X%,186)
6550     J%=4+I%*8
```

```
6560    IF I%>=2 THEN J%=J%+1
6570    IF I%>=5 THEN J%=J%+1
6580    IF I%>=7 THEN J%=J%+1
6590    LOCATE 25,J%: PRINT USING "#";I%;
6600 NEXT I%
6610 LOCATE 25,39: PRINT "r (fm)";                    'horizontal legend
6620 FOR J%=1 TO 90                                   'plot rho and error
6630    RAD=.1*J%                                      ' every 0.1 fm
6640    IF RAD>RMAX THEN GOTO 6740
6650    X%=30+600*RAD/9
6660    Y%=189-89*(RHO(J%)+DRHO(J%))/.08
6670    IF Y%>189 THEN Y%=189
6680    IF Y%<0 THEN Y%=0
6690    Y1%=189-89*(RHO(J%)-DRHO(J%))/.08
6700    IF Y1%>189 THEN Y1%=189
6710    IF Y1%<0 THEN Y1%=0
6720    LINE (X%,Y%)-(X%,Y1%)
6730 NEXT J%
6740 LOCATE 14,50
6750 PRINT "Plot of rho in fm^-3 for "+TARGET$;         'plot title
6760 LOCATE 15,64: PRINT USING "Sines used=  ##";NBASIS%;
6770 LOCATE 16,64: PRINT USING "Rmax=  ##.## fm";RMAX;
6780 LOCATE 17,64: PRINT USING "Qmax=#.### fm^-1";QMAX
6790 RETURN
6800 '
7000 '***********************************************************************
7010 'subroutine to display the quantitative data on the fit
7020 'input  variables: none
7030 'output variables: none
7040 'global variables: A#,CHISQ,CZERO,ITER%,NBASIS%,NDOF%,RHO,RMAX,TARGET$
7050 'local  variables: DELTAC,J%,N%
7060 '***********************************************************************
7070 CLS
7080 LOCATE 1,22: PRINT "Fit to data for "+TARGET$;
7090 LOCATE 1,43: PRINT USING " Iteration ##";ITER%;
7100 LOCATE 2,26
7110 PRINT USING "Chisq=+#.###^^^^ for ## DoF";CHISQ,NDOF%
7120 LOCATE 3,13: PRINT USING "Number of sines=##";NBASIS%;
7130 LOCATE 3,33: PRINT USING "Rmax=##.### fm  Qmax=#.### fm^-1";RMAX,QMAX
7140 LOCATE 4,16: PRINT "The expansion coefficients of the density are:"
7150 FOR N%=1 TO NBASIS%
7160    IF N% MOD 2 = 1 THEN LOCATE (N%-1)/2+5,4 ELSE LOCATE N%/2+4,44
7170    IF A#(N%,N%)>0 THEN DELTAC=SQR(A#(N%,N%)) ELSE DELTAC=0
7180    PRINT USING "C(##)=+#.####^^^^ +- +#.####^^^^";N%,CZERO(N%),DELTAC
7190 NEXT N%
7200 LOCATE 14,25: PRINT "The charge density in fm^-3 is:";
7210 FOR J%=1 TO 9
7220    LOCATE 14+J%,1: PRINT USING "r=#.# to #.# fm";.2+J%-1,J%
7230    FOR I%=1 TO 5
7240       LOCATE 14+J%,12*I%+7: PRINT USING "+#.####^^^^";RHO((J%-1)*10+2*I%);
7250    NEXT I%
7260 NEXT J%
7270 LOCATE 25,33: PRINT "Press e to end";
7280 RETURN
7290 '
8000 '***********************************************************************
8010 'subroutine to solve for the new C's
8020 'input  variables: none
```

```
8030 'output variables: A#,BVECTOR,CZERO
8040 'global variables: CCHI,CHIN,DSIGE,IMFT,JTABLE,KBEAM,N%,NBASIS%,NLEG%,
8050 '                  NPOINTS%,REFT,RMAX,SCHI,SIGE,SIGT,THETA
8060 'local  variables: I%,ILEG%,IMW,M%,REW,SN,W
8070 '****************************************************************************
8080 FOR N%=1 TO NBASIS%                              'zero the B vector and A matrix
8090     BVECTOR(N%)=0
8100     FOR M%=1 TO N%
8110         A#(N%,M%)=0
8120     NEXT M%
8130 NEXT N%
8140 '
8150 FOR I%=1 TO NPOINTS%                             'calculate A matrix
8160     IF QEFF(I%)>QMAX THEN GOTO 8340
8170     FOR N%=1 TO NBASIS%                          'calculate real and imag W vector
8180         REW=0: IMW=0
8190         FOR ILEG%=1 TO NLEG%
8200             REW=REW+JTABLE(ILEG%,I%)*CCHI(ILEG%)*CHIN(ILEG%,N%)
8210             IMW=IMW+JTABLE(ILEG%,I%)*SCHI(ILEG%)*CHIN(ILEG%,N%)
8220         NEXT ILEG%
8230         REW=2*REW*KBEAM*RMAX^2
8240         IMW=2*IMW*KBEAM*RMAX^2
8250         W(N%)=2*COS(THETA(I%)/2)^2*(REFT(I%)*REW+IMFT(I%)*IMW)*10
8260     NEXT N%
8270 '
8280     FOR N%=1 TO NBASIS%                          'this point's contribution to A,B
8290         FOR M%=1 TO N%
8300             A#(N%,M%)=A#(N%,M%)+CDBL(W(N%)*W(M%)/DSIGE(I%)^2)
8310         NEXT M%
8320         BVECTOR(N%)=BVECTOR(N%)+(SIGE(I%)-SIGT(I%))*W(N%)/DSIGE(I%)^2
8330     NEXT N%
8340 NEXT I%
8350 '
8360 FOR N%=1 TO NBASIS%                              'extend the A matrix and scale
8370     SN=RMAX^2/N%/PI
8380     IF N% MOD 2 =1 THEN SN=-SN
8390     A#(N%,NBASIS%+1)=SN*A#(1,1)
8400     A#(NBASIS%+1,N%)=A#(N%,NBASIS%+1)            'A is symmetric
8410     FOR M%=1 TO N%
8420         A#(M%,N%)=A#(N%,M%)
8430     NEXT M%
8440 NEXT N%
8450 A#(NBASIS%+1,NBASIS%+1)=0
8460 '
8470 N%=NBASIS%+1                                     'invert A
8480 GOSUB 11000
8490 '
8500 FOR N%=1 TO NBASIS%                              'multiply B by A to find new C's
8510     SUM=0
8520     FOR M%=1 TO NBASIS%
8530         SUM=SUM+A#(N%,M%)*BVECTOR(M%)
8540     NEXT M%
8550     CZERO(N%)=CZERO(N%)+SUM
8560 NEXT N%
8570 RETURN
8580 '
9000 '****************************************************************************
9010 'subroutine to compute J0 and J1 by polynomial approximation
```

```
9020 'input   variables:   X,ORDER%
9030 'output  variables:   BESSJ0 or BESSJ1
9040 'global  variables:   none
9050 'local   variables:   TEMP,TEMP2,Y,Y2
9060 '***********************************************************
9070 Y=X/3
9080 '
9090 IF ABS(X)>3 THEN GOTO 9240                                    'X<3 region
9100    Y2=Y*Y
9110    IF ORDER%=1 THEN GOTO 9180
9120       TEMP=-.0039444+.00021*Y2: TEMP=.0444479+Y2*TEMP          'J0
9130       TEMP=-.3163866+Y2*TEMP:    TEMP=1.2656208#+Y2*TEMP
9140       TEMP=-2.2499997#+Y2*TEMP
9150       BESSJ0=1+Y2*TEMP
9160       RETURN
9170    '
9180       TEMP=-3.1761E-04+1.109E-05*Y2: TEMP=4.43319E-03+Y2*TEMP  'J1
9190       TEMP=-3.954289E-02+Y2*TEMP:    TEMP=.21093573#+Y2*TEMP
9200       TEMP=-.56249985#+Y2*TEMP
9210       BESSJ1=X*(.5+Y2*TEMP)
9220       RETURN
9230    '
9240    Y=1/Y                                                       'X>3 region
9250    IF ORDER%=1 THEN GOTO 9350
9260       TEMP=-7.2805E-04+1.4476E-04*Y:  TEMP=1.37237E-03+TEMP*Y  'J0
9270       TEMP=-9.512E-05+TEMP*Y:         TEMP=-.0055274+TEMP*Y:
9280       TEMP=-7.7E-07+TEMP*Y:           TEMP=.79788456#+TEMP*Y
9290       TEMP2=-2.9333E-04+1.3558E-04*Y: TEMP2=-5.4125E-04+TEMP2*Y
9300       TEMP2=2.62573E-03+TEMP2*Y:      TEMP2=-3.954E-05+TEMP2*Y
9310       TEMP2=-4.166397E-02+TEMP2*Y:    TEMP2=X-.78539816#+TEMP2*Y
9320       BESSJ0=TEMP*COS(TEMP2)/SQR(X)
9330       RETURN
9340    '
9350    TEMP=1.13653E-03-2.0033E-04*Y:  TEMP=-2.49511E-03+TEMP*Y    'J1
9360    TEMP=1.7105E-04+TEMP*Y:         TEMP=1.659667E-02+TEMP*Y
9370    TEMP=1.56E-06+TEMP*Y:           TEMP=.79788456#+TEMP*Y
9380    TEMP2=7.9824E-04-2.9166E-04*Y:  TEMP2=7.4348E-04+TEMP2*Y
9390    TEMP2=-6.37879E-03+TEMP2*Y:     TEMP2=.0000565+TEMP2*Y
9400    TEMP2=.12499612#+TEMP2*Y:       TEMP2=X-2.35619449#+TEMP2*Y
9410    BESSJ1=TEMP*COS(TEMP2)/SQR(X)
9420    RETURN
9430    '
10000 '***********************************************************
10010 'subroutine to establish Gauss-Legendre weights and abscissae for 20 pts.
10020 'input   variables: none
10030 'output  variables: WLEG,XLEG
10040 'global  variables: none
10050 'local   variables: ILEG%
10060 '***********************************************************
10070 XLEG(1)=.993128599185094#  : WLEG(1)=.017614007139152#
10080 XLEG(2)=.963971927277913#  : WLEG(2)=.040601429800386#
10090 XLEG(3)=.912234428251325#  : WLEG(3)=.062672048334109#
10100 XLEG(4)=.839116971822218#  : WLEG(4)=.083276741576704#
10110 XLEG(5)=.74633190646015#   : WLEG(5)=.10193011981724#
10120 XLEG(6)=.636053680726515#  : WLEG(6)=.118194531961518#
10130 XLEG(7)=.510867001950827#  : WLEG(7)=.131688638449176#
10140 XLEG(8)=.373706088715419#  : WLEG(8)=.142096109318382#
10150 XLEG(9)=.227785851141645#  : WLEG(9)=.149172986472603#
```

```
10160 XLEG(10)=.076526521133497# : WLEG(10)=.152753387130725#
10170 FOR ILEG%=1 TO 10                              'since the weights and abscissa
10180     XLEG(21-ILEG%)=-XLEG(ILEG%)                ' are even and odd functions
10190     WLEG(21-ILEG%)=WLEG(ILEG%)                 ' functions respectively, the
10200 NEXT ILEG%                                     ' other half of the array is
10210 RETURN                                         ' trivial to fill out
10220 '
11000 '***********************************************************************
11010 'subroutine to invert a symmetric matrix A#(N%,N%) by Gauss-Jordan elim.
11020 ' On output, A# contains the inverse of the input matrix.
11030 ' Main code must dimension A# and auxiliary arrays R%(N%),P#(N%),Q#(N%).
11040 'input  variables: A#,N%
11050 'output variables: A#
11060 'global variables: P#,Q#,R%
11070 'local  variables: BIG,I%,J%,K%,PIVOT1#,TEST
11080 '***********************************************************************
11090 FOR I%=1 TO N%                                 'R%=0 tells which rows
11100     R%(I%)=1                                    ' already zeroed
11110 NEXT I%
11120 '
11130 FOR I%=1 TO N%                                 'loop over rows
11140     K%=0: BIG=0                                 'pivot is largest diag
11150     FOR J%=1 TO N%                              'in rows not yet zeroed
11160        IF R%(J%)=0 THEN GOTO 11200
11170          TEST=ABS(A#(J%,J%))
11180          IF TEST<=BIG THEN GOTO 11200
11190            BIG=TEST: K%=J%
11200     NEXT J%
11210 '
11220     IF K%=0 THEN GOTO 11580                     'all diags=0;A# singulr
11230     IF I%=1 THEN PIVOT1#=A#(K%,K%)              'largest diag of A#
11240     IF ABS(A#(K%,K%)/PIVOT1#)<1E-14 THEN GOTO 11580 'ill-conditioned A#
11250 '
11260     R%(K%)=0: Q#(K%)=1/A#(K%,K%)
11270     P#(K%)=1: A#(K%,K%)=0                       'begin zeroing this row
11280     IF K%=1 THEN GOTO 11360
11290       FOR J%=1 TO K%-1                          'elements above diag
11300         P#(J%)=A#(J%,K%)
11310         Q#(J%)=A#(J%,K%)*Q#(K%)
11320         IF R%(J%)=1 THEN Q#(J%)=-Q#(J%)
11330         A#(J%,K%)=0
11340       NEXT J%
11350 '
11360     IF K%=N% THEN GOTO 11440
11370       FOR J%=K%+1 TO N%                         'elements right of diag
11380         P#(J%)=A#(K%,J%)
11390         Q#(J%)=-A#(K%,J%)*Q#(K%)
11400         IF R%(J%)=0 THEN P#(J%)=-P#(J%)
11410         A#(K%,J%)=0
11420       NEXT J%
11430 '
11440     FOR J%=1 TO N%                              'transform all of A#
11450       FOR K%=J% TO N%
11460         A#(J%,K%)=A#(J%,K%)+P#(J%)*Q#(K%)
11470       NEXT K%
11480     NEXT J%
11490 NEXT I%
11500 '
```

```
11510 FOR J%=2 TO N%                                    'symmetrize A#^-1
11520    FOR K%=1 TO J%-1
11530       A#(J%,K%)=A#(K%,J%)
11540    NEXT K%
11550 NEXT J%
11560 '
11570 RETURN
11580 BEEP: PRINT "FAILURE IN MATRIX INVERSION";
11590 END
11600 '
12000 '****************************************************************
12010 'subroutine to display the header screen
12020 'input   variables: none
12030 'output variables: GRAPHICS%,MONO%
12040 'global variables: none
12050 'local   variables: G$,M$,ROW%
12060 '****************************************************************
12070 SCREEN 0: CLS: KEY OFF                            'program begins in text mode
12080 '
12090 LOCATE 1,30: COLOR 15                             'display book title
12100 PRINT "COMPUTATIONAL PHYSICS";
12110 LOCATE 2,40: COLOR 7
12120 PRINT "by";
12130 LOCATE 3,33: PRINT "Steven E. Koonin";
12140 LOCATE 4,14
12150 PRINT "Copyright 1985, Benjamin/Cummings Publishing Company";
12160 '
12170 LOCATE 5,10                                       'draw the box
12180 PRINT CHR$(201)+STRING$(59,205)+CHR$(187);
12190 FOR ROW%=6 TO 19
12200    LOCATE ROW%,10: PRINT CHR$(186);
12210    LOCATE ROW%,70: PRINT CHR$(186);
12220 NEXT ROW%
12230 LOCATE 20,10
12240 PRINT CHR$(200)+STRING$(59,205)+CHR$(188);
12250 '
12260 COLOR 15                                          'print title, etc.
12270 LOCATE  7,36:  PRINT "EXAMPLE 5";
12280 COLOR 7
12290 LOCATE 10,22:  PRINT "Determining nuclear charge densities";
12300 LOCATE 13,36:  PRINT "**********"
12310 LOCATE 15,25:  PRINT "Type 'e' to end while running."
12320 LOCATE 16,22:  PRINT "Type ctrl-break to stop at a prompt."
12330 '
12340 LOCATE 19,16: BEEP                                'get screen configuration
12350 INPUT "Does your computer have graphics capability (y/n)";G$
12360 IF LEFT$(G$,1)="y" OR LEFT$(G$,1)="n" GOTO 12390
12370    LOCATE 19,12:  PRINT SPACE$(58):  BEEP
12380    LOCATE 18,35:  PRINT "Try again,":  GOTO 12340
12390 IF LEFT$(G$,1)="y" GOTO 12420
12400    GRAPHICS%=0:   MONO%=-1
12410    GOTO 12510
12420 GRAPHICS%=-1
12430 LOCATE 18,15:  PRINT SPACE$(55)
12440 LOCATE 19,15:  PRINT SPACE$(55)
12450 LOCATE 19,15: BEEP
12460 INPUT "Do you also have a separate screen for text (y/n)";M$
12470 IF LEFT$(M$,1)="y" OR LEFT$(M$,1)="n" GOTO 12490  'error trapping
```

```
12480    LOCATE 18,35:  PRINT "Try again,":   GOTO 12440
12490 IF LEFT$(M$,1)="y" THEN MONO%=-1 ELSE MONO%=0
12500 '
12510 LOCATE 21,10
12520 PRINT "This program analyzes elastic electron-nucleus cross sections";
12530 LOCATE 22,17
12540 PRINT"to determine the charge density of the nucleus.";
12550 LOCATE 23,15
12560 PRINT "The density is expanded as a series of sine functions";
12570 LOCATE 24,25
12580 PRINT "confined within a radius Rmax";
12590 LOCATE 25,31,1:   BEEP
12600 PRINT "Type c to continue";
12610 IF INKEY$<>"c" GOTO 12610
12620 LOCATE ,,0
12630 '
12640 IF MONO% AND GRAPHICS% THEN GOSUB 14000              'switch to text screen
12650 '
12660 RETURN
12670 '
13000 '********************************************************************
13010 'subroutine to switch from mono to graphics screen
13020 'input  variables: none
13030 'output variables: none
13040 'global variables: none
13050 'local  variables: none
13060 '********************************************************************
13070 DEF SEG=0
13080 POKE &H410,  (PEEK(&H410) AND &HCF) OR &H10
13090 SCREEN 0:  WIDTH 80: LOCATE ,,0: SCREEN 2,,0,0
13100 RETURN
13110 '
14000 '********************************************************************
14010 'subroutine to switch from graphics to mono screen
14020 'input  variables: none
14030 'output variables: none
14040 'global variables: none
14050 'local  variables: none
14060 '********************************************************************
14070 DEF SEG=0
14080 POKE &H410,  (PEEK(&H410) OR &H30)
14090 SCREEN 0:  WIDTH 80:  LOCATE ,,0
14100 RETURN
14110 '
15000 '********************************************************************
15010 'data for the nucleus 40Ca at a beam energy of 400.0 MeV
15020 '    each line after the first lists the lab angle (in degrees),
15030 '    the cross section, and the error in the cross section (in mb)
15040 '    for two data points
15050 '********************************************************************
15060 DATA 40Ca,20,40,400.00,100
15070 DATA  15.129,0.129E+02,0.388E+00, 16.372,0.750E+01,0.225E+00
15080 DATA  17.612,0.448E+01,0.151E+00, 18.849,0.264E+01,0.846E-01
15090 DATA  20.083,0.148E+01,0.444E-01, 21.312,0.861E+00,0.258E-01
15100 DATA  22.537,0.486E+00,0.146E-01, 23.752,0.261E+00,0.783E-02
15110 DATA  24.925,0.150E+00,0.451E-02, 26.179,0.713E-01,0.214E-02
15120 DATA  27.391,0.398E-01,0.128E-02, 28.211,0.239E-01,0.309E-02
15130 DATA  28.543,0.176E-01,0.527E-03, 28.758,0.162E-01,0.518E-03
```

```
15140 DATA   29.786,0.875E-02,0.263E-03,  30.010,0.694E-02,0.225E-03
15150 DATA   30.962,0.365E-02,0.110E-03,  31.264,0.331E-02,0.112E-03
15160 DATA   32.023,0.261E-02,0.403E-03,  32.519,0.186E-02,0.627E-04
15170 DATA   33.285,0.146E-02,0.438E-04,  33.939,0.152E-02,0.108E-03
15180 DATA   34.488,0.133E-02,0.400E-04,  34.757,0.125E-02,0.527E-04
15190 DATA   35.615,0.141E-02,0.423E-04,  35.863,0.137E-02,0.945E-04
15200 DATA   36.797,0.138E-02,0.415E-04,  37.277,0.143E-02,0.544E-04
15210 DATA   37.794,0.134E-02,0.950E-04,  39.081,0.118E-02,0.353E-04
15220 DATA   39.734,0.117E-02,0.824E-04,  40.182,0.114E-02,0.342E-04
15230 DATA   41.347,0.935E-03,0.287E-04,  41.683,0.878E-03,0.611E-04
15240 DATA   42.417,0.778E-03,0.233E-04,  42.696,0.764E-03,0.268E-04
15250 DATA   43.516,0.609E-03,0.183E-04,  43.955,0.585E-03,0.175E-04
15260 DATA   44.635,0.452E-03,0.136E-04,  45.229,0.432E-03,0.161E-04
15270 DATA   45.611,0.346E-03,0.310E-04,  46.518,0.298E-03,0.134E-04
15280 DATA   46.803,0.257E-03,0.770E-05,  47.591,0.213E-03,0.213E-04
15290 DATA   47.797,0.204E-03,0.824E-05,  48.870,0.123E-03,0.408E-05
15300 DATA   49.078,0.127E-03,0.521E-05,  49.583,0.104E-03,0.105E-04
15310 DATA   49.933,0.826E-04,0.248E-05,  50.413,0.768E-04,0.316E-05
15320 DATA   50.974,0.562E-04,0.180E-05,  51.588,0.438E-04,0.625E-05
15330 DATA   51.986,0.353E-04,0.119E-05,  52.654,0.243E-04,0.120E-05
15340 DATA   52.976,0.214E-04,0.642E-06,  53.605,0.164E-04,0.273E-05
15350 DATA   53.933,0.123E-04,0.516E-05,  54.308,0.109E-04,0.476E-06
15360 DATA   54.936,0.643E-05,0.228E-06,  55.539,0.509E-05,0.229E-06
15370 DATA   56.902,0.233E-05,0.112E-06,  57.684,0.154E-05,0.332E-06
15380 DATA   58.061,0.128E-05,0.536E-07,  59.458,0.121E-05,0.471E-07
15390 DATA   59.747,0.120E-05,0.179E-06,  60.453,0.143E-05,0.545E-07
15400 DATA   61.826,0.191E-05,0.171E-06,  63.423,0.213E-05,0.926E-07
15410 DATA   66.041,0.226E-05,0.181E-06,  68.386,0.191E-05,0.654E-07
15420 DATA   68.734,0.172E-05,0.829E-07,  70.338,0.148E-05,0.134E-06
15430 DATA   71.067,0.123E-05,0.532E-07,  71.588,0.120E-05,0.418E-07
15440 DATA   73.768,0.848E-06,0.461E-07,  74.316,0.664E-06,0.253E-07
15450 DATA   74.729,0.601E-06,0.837E-07,  76.491,0.473E-06,0.250E-07
15460 DATA   77.067,0.394E-06,0.118E-07,  79.227,0.231E-06,0.356E-07
15470 DATA   79.845,0.175E-06,0.524E-08,  80.623,0.142E-06,0.122E-07
15480 DATA   82.650,0.711E-07,0.335E-08,  83.849,0.474E-07,0.137E-07
15490 DATA   84.817,0.309E-07,0.408E-08,  85.486,0.248E-07,0.144E-08
15500 DATA   88.355,0.721E-08,0.467E-09,  89.083,0.611E-08,0.186E-08
15510 DATA   91.011,0.214E-08,0.557E-09,  93.501,0.138E-08,0.578E-09
15520 DATA   95.671,0.668E-09,0.650E-10,  96.041,0.584E-09,0.229E-09
15530 DATA   98.689,0.638E-09,0.334E-09,101.292,0.621E-09,0.279E-09
15540 DATA  103.398,0.828E-09,0.832E-10,104.015,0.915E-09,0.450E-09
15550 DATA  105.550,0.882E-09,0.521E-09,109.695,0.561E-09,0.234E-09
15560 DATA  116.507,0.836E-10,0.215E-10,123.719,0.511E-11,0.299E-11
15570 '
16000 '********************************************************************
16010 'data for the nucleus 58Ni at a beam energy of 449.8 MeV
16020 '********************************************************************
16030 DATA   58Ni,28,58,449.8,82
16040 DATA   14.743,0.125E+02,0.258E+00,  16.073,0.609E+01,0.122E+00
16050 DATA   17.399,0.310E+01,0.620E-01,  18.723,0.136E+01,0.273E-01
16060 DATA   20.044,0.587E+00,0.117E-01,  21.361,0.246E+00,0.492E-02
16070 DATA   22.674,0.986E-01,0.197E-02,  23.983,0.391E-01,0.782E-03
16080 DATA   25.288,0.202E-01,0.405E-03,  26.588,0.144E-01,0.302E-03
16090 DATA   27.884,0.131E-01,0.282E-03,  29.174,0.124E-01,0.278E-03
16100 DATA   30.459,0.113E-01,0.228E-03,  31.738,0.918E-02,0.184E-03
16110 DATA   33.012,0.722E-02,0.144E-02,  34.279,0.527E-02,0.105E-03
16120 DATA   35.539,0.347E-02,0.695E-04,  36.793,0.226E-02,0.452E-04
16130 DATA   37.117,0.199E-02,0.597E-04,  38.039,0.133E-02,0.267E-04
```

```
16140 DATA   39.278,0.760E-03,0.152E-04,  40.510,0.395E-03,0.791E-05
16150 DATA   41.733,0.201E-03,0.401E-05,  42.834,0.110E-03,0.330E-05
16160 DATA   42.948,0.823E-04,0.167E-05,  44.154,0.374E-04,0.130E-05
16170 DATA   44.963,0.254E-04,0.759E-06,  45.351,0.193E-04,0.133E-05
16180 DATA   46.538,0.168E-04,0.156E-05,  47.715,0.221E-04,0.197E-05
16190 DATA   47.971,0.195E-04,0.582E-06,  48.883,0.241E-04,0.173E-05
16200 DATA   49.870,0.251E-04,0.753E-06,  49.920,0.249E-04,0.753E-06
16210 DATA   49.950,0.249E-04,0.753E-06,  49.980,0.241E-04,0.753E-06
16220 DATA   50.000,0.246E-04,0.753E-06,  50.020,0.253E-04,0.753E-06
16230 DATA   50.040,0.256E-04,0.154E-05,  50.120,0.242E-04,0.753E-06
16240 DATA   51.185,0.231E-04,0.146E-05,  51.969,0.231E-04,0.753E-06
16250 DATA   52.149,0.238E-04,0.753E-06,  52.320,0.232E-04,0.128E-05
16260 DATA   53.442,0.178E-04,0.117E-05,  53.918,0.192E-04,0.577E-06
16270 DATA   53.968,0.182E-04,0.544E-06,  54.118,0.190E-04,0.577E-06
16280 DATA   54.553,0.162E-04,0.101E-05,  55.967,0.129E-04,0.390E-06
16290 DATA   56.736,0.104E-04,0.489E-06,  58.166,0.786E-05,0.237E-06
16300 DATA   58.865,0.563E-05,0.367E-06,  59.965,0.425E-05,0.128E-06
16310 DATA   60.937,0.316E-05,0.178E-06,  62.004,0.197E-05,0.596E-07
16320 DATA   62.948,0.143E-05,0.197E-06,  63.973,0.856E-06,0.299E-07
16330 DATA   64.894,0.516E-06,0.538E-07,  66.013,0.274E-06,0.110E-07
16340 DATA   66.772,0.201E-06,0.349E-07,  66.972,0.142E-06,0.645E-08
16350 DATA   67.002,0.170E-06,0.135E-07,  67.972,0.691E-07,0.449E-08
16360 DATA   68.577,0.650E-07,0.211E-07,  69.971,0.387E-07,0.290E-08
16370 DATA   70.305,0.499E-07,0.212E-07,  70.770,0.419E-07,0.272E-08
16380 DATA   71.952,0.364E-07,0.119E-07,  72.070,0.321E-07,0.321E-08
16390 DATA   72.969,0.369E-07,0.277E-08,  76.018,0.429E-07,0.343E-08
16400 DATA   78.617,0.277E-07,0.222E-08,  80.696,0.217E-07,0.196E-08
16410 DATA   82.925,0.113E-07,0.136E-08,  85.974,0.318E-08,0.474E-09
16420 DATA   89.702,0.990E-09,0.203E-09,  91.971,0.117E-09,0.698E-10
16430 DATA   96.879,0.190E-09,0.614E-10,100.777,0.302E-09,0.108E-09
16440 DATA  104.485,0.362E-09,0.105E-09,113.789,0.909E-10,0.455E-10
16450 '
17000 '*****************************************************************
17010 'data for the nucleus 208Pb at a beam energy of 502.0 MeV
17020 '*****************************************************************
17030 DATA 208Pb,82,208,502.00,93
17040 DATA    8.368,0.165E+04,0.246E+02,   9.952,0.457E+03,0.106E+02
17050 DATA   10.436,0.284E+03,0.717E+01,  11.431,0.136E+03,0.446E+01
17060 DATA   12.496,0.573E+02,0.146E+01,  13.463,0.288E+02,0.279E+00
17070 DATA   13.984,0.200E+02,0.444E+00,  14.544,0.152E+02,0.390E+00
17080 DATA   15.040,0.113E+02,0.262E+00,  15.924,0.749E+01,0.104E+00
17090 DATA   16.579,0.565E+01,0.130E+00,  17.070,0.486E+01,0.153E+00
17100 DATA   17.503,0.392E+01,0.889E-01,  18.070,0.307E+01,0.991E-01
17110 DATA   18.446,0.257E+01,0.236E-01,  19.070,0.191E+01,0.612E-01
17120 DATA   19.602,0.153E+01,0.476E-01,  20.070,0.116E+01,0.215E+00
17130 DATA   20.596,0.861E+00,0.196E-01,  21.070,0.640E+00,0.161E-01
17140 DATA   21.590,0.508E+00,0.157E-01,  21.918,0.370E+00,0.396E-02
17150 DATA   22.579,0.245E+00,0.535E-02,  23.070,0.178E+00,0.444E-02
17160 DATA   23.563,0.132E+00,0.409E-02,  24.070,0.875E-01,0.353E-02
17170 DATA   24.439,0.686E-01,0.679E-03,  25.070,0.541E-01,0.166E-02
17180 DATA   25.514,0.440E-01,0.973E-03,  26.070,0.371E-01,0.109E-02
17190 DATA   26.461,0.331E-01,0.325E-03,  27.070,0.311E-01,0.817E-03
17200 DATA   27.441,0.289E-01,0.652E-03,  28.071,0.264E-01,0.708E-03
17210 DATA   28.934,0.242E-01,0.568E-03,  29.266,0.197E-01,0.187E-03
17220 DATA   30.071,0.155E-01,0.514E-03,  31.071,0.106E-01,0.304E-03
17230 DATA   32.071,0.654E-02,0.176E-03,  33.054,0.382E-02,0.851E-04
17240 DATA   33.962,0.202E-02,0.484E-04,  34.415,0.149E-02,0.195E-04
17250 DATA   34.862,0.115E-02,0.312E-04,  35.753,0.703E-03,0.215E-04
```

```
17260 DATA  36.071,0.649E-03,0.170E-04,  36.635,0.537E-03,0.816E-04
17270 DATA  37.071,0.509E-03,0.146E-04,  37.508,0.510E-03,0.131E-04
17280 DATA  37.975,0.489E-03,0.146E-04,  38.371,0.480E-03,0.181E-04
17290 DATA  38.961,0.476E-03,0.190E-04,  39.588,0.408E-03,0.130E-04
17300 DATA  40.011,0.410E-03,0.164E-04,  40.899,0.331E-03,0.909E-05
17310 DATA  41.721,0.288E-03,0.977E-05,  42.062,0.261E-03,0.180E-04
17320 DATA  42.952,0.159E-03,0.478E-05,  43.330,0.130E-03,0.477E-05
17330 DATA  44.062,0.878E-04,0.269E-05,  44.961,0.540E-04,0.162E-05
17340 DATA  45.880,0.290E-04,0.870E-06,  46.407,0.227E-04,0.187E-05
17350 DATA  46.930,0.172E-04,0.517E-06,  48.063,0.128E-04,0.495E-06
17360 DATA  48.948,0.129E-04,0.388E-06,  50.013,0.133E-04,0.531E-06
17370 DATA  50.628,0.138E-04,0.156E-05,  50.957,0.137E-04,0.410E-06
17380 DATA  52.013,0.112E-04,0.448E-06,  52.063,0.115e-04,0.433e-06
17390 DATA  52.956,0.905E-05,0.272E-06,  54.064,0.620E-05,0.332E-06
17400 DATA  54.974,0.450E-05,0.149E-06,  56.064,0.273E-05,0.166E-06
17410 DATA  56.963,0.165E-05,0.676E-07,  58.064,0.804E-06,0.777E-07
17420 DATA  58.982,0.568E-06,0.244E-07,  59.961,0.409E-06,0.204E-07
17430 DATA  60.971,0.357E-06,0.186E-07,  62.065,0.285E-06,0.287E-07
17440 DATA  62.949,0.288E-06,0.178E-07,  64.918,0.201E-06,0.141E-07
17450 DATA  66.017,0.118E-06,0.946E-08,  66.977,0.835E-07,0.752E-08
17460 DATA  68.965,0.365E-07,0.437E-08,  70.914,0.987E-08,0.139E-08
17470 DATA  72.933,0.615E-08,0.117E-08,  74.952,0.547E-08,0.126E-08
17480 DATA  76.940,0.367E-08,0.845E-09,  78.939,0.233E-08,0.562E-09
17490 DATA  80.928,0.495E-09,0.223E-09,  83.013,0.379E-09,0.190E-09
17500 DATA  87.015,0.115E-10,0.115E-09
17510 '
```

B.6 Example 6

This program solves Laplace's equation in two dimensions on a uniform rectangular lattice by Gauss-Seidel iteration. The dimensions of the lattice and initial value of the potential on its border are defined in subroutine 2000 and a value of the relaxation parameter is defined (subroutine 3000). Dirichlet boundary conditions can then be entered from the keyboard at each point on the lattice (subroutine 6000) and, if desired, relaxation sweeps can be restricted to a sub-lattice (subroutine 9000). The relaxation iterations then proceed (loop 370-440), with Eq. (6.17) being solved successively at each lattice point (subroutine 13000). After each sweep, the lattice is displayed as an array of colored characters (subroutine 7000). Also shown are the current energy calculated from Eq. (6.7) and its change from the previous iteration (lines 13240, 13290), as well as the maximum fractional change in the potential anywhere in the lattice during the current iteration.

To demonstrate this program, input a size of 15 horizontal points, 15 vertical points, a value of 1 for the potential, P, on the boundary, a relaxation parameter OMEGA of 1.5, use the cursor to place a boundary condition of 9 somewhere near the center of the lattice, and request no sub-lattice. Iterations will then show the potential developing in the pattern expected.

```
10  ' ***********************************************************************
20  'Example 6:   Solving Laplace's equation in two dimensions
21  'COMPUTATIONAL PHYSICS by Steven E. Koonin
22  'Copyright 1985, Benjamin/Cummings Publishing Company
30  ' ***********************************************************************
40  GOSUB 15000                                      'display header screen
50  '
60  FIRST%=-1                                        'define constants, functions
70  DIM INTERIOR%(10)                                'array for color of interior
80  DIM BOUNDARY%(10)                                'array for color of boundary
90  DIM BUFFER%(24,80)                               'gives position in screen buff
100 DIM P(24,80)                                     'value of potential
110 OMEGA%=-1                                        'flag to input omega
120 LATTICE%=-1                                      'flag to input lattice size
130 LSAVE%=0                                         'flag to save lattice size
140 BC%=0:  BCSAVE%=0                                'keep size, change bound cond
150 PSAVE%=0                                         'keep old Phi values
160 CLR%=-1                                          'flag to clear screen
170 '
180 IF FIRST% THEN GOSUB 1000                        'print cautionary message
190 IF LATTICE% AND NOT BCSAVE% THEN GOSUB 2000      'input lattice size, INIT%
200 IF OMEGA% THEN GOSUB 3000                        'input OMEGA
210 IF LATTICE% THEN GOSUB 4000                      'initialize P array
220 IF FIRST% THEN GOSUB 5000                        'display instruction for input
230 IF LATTICE% OR BC% THEN GOSUB 6000              'input P from screen
240 IF SUBLATTICE% THEN GOSUB 9000                  'input sublattice from screen
250 GOSUB 11000                                      'make initial guess for P
260 GOSUB 12000                                      'display commands
270 '
```

```
280 BC%=0:       OMEGA%=0: LATTICE%=0:  LSAVE%=0  'reset all flags
290 PSAVE%=0:  CLR%=0:   SUBLATTICE%=0:  BCSAVE%=0
300 '
310 LOCATE ,,0:    ITERATION%=0                  'iteration count begins from 0
320 K$=INKEY$                                    'accept keyboard input
330 IF NOT SUBL% THEN GOTO 370                   'if there is a sublattice,then
340   X1%=SUBX1%:  X2%=SUBX2%                     ' the limits on X and Y are not
350   Y1%=SUBY1%:  Y2%=SUBY2%                     ' those of the whole lattice
360 '
370 WHILE K$<>"e"                                 'continue loop until e is typed
380   ITERATION%=ITERATION%+1
390   OLDE=ENERGY:   ENERGY=0                     'save previous energy
400   OLDDELP=DELTAPMAX:  DELTAPMAX=0             'save previous max delta P
410   GOSUB 13000                                 'relax the lattice
420 '
430   K$=INKEY$                                   'accept keyboard input
440 WEND
450 '
460 IF NOT SUBL% THEN GOTO 510                    'if there is a sublattice,
470   X1%=XFIRST%:  X2%=XLAST%                     ' typing e once will begin
480   Y1%=YFIRST%:  Y2%=YLAST%                     ' the loop for the entire
490   LOCATE 25,59:  PRINT SPACE$(10);            ' lattice
500   SUBL%=0:  GOTO 310
510 GOSUB 14000                                   'display options
520 GOTO 180                                      'begin again
530 '
1000 '***************************************************************************
1010 'subroutine to print cautionary message about POKEing
1020 'input   variables:   none
1030 'output variables:   BOUNDARY%(I%),  INTERIOR%(I%),  POKING%
1040 'global variables:   GRAPHICS%
1050 'local   variables:   ANSWER$, OPT$
1060 '***************************************************************************
1070 CLS
1080 '
1090 LOCATE 1,34,0
1100 PRINT "C A U T I O N"
1110 LOCATE 3,9
1120 PRINT "For efficient screen display this program uses the POKE command";
1130 LOCATE 5,9
1140 PRINT "which places data directly into memory.  The location given in";
1150 LOCATE 7,8
1160 PRINT "POKE (syntax:  POKE location,data) is an offset from the current";
1170 LOCATE 9,8
1180 PRINT "segment as given by the DEF SEG statement.  This program sets the";
1190 LOCATE 11,10
1200 PRINT "segment to the beginning of either the graphics or monochrome";
1210 LOCATE 13,8
1220 PRINT "screen buffer for the IBM PC.  If your memory map isn't the same";
1230 LOCATE 15,8
1240 PRINT "as an IBM, POKE may write over important data in RAM.  Check your";
1250 LOCATE 17,9
1260 PRINT "BASIC manual (DEF SEG statement) for the segment of the screen";
1270 LOCATE 19,8
1280 PRINT "buffer and compare with lines 1450,1540 in this program to ensure";
1290 LOCATE 21,33
1300 PRINT "smooth running.";
1310 '
```

```
1320 LOCATE 24,1                                          'input method of screen display
1330 PRINT "Do you wish to use 1)the POKE method, or 2)the slower and sure";
1340 INPUT; " LOCATE method";OPT$
1350 IF OPT$="1" OR OPT$="2" THEN GOTO 1400
1360    LOCATE 25,24:  BEEP
1370    PRINT "Try again, the response is 1 or 2";
1380    LOCATE 24,1:  PRINT SPACES$(79);
1390    GOTO 1320
1400 IF OPT$="1" THEN POKING%=-1 ELSE POKING%=0
1410 IF NOT POKING% THEN RETURN
1420 '
1430 IF NOT GRAPHICS% THEN GOTO 1540
1440    GOSUB 16000                                       'switch screens
1450    DEF SEG=&HB800                                     'set memory
1460    RESTORE 1510
1470    FOR I%=0 TO 9                                      'set up array for colors
1480       READ INTERIOR%(I%)
1490       BOUNDARY%(I%)=INTERIOR%(I%)+96
1500    NEXT I%
1510    DATA 4,5,1,3,2,12,13,9,11,10
1520    RETURN
1530 '
1540 DEF SEG=&HB000                                        'set memory
1550 FOR I%=0 TO 9                                         'set up arrays for colors
1560    INTERIOR%(I%)=2                                     ' (viz. boldface or regular
1570    BOUNDARY%(I%)=11                                    ' type)
1580 NEXT I%
1590 RETURN
1600 '
2000 '****************************************************************************
2010 'subroutine to input Nx, Ny, INIT%
2020 'input     variables:  LATTICE%, LSAVE%
2030 'output variables:  INIT%, STP%, XFIRST%, XLAST%, YFIRST%, YLAST%
2040 'global variables:  none
2050 'local    variables:  NX%, NY%
2060 '****************************************************************************
2070 CLS
2080 IF LSAVE% THEN GOTO 2380
2090 '
2100 LOCATE 3,19                                           'display introduction
2110 PRINT "This program solves Laplace's equation on a";
2120 LOCATE 4,19
2130 PRINT "uniform rectangular lattice of unit spacing";
2140 '
2150 LOCATE 6,18                                           'input NX
2160 INPUT; "Enter the number of horizontal points (<=79)";NX%
2170 IF NX%>0 AND NX%<80 GOTO 2230
2180    LOCATE 7,29:  BEEP
2190    PRINT "Try again, 0 < Nx <= 79";
2200    LOCATE 6,1:  PRINT SPACE$(80);
2210    GOTO 2150
2220 '
2230 LOCATE 9,19                                           'input NY
2240 INPUT; "Enter the number of vertical points (<=23)";NY%
2250 IF NY%>0 AND NY%<=23 THEN GOTO 2310
2260    LOCATE 10,29:  BEEP
2270    PRINT "Try again, 0 < Ny <=24";
2280    LOCATE 9,1:   PRINT SPACE$(80);
```

```
2290    GOTO 2230
2300  '
2310 YFIRST%=(24-NY%)/2:   YLAST%=YFIRST%+NY%-1          'center the lattice
2320 IF NX%>39 GOTO 2350                                 'if possible, let the
2330    XFIRST%=(79-2*NX%)/2:   XLAST%=XFIRST%+2*(NX%-1) ' columns be separated
2340    STP%=2:   GOTO 2380                              ' by a space
2350 XFIRST%=(81-NX%)/2:   XLAST%=XFIRST%+NX%-1
2360 STP%=1
2370  '
2380 LOCATE 12,13                                        'input bound value of P
2390 INPUT; "Enter the initial value for P along the boundary (1-35)";INIT%
2400 IF INIT%>=1 AND INIT%<=35 THEN GOTO 2460
2410    LOCATE 13,36:   BEEP
2420    PRINT "Try again";
2430    LOCATE 12,1:   PRINT SPACE$(79);
2440    GOTO 2380
2450  '
2460 RETURN
2470  '
3000  '*******************************************************************
3010 'subroutine to input OMEGA
3020 'input  variables:   BC%, FIRST%, LATTICE%
3030 'output variables:   OMEGA
3040 'global variables:   none
3050 'local  variables:   none
3060  '*******************************************************************
3070 IF NOT LATTICE% THEN CLS                            'clear screen
3080  '
3090 IF FIRST% GOTO 3120                                 'input OMEGA
3100 LOCATE 19,24
3110 PRINT USING "Your last value of Omega was #.##";OMEGA;
3120 LOCATE 18,11
3130 PRINT "Omega is the parameter which controls the rate of relaxation";
3140 LOCATE 20,33
3150 PRINT "0 < OMEGA <= 2";
3160 LOCATE 22,36
3170 INPUT; "OMEGA = ";OMEGA
3180 IF OMEGA>0 AND OMEGA<=2 GOTO 3240
3190    LOCATE 23,36:   BEEP
3200    PRINT "Try again"
3210    LOCATE 22,1:   PRINT SPACE$(79);
3220    GOTO 3160
3230  '
3240 RETURN
3250  '
4000  '*******************************************************************
4010 'subroutine to input initial values in P and BUFFER% arrays
4020 'input  variables:   BCSAVE%, INIT%, STP%, XFIRST%, XLAST%, YFIRST%, YLAST%
4030 'output variables:   BUFFER%(I%,J%), P(I%,J%)
4040 'global variables:   none
4050 'local  variables:   COL%, ROW%
4060  '*******************************************************************
4070 IF NOT BCSAVE% THEN GOTO 4150                       'if saving bound. cond.
4080 FOR ROW%=YFIRST% TO YLAST%                          ' then release all non
4090    FOR COL%=XFIRST% TO XLAST% STEP STP%             ' boundary points
4100       IF P(ROW%,COL%)>=0 THEN P(ROW%,COL%)=0
4110    NEXT COL%
4120 NEXT ROW%
```

```
4130  RETURN
4140  '
4150  FOR ROW%=YFIRST% TO YLAST%
4160      FOR COL%=XFIRST% TO XLAST% STEP STP%          'location in
4170          BUFFER%(ROW%,COL%)=((ROW%-1)*80+(COL%-1))*2  ' screen buffer
4180          P(ROW%,COL%)=0                            'relase all pts
4190      NEXT COL%
4200  NEXT ROW%
4210  '
4220  FOR ROW%=YFIRST% TO YLAST%                        'set all points on the edge of
4230      P(ROW%,XFIRST%)=-INIT%                        ' of the lattice to the initial
4240      P(ROW%,XLAST%)=-INIT%                         ' value for P
4250  NEXT ROW%
4260  '
4270  FOR COL%=XFIRST% TO XLAST% STEP STP%             'a negative value for P
4280      P(YFIRST%,COL%)=-INIT%                        ' indicates that it is a
4290      P(YLAST%,COL%)=-INIT%                         ' boundary point
4300  NEXT COL%
4310  '
4320  RETURN
4330  '
5000  '***************************************************************
5010  'subroutine to display instructions for altering boundary conditions
5020  'input  variables:  none
5030  'output variables:  none
5040  'global variables:  none
5050  'local  variables:  none
5060  '***************************************************************
5070  CLS
5080  '
5090  LOCATE 1,8,0
5100  PRINT "The initial boundary conditions will be displayed on the screen";
5110  LOCATE 3,17
5120  PRINT "and you will be able to alter them as follows:";
5130  LOCATE 5,27
5140  PRINT "Use "+CHR$(24)+CHR$(25)+CHR$(26)+CHR$(27)+" to move the cursor";
5150  LOCATE 7,18
5160  PRINT "At any point you may enter characters 1-9,a-z";
5170  LOCATE 9,14
5180  PRINT "corresponding to Dirichlet boundary values from 1-35";
5190  LOCATE 11,10
5200  PRINT "Entering . at any point will release it from boundary conditions";
5210  LOCATE 13,24
5220  PRINT "Type return when you are finished";
5230  LOCATE 16,3
5240  PRINT "Then by placing a < in the upper left hand and ";
5250  PRINT "a > in the lower right hand";
5260  LOCATE 18,8
5270  PRINT "corners you may specify a sublattice which will be relaxed first,";
5280  LOCATE 20,11
5290  PRINT "thus speeding the relaxation process for the whole lattice.";
5300  LOCATE 22,14
5310  PRINT "Type . to erase a mistake, type return when finished.";
5320  '
5330  LOCATE 25,31,1
5340  PRINT "Type c to continue";
5350  IF INKEY$<>"c" THEN GOTO 5350
5360  LOCATE ,,0
```

```
5370 '
5380 RETURN
5390 '
6000 '***********************************************************
6010 'subroutine to input boundary conditions from the screen
6020 'input   variables:  PSAVE%, STP%, XFIRST%, XLAST%, YFIRST%, YLAST%
6030 'output  variables:  SUBL%, SUBLATTICE%, SUBSAVE%, X1%,X2%, Y1%, Y2%
6040 'global  variables:  BOUNDARY%(I%), INTERIOR%(I%), POKING%
6050 'local   variables:  ASKII%, ATT%, COL%, OPT$, P%, ROW%
6060 '***********************************************************
6070 X1%=XFIRST%:   X2%=XLAST%                  'set values so that the entire
6080 Y1%=YFIRST%:   Y2%=YLAST%                  ' lattice will be displayed
6090 IF NOT PSAVE% THEN GOSUB 7000 ELSE GOSUB 8000    'display lattice
6100 '
6110 LOCATE 24,16                               'display instructions
6120 PRINT "enter values 1-9,a-z;  type return when finished";
6130 '
6140 COL%=XFIRST%:  ROW%=YFIRST%               'start cursor in upper right
6150 '
6160 LOCATE ROW%,COL%,1                        'place cursor at current pos
6170     COLOR 15                              'attribute for boundary
6180     K$=INKEY$                             'wait for keyboard input
6190     IF K$="" THEN GOTO 6180
6200     LOCATE 25,34: PRINT SPACE$(13);       'erase old error message
6210 '
6220     IF LEN(K$)<2 THEN GOTO 6420           'check for two character
6230       K$=RIGHT$(K$,1)                     ' input indicating cursors
6240 '
6250       IF K$<>CHR$(72) THEN GOTO 6290      'up arrow
6260         ROW%=ROW%-1
6270         IF ROW%<YFIRST% THEN ROW%=YFIRST%
6280         GOTO 6160
6290       IF K$<>CHR$(80) THEN GOTO 6330      'down arrow
6300         ROW%=ROW%+1
6310         IF ROW%>YLAST% THEN ROW%=YLAST%
6320         GOTO 6160
6330       IF K$<>CHR$(77) THEN GOTO 6370      'left arrow
6340         COL%=COL%+STP%
6350         IF COL%>XLAST% THEN COL%=XLAST%
6360         GOTO 6160
6370       IF K$<>CHR$(75) THEN GOTO 6610      'right arrow
6380         COL%=COL%-STP%
6390         IF COL%<XFIRST% THEN COL%=XFIRST%
6400         GOTO 6160
6410 '
6420     IF K$=CHR$(13) THEN GOTO 6640         'typing return exits
6430 '
6440     IF K$<>CHR$(46) THEN GOTO 6490        'release point if . entered
6450       P(ROW%,COL%)=0                      'P=0 indicates released point
6460       ASKII%=46                           '. is ASCII character 46
6470           COLOR 7:  ATT%=INTERIOR%(0)     'get correct attribute
6480           GOTO 6560
6490     ASKII%=ASC(K$)                        'find ASCII code for input
6500     IF ASKII%<49 OR ASKII%>122 GOTO 6610  'if code in this
6510     IF ASKII%>57 AND ASKII%<97 GOTO 6610  ' range, not 1-9,a-z
6520       IF ASKII%>96 THEN P%=ASKII%-87 ELSE P%=ASKII%-48
6530       P(ROW%,COL%)=-P%                    'translate ASCII to P value
6540       ATT%=BOUNDARY%(P% MOD 10)           'get correct attribute
```

```
6550 '
6560      IF NOT POKING% THEN LOCATE ROW%,COL%:  PRINT CHR$(ASKII%);
6570      IF POKING% THEN POKE BUFFER%(ROW%,COL%),ASKII%
6580      IF POKING% THEN POKE BUFFER%(ROW%,COL%)+1,ATT%
6590      GOTO 6160
6600 '
6610    LOCATE 25,34: COLOR 7:  PRINT "Illegal input";
6620 GOTO 6160
6630 '
6640 COLOR 7: LOCATE 24,1:  PRINT SPACE$(79);          'do you want a sublattice
6650 LOCATE 25,1:  PRINT SPACE$(79);
6660 LOCATE 24,19
6670 INPUT; "Do you wish to specify a sub lattice (y/n)";OPT$
6680 IF OPT$="y" OR OPT$="n" THEN GOTO 6730
6690    LOCATE 25,36:  BEEP
6700    PRINT "Try again"
6710    LOCATE 24,1:  PRINT SPACE$(79);
6720    GOTO 6660
6730 IF OPT$="y" THEN SUBL%=-1 ELSE SUBL%=0          'flag to relax sublattice
6740 IF OPT$="y" THEN SUBLATTICE%=-1 ELSE SUBLATTICE%=0  'flag to enter sublat
6750 IF OPT$="y" THEN SUBSAVE%=-1 ELSE SUBSAVE%=0     'flag to redo sublattice
6760 '                                              if only Omega is changed
6770 RETURN
6780 '
7000 '**************************************************************************
7010 'subroutine to display lattice on the screen
7020 'input   variables:  BUFFER%(I%,J%), CLR%, P(I%,J%), STP%,X1%, X2%, Y1%,Y2%
7030 'output variables:   none
7040 'global variables:   BOUNDARY%(I%), INTERIOR%(I%), POKING%
7050 'local  variables:   ASKII%, ATT%, COL%, MP%, NEG%, P%, ROW%
7060 '**************************************************************************
7070 IF CLR% THEN CLS
7080 LOCATE ,,0
7090 IF NOT POKING% THEN GOTO 7250
7100 '
7110 FOR ROW%=Y1% TO Y2%                            'the attribute is different for
7120    FOR COL%=X1% TO X2% STEP STP%               ' bound and non-bound points so
7130       P%=ABS(P(ROW%,COL%))                     ' that they look different on
7140       IF P%<10 THEN ASKII%=P%+48 ELSE ASKII%=P%+87     ' the screen
7150       IF P(ROW%,COL%)=0 THEN ASKII%=46
7160       MP%=P% MOD 10:  IF P(ROW%,COL%)<0 THEN NEG%=-1 ELSE NEG%=0
7170       IF NEG% THEN ATT%=BOUNDARY%(MP%) ELSE ATT%=INTERIOR%(MP%)
7180       POKE BUFFER%(ROW%,COL%)+1,ATT%
7190       POKE BUFFER%(ROW%,COL%),ASKII%
7200    NEXT COL%
7210 NEXT ROW%
7220 '
7230 RETURN
7240 '
7250 FOR ROW%=Y1% TO Y2%                            'the color is different for
7260    FOR COL%=X1% TO X2% STEP STP%               ' bound. and non-bound. pts
7270       P%=ABS(P(ROW%,COL%))
7280       IF P%<10 THEN ASKII%=P%+48 ELSE ASKII%=P%+87
7290       IF P(ROW%,COL%)=0 THEN ASKII%=46
7300       IF P(ROW%,COL%)<0 THEN COLOR 15 ELSE COLOR 7
7310       LOCATE ROW%,COL%,0:  PRINT CHR$(ASKII%);
7320    NEXT COL%
7330 NEXT ROW%
```

```
7340 '
7350 COLOR 7:   RETURN
7360 '
8000 '*********************************************************************
8010 'subroutine to put lattice on screen when P is to be saved
8020 'input   variables:   BUFFER%(I%,J%), CLR%, P(I%,J%), STP%, XFIRST%, XLAST%,
8030 '                     YFIRST%, YLAST%
8040 'output variables:    none
8050 'global variables:    BOUNDARY%(I%), INTERIOR%(I%), POKING%
8060 'local   variables:   ASKII%, ATT%, COL%, MP%, P%, ROW%
8070 '*********************************************************************
8080 IF CLR% THEN CLS
8090 LOCATE ,,0
8100 IF NOT POKING% THEN GOTO 8250
8110 '
8120 FOR ROW%=YFIRST% TO YLAST%                      'non-bound pts will be displayed
8130    FOR COL%=XFIRST% TO XLAST% STEP STP%      ' as ., but the value in P will
8140       IF P(ROW%,COL%)<0 THEN P%=-P(ROW%,COL%)          ' not be lost
8150       IF P%<10 THEN ASKII%=P%+48 ELSE ASKII%=P%+87
8160       IF P(ROW%,COL%)>0 THEN ASKII%=46
8170       MP%=P% MOD 10
8180       IF ASKII%=46 THEN ATT%=INTERIOR%(0) ELSE ATT%=BOUNDARY%(MP%)
8190       POKE BUFFER%(ROW%,COL%),ASKII%
8200       POKE BUFFER%(ROW%,COL%)+1,ATT%
8210    NEXT COL%
8220 NEXT ROW%
8230 RETURN
8240 '
8250 FOR ROW%=YFIRST% TO YLAST%
8260    FOR COL%=XFIRST% TO XLAST% STEP STP%
8270       IF P(ROW%,COL%)<0 THEN P%=-P(ROW%,COL%)
8280       IF P%<10 THEN ASKII%=P%+48 ELSE ASKII%=P%+87
8290       IF P(ROW%,COL%)>0 THEN ASKII%=46
8300       IF ASKII%=46 THEN COLOR 7 ELSE COLOR 15
8310       LOCATE ROW%,COL%:   PRINT CHR$(ASKII%);
8320    NEXT COL%
8330 NEXT ROW%
8340 COLOR 7:   RETURN
8350 '
9000 '*********************************************************************
9010 'subroutine to input parameters of sublattice
9020 'input   variables:   BUFFER%(I%,J%), P(I%,J%), STP%, XFIRST%, XLAST%,
9030 '                     YFIRST%, YLAST%
9040 'output variables:    SUBX1%, SUBX2%, SUBY1%, SUBY2%
9050 'global variables:    BOUNDARY%(I%), INTERIOR%(I%), POKING%
9060 'local   variables:   ASKII%, ATTRIBUTE%, BOTH%, COL%, CORRECT%, DELX%,
9070 '                     DELY%, K$, ONEDIM%, P%, ROW%, SWITCHED%, SUB1%, SUB2%
9080 '                     TYPEOVER%
9090 '*********************************************************************
9100 SUB1%=0:   SUB2%=0                             'set flags to zero
9110 '
9120 LOCATE 24,1:   PRINT SPACE$(79);               'clear bottom of screen
9130 LOCATE 25,1:   PRINT SPACE$(79);
9140 LOCATE 24,3                                    'display instructions
9150 PRINT "Enter < in upper left and > in lower right corners to specify a";
9160 PRINT " sublattice.";
9170 LOCATE 25,13
9180 PRINT "Type . to correct a mistake, type return when finished.";
```

```
9190 '
9200 COL%=XFIRST%:  ROW%=YFIRST%                    'start cursor in upper right
9210 '
9220 LOCATE ROW%,COL%,1                             'place cursor at current pos
9230    K$=INKEY$                                   'wait for keyboard input
9240    IF K$="" THEN GOTO 9230
9250    '
9260    IF LEN(K$)<2 THEN GOTO 9460                 'check for two character
9270     K$=RIGHT$(K$,1)                            ' input indicating cursors
9280         '
9290      IF K$<>CHR$(72) THEN GOTO 9330            'up arrow
9300        ROW%=ROW%-1
9310         IF ROW%<YFIRST% THEN ROW%=YFIRST%
9320        GOTO 9220
9330      IF K$<>CHR$(80) THEN GOTO 9370            'down arrow
9340        ROW%=ROW%+1
9350         IF ROW%>YLAST% THEN ROW%=YLAST%
9360        GOTO 9220
9370      IF K$<>CHR$(77) THEN GOTO 9410            'left arrow
9380        COL%=COL%+STP%
9390         IF COL%>XLAST% THEN COL%=XLAST%
9400        GOTO 9220
9410      IF K$<>CHR$(75) THEN GOTO 6610            'right arrow
9420        COL%=COL%-STP%
9430         IF COL%<XFIRST% THEN COL%=XFIRST%
9440        GOTO 9220
9450 '
9460    IF K$<>CHR$(13) THEN 9620                   'typing return exits
9470      BOTH%=SUB1% AND SUB2%                     'make sure that the sublattice
9480      DELX%=SUBX2%-SUBX1%                       ' parameters are sensible
9490      DELY%=SUBY2%-SUBY1%
9500      ONEDIM%=BOTH% AND (DELX%=0 OR DELY%=0)
9510      SWITCHED%=BOTH% AND (DELX%<0 OR DELY%<0) AND NOT ONEDIM%
9520      CORRECT%=BOTH% AND (DELX%>0) AND (DELY%>0)
9530      IF CORRECT% THEN LOCATE ,,0: COLOR 7:  RETURN
9540      LOCATE 25,1:  PRINT SPACE$(79);
9550      LOCATE 25,29:  BEEP
9560      IF NOT SUB2% THEN PRINT "Specify lower right hand corner.";
9570      IF NOT SUB1% THEN PRINT "Specify upper left hand corner.";
9580      IF ONEDIM% THEN PRINT "Sublattice must be 2 dimensional.";
9590      IF SWITCHED% THEN PRINT "Try again; corners are switched";
9600      GOTO 9200
9610 '
9620    IF K$<>CHR$(60) THEN GOTO 9740             'upper left corner
9630      IF SUB1% THEN GOTO 9700
9640        TYPEOVER%=(SUB2% AND COL%=SUBX2% AND ROW%=SUBY2%)
9650         IF TYPEOVER% THEN SUB2%=0
9660        SUB1%=-1                                'set flags
9670        SUBX1%=COL%                             'input sublattice parameters
9680        SUBY1%=ROW%
9690        LOCATE ROW%,COL%:  PRINT K$;:  GOTO 9220
9700      LOCATE 25,1:  PRINT SPACE$(79);
9710      LOCATE 25,16:  BEEP
9720      PRINT "Upper left already specified, type . to correct";
9730      GOTO 9220
9740    IF K$<>CHR$(62) THEN GOTO 9870             'lower right corner
9750      IF SUB2% THEN GOTO 9820
9760        TYPEOVER%=(SUB1% AND COL%=SUBX1% AND ROW%=SUBY1%)
```

```
9770        IF TYPEOVER% THEN SUB1%=0
9780        SUB2%=-1                                'set flags
9790        SUBX2%=COL%                             'input parameters
9800        SUBY2%=ROW%
9810        LOCATE ROW%,COL%:  PRINT K$;: GOTO 9220
9820      LOCATE 25,1:  PRINT SPACE$(79);
9830      LOCATE 25,16:  BEEP
9840      PRINT "Lower right already specified, type . to correct";
9850      GOTO 9220
9860 '
9870      IF K$<>CHR$(46) THEN GOTO 9220           'typing . corrects a mistake
9880      IF COL%=SUBX1% AND ROW%=SUBY1% THEN SUB1%=0
9890      IF COL%=SUBX2% AND ROW%=SUBY2% THEN SUB2%=0
9900      IF P(ROW%,COL%)<0 THEN GOTO 9930
9910        ASKII%=46:  COLOR 7:  ATTRIBUTE%=INTERIOR%(0)
9920        GOTO 9970
9930      P%=ABS(P(ROW%,COL%))
9940      IF P%<10 THEN ASKII%=P%+48 ELSE ASKII%=P%+87
9950      COLOR 7:  ATTRIBUTE%=BOUNDARY%(P% MOD 10)
9960 '
9970      IF NOT POKING% THEN LOCATE ROW%,COL%:  PRINT CHR$(ASKII%);
9980      IF POKING% THEN POKE BUFFER%(ROW%,COL%),ASKII%
9990      IF POKING% THEN POKE BUFFER%(ROW%,COL%)+1,ATTRIBUTE%
10000     GOTO 9220
10010 '
11000 '************************************************************
11010 'subroutine to guess initial values
11020 'input  variables:  INIT%, P(I%,J%), STP%, XFIRST%, XLAST%, YFIRST%
11030 '                   YLAST%
11040 'output variables:  P(I%,J%)
11050 'global variables:  none
11060 'local  variables:  COL%, ROW%
11070 '************************************************************
11080 IF PSAVE% THEN GOTO 11180
11090 '
11100   FOR ROW%=YFIRST% TO YLAST%                 'set all non-bound pts to the
11110     FOR COL%=XFIRST% TO XLAST% STEP STP%     ' initial value for the bound
11120       IF P(ROW%,COL%)<0 THEN GOTO 11140
11130         P(ROW%,COL%)=INIT%
11140     NEXT COL%
11150   NEXT ROW%
11160   RETURN
11170 '
11180 FOR ROW%=YFIRST% TO YLAST%                   'if saving last PHI, set all
11190   FOR COL%=XFIRST% TO XLAST% STEP STP%       ' released points to the init
11200     IF P(ROW%,COL%)<>0 THEN GOTO 11220       ' value for the bound
11210       P(ROW%,COL%)=INIT%
11220   NEXT COL%
11230 NEXT ROW%
11240 '
11250 RETURN
11260 '
12000 '************************************************************
12010 'subroutine to prepare the screen for relaxation loop
12020 'input  variables:  FIRST%, SUBL%
12030 'output variables:  FIRST%
12040 'global variables:  none
12050 'local  variables:  none
```

```
12060 '**********************************************************************
12070 CLS:  COLOR 7:  LOCATE ,,0
12080 IF NOT FIRST% THEN GOTO 12320
12090 '
12100 LOCATE 5,16                                        'display instructions
12110 PRINT "Typing e will end the current relaxation process.";
12120 LOCATE 8,11
12130 PRINT "If you have specified a sub lattice, typing e once will end";
12140 LOCATE 11,11
12150 PRINT "relaxation of the sub lattice and begin the process for the";
12160 LOCATE 14,11
12170 PRINT "entire lattice.  Typing e once more will end the relaxation";
12180 LOCATE 17,31
12190 PRINT "process altogether.";
12200 LOCATE 20,17
12210 PRINT "The energy printed is the value per unit area.";
12220 LOCATE 22,4
12230 PRINT "Delta P is the maximum percentage change in Phi from previous";
12240 PRINT " iteration.";
12250 '
12260 LOCATE 25,31,1
12270 PRINT "Type c to continue";
12280 IF INKEY$<>"c" THEN GOTO 12280
12290 LOCATE ,,0
12300 FIRST%=0
12310 '
12320 GOSUB 7000                                         'display lattice
12330 '
12340 LOCATE 24,1                                         'prepare for output
12350 PRINT "iteration =   0";
12360 LOCATE 24,27
12370 PRINT "delta P= 00.000000";
12380 LOCATE 24,57
12390 PRINT "last delta P= 00.000000";
12400 LOCATE 25,1
12410 PRINT "energy = 000.000000";
12420 LOCATE 25,24
12430 PRINT "change in energy = +00.000000";
12440 IF SUBL% THEN LOCATE 25,59:  PRINT "sublattice";
12450 LOCATE 25,75: COLOR 15
12460 PRINT "e:end";: COLOR 7
12470 '
12480 RETURN
12490 '
13000 '**********************************************************************
13010 'subroutine to relax the lattice
13020 'input  variables:  DELTAPMAX, ENERGY, ITERATION%, OLDE, OLDDELP,
13030 '                   P(I%,J%), X1%, X2%, Y1%, Y2%
13040 'output variables:  DELTAPMAX, ENERGY
13050 'global variables:  NX%, NY%, OMEGA, STP%
13060 'local  variables:  A, B, C, COL%, D, DELENERGY, DELTAP, PNEW, POLD, ROW%
13070 '**********************************************************************
13080 FOR ROW%=Y1% TO Y2%                        'loop over all lattice or
13090    FOR COL%=X1% TO X2% STEP STP%           ' sublattice points
13100       IF ROW%=Y1% AND COL%=X1% THEN GOTO 13280
13110       B=ABS(P(ROW%-1,COL%))
13120       D=ABS(P(ROW%,COL%-STP%))             'don't relax a boundary point
13130       IF P(ROW%,COL%)<0 THEN PNEW=ABS(P(ROW%,COL%)): GOTO 13250
```

```
13140 '
13150         POLD=P(ROW%,COL%)                        'relax the point
13160         A=ABS(P(ROW%+1,COL%))
13170         C=ABS(P(ROW%,COL%+STP%))
13180         PNEW=(1-OMEGA)*POLD+OMEGA*.25*(A+B+C+D)
13190         P(ROW%,COL%)=PNEW
13200         IF PNEW<1 THEN PNEW=1:  P(ROW%,COL%)=1   'avoid neg values
13210 '
13220         DELTAP=ABS(POLD-PNEW)/POLD               'check for max change in P
13230         IF DELTAP>DELTAPMAX THEN DELTAPMAX=DELTAP
13240 '
13250         IF COL%=X1% THEN ENERGY=ENERGY+(PNEW-B)^2:   GOTO 13280
13260         IF ROW%=Y1% THEN ENERGY=ENERGY+(PNEW-D)^2:   GOTO 13280
13270         ENERGY=ENERGY+(PNEW-B)^2+(PNEW-D)^2'term in the energy
13280     NEXT COL%
13290 NEXT ROW%
13300 '
13310 '
13320 ENERGY=ENERGY/(2*NX%*NY%)                        'energy per unit area
13330 DELENERGY=ENERGY-OLDE
13340 IF ITERATION%=1 THEN DELENERGY=0:   OLDDELP=0
13350 '
13360 GOSUB 7000                                       'put new P on screen
13370 LOCATE 24,13,0:   PRINT USING "###";ITERATION%;
13380 LOCATE 24,36,0:   PRINT USING "##.#####";DELTAPMAX;
13390 LOCATE 24,71,0:   PRINT USING "##.#####";OLDDELP;
13400 LOCATE 25,10,0:   PRINT USING "###.#####";ENERGY;
13410 LOCATE 25,43,0:   PRINT USING "+##.#####";DELENERGY;
13420 '
13430 RETURN
13440 '
14000 '*********************************************************************
14010 'subroutine to display options
14020 'input  variables:   SUBSAVE%
14030 'output variables:   BC%, BCSAVE%, CLR%, LATTICE%, LSAVE%, OMEGA%,
14040 '                    PSAVE%, SUBL%, SUBSAVE%
14050 'global variables:   none
14060 'local  variables:   A$, B$, C$, D$, E$
14070 '*********************************************************************
14080 CLS:  CLR%=-1
14090 '
14100 LOCATE 1,28                                      'end command
14110 INPUT; "Do you wish to end (y/n)";A$
14120 IF A$="y" OR A$="n" THEN GOTO 14170
14130    LOCATE 2,36:   BEEP
14140    PRINT "Try again";
14150    LOCATE 1,1:    PRINT SPACE$(79);
14160    GOTO 14100
14170 IF A$="y" THEN END
14180 '
14190 LOCATE 5,24                                      'change Omega command
14200 INPUT; "Do you wish to change Omega (y/n)";B$
14210 IF B$="y" OR B$="n" THEN GOTO 14260
14220    LOCATE 6,36:   BEEP
14230    PRINT "Try again";
14240    LOCATE 5,1:    PRINT SPACE$(79);
14250    GOTO 14190
14260 IF B$="y" THEN OMEGA%=-1 ELSE OMEGA%=0           'set Omega flag
```

```
14270 '
14280 LOCATE 9,18                                       'change lattice size command
14290 INPUT; "Do you wish to change the lattice size (y/n)";C$
14300 IF C$="y" OR C$="n" THEN GOTO 14350
14310   LOCATE 10,36:  BEEP
14320   PRINT "Try again";
14330   LOCATE 9,1:  PRINT SPACE$(79);
14340   GOTO 14280
14350 IF C$="y" THEN LATTICE%=-1 ELSE LATTICE%=0 'set lattice flag
14360 IF LATTICE% THEN SUBSAVE%=0:  RETURN
14370 '
14380 LOCATE 13,34:  PRINT "Do you wish to";      'command to change bound. cond
14390 LOCATE 15,1
14400 PRINT "1)preserve the boundary conditions and start again";
14410 PRINT " (useful for trying new OMEGA)";
14420 LOCATE 16,18
14430 PRINT "2)change boundary conditions and start again";
14440 LOCATE 17,3
14450 PRINT "3)change boundary conditions (starting from current b.c.)";
14460 PRINT " and keep current PHI";
14470 LOCATE 19,29
14480 INPUT; "Enter your choice (1-3)";D$
14490 IF D$="1" OR D$="2" OR D$="3" THEN GOTO 14540
14500   LOCATE 20,36:  BEEP
14510   PRINT "Try again";
14520   LOCATE 19,1:  PRINT SPACE$(79); ·
14530   GOTO 14470
14540 IF D$="1" THEN BC%=0: PSAVE%=0:  IF SUBSAVE% THEN SUBL%=-1:  RETURN
14550 IF D$="1" THEN RETURN
14560 IF D$="3" THEN BC%=-1: PSAVE%=-1:  SUBSAVE%=0:  RETURN
14570 '
14580 LOCATE 23,11
14590 INPUT; "Do you wish to 1)begin from current b.c. or 2)start fresh";E$
14600 IF E$="1" OR E$="2" THEN GOTO 14640
14610   BEEP:  LOCATE 23,1
14620   PRINT SPACE$(80);
14630   GOTO 14580
14640 IF E$="1" THEN LATTICE%=-1:  BCSAVE%=-1:  PSAVE%=0:  SUBSAVE%=0
14650 IF E$="2" THEN LATTICE%=-1:  LSAVE%=-1:  PSAVE%=0:  SUBSAVE%=0
14660 RETURN
14670 '
15000 '*********************************************************************
15010 'subroutine to display the header screen
15020 'input  variables: none
15030 'output variables: GRAPHICS%, MONO%
15040 'global variables: none
15050 'local  variables: G$, M$, ROW%
15060 '*********************************************************************
15070 SCREEN 0: CLS: KEY OFF                          'program begins in text mode
15080 '
15090 LOCATE 1,30: COLOR 15                           'display book title
15100 PRINT "COMPUTATIONAL PHYSICS";
15110 LOCATE 2,40: COLOR 7
15120 PRINT "by";
15130 LOCATE 3,33: PRINT "Steven E. Koonin";
15140 LOCATE 4,14
15150 PRINT "Copyright 1985, Benjamin/Cummings Publishing Company";
15160 '
```

```
15170 LOCATE 5,10                                        'draw the box
15180 PRINT CHR$(201)+STRING$(59,205)+CHR$(187);
15190 FOR ROW%=6 TO 19
15200     LOCATE ROW%,10: PRINT CHR$(186);
15210     LOCATE ROW%,70: PRINT CHR$(186);
15220 NEXT ROW%
15230 LOCATE 20,10
15240 PRINT CHR$(200)+STRING$(59,205)+CHR$(188);
15250 '
15260 COLOR 15                                           'print title, etc.
15270 LOCATE  7,36:  PRINT "EXAMPLE 6";
15280 COLOR 7
15290 LOCATE 10,22:  PRINT "Laplace's  equation in two dimensions"
15300 LOCATE 13,36:  PRINT "**********"
15310 LOCATE 15,26:  PRINT "Type 'e' to end while running"
15320 LOCATE 16,23:  PRINT "Type ctrl-break to stop at a prompt"
15330 '
15340 LOCATE 19,16: BEEP                                 'get screen configuration
15350 INPUT "Does your computer have graphics capability (y/n)";G$
15360 IF LEFT$(G$,1)="y" OR LEFT$(G$,1)="n" GOTO 15390
15370     LOCATE 19,12:  PRINT SPACES$(58):  BEEP
15380     LOCATE 18,35:  PRINT "Try again,":  GOTO 15340
15390 IF LEFT$(G$,1)="y" GOTO 15420
15400     GRAPHICS%=0:  MONO%=-1
15410     RETURN
15420 GRAPHICS%=-1
15430 LOCATE 18,15:  PRINT SPACES$(55)
15440 LOCATE 19,15:  PRINT SPACES$(55)
15450 LOCATE 19,15: BEEP
15460 INPUT "Do you also have a separate screen for text (y/n)";M$
15470 IF LEFT$(M$,1)="y" OR LEFT$(M$,1)="n" GOTO 15490    'error trapping
15480     LOCATE 18,35:  PRINT "Try again,":  GOTO 15440
15490 IF LEFT$(M$,1)="y" THEN MONO%=-1 ELSE MONO%=0
15500 '
15510 LOCATE 21,14
15520 PRINT "This program uses the iterative relaxation method to";
15530 LOCATE 22,17
15540 PRINT "solve an elliptic partial differential equation"
15550 LOCATE 23,8
15560 PRINT "(viz. Laplace's equation) with user defined boundary conditions";
15570 LOCATE 25,31,1:  BEEP
15580 PRINT "Type c to continue";
15590 IF INKEY$<>"c" THEN GOTO 15590
15600 LOCATE ,,0
15610 RETURN
15620 '
16000 '*************************************************************************
16010 'subroutine to switch from mono to graphics screen
16020 'input  variables: none
16030 'output variables: none
16040 'global variables: none
16050 'local  variables: none
16060 '*************************************************************************
16070 DEF SEG=0
16080 POKE &H410,  (PEEK(&H410) AND &HCF) OR &H10
16090 SCREEN 1,0,0,0: SCREEN 0: WIDTH 80: LOCATE ,,1,6,7
16100 RETURN
16110 '
```

B.7 Example 7

This program solves the time-dependent Schroedinger equation for a particle moving in one dimension. The potential is defined by subroutines 1000, 3000, 4000, and 5000. Various analytical forms are possible (square-well, gaussian barrier, potential step, or parabolic well), as defined in subroutine 4000, and any of these can be altered pointwise from the keyboard (subroutine 5000). The overall scale of the potential (VZERO) can also be specified. After defining the potential, the initial wavepacket can be chosen to have either a Gaussian or Lorentzian shape (subroutine 7000), with specified width and average momentum and position. The time evolution then proceeds (loop 330-390). Here, the basic task is to find the new wavefunction at each time step by solving Eqs. (7.31,7.33) in subroutine 16000 using the matrix inversion scheme (7.12-7.16). For this purpose, the α and γ coefficients are computed once (in subroutine 9000), before the time evolution begins. At each time step, the normalization of the packet and various expectation values are computed (subroutine 11000) and a plot of the wavefunction and potential is drawn (subroutine 14000 or 15000). Upon interrupting the time evolution, the potential, initial wavepacket, or time step can be changed to do another run.

For a typical output from this program, select the analytical form of a square barrier with height, width, and centroid

$$Vzero=0.1, \quad A=10, \quad X0=0,$$

and choose 0 as the value of X which separates left from right. Then select a Gaussian wavepacket with average wavenumber, width, and centroid given by

$$K0=0.5, \quad SIGMA=15, \quad X0=-40.$$

Finally, choose a time step

$$DT=0.5$$

and then watch the evolution unfold.

```
10  '******************************************************************
20  'Example 7: The time-dependent Schrodinger Equation
21  'COMPUTATIONAL PHYSICS by Steven E. Koonin
22  'Copyright 1985, Benjamin/Cummings Publishing Company
30  '******************************************************************
40  GOSUB 18000                                    'display header screen
50  '
60  FIRST%=-1:   ONCE%=-1                          'define constants, functions
70  NPTS%=160                                      'following arrays dim to NPTS%
80  DIM V(160)                                     'potential array
90  DIM V%(160),V1%(160),V2%(160),V3%(160)         'potential arrays for graphing
100 DIM AZR(160)                                   'real part of matrix diagonal
110 DIM GAMR(160),GAMI(160)                        'real and imaginary parts for
```

```
120 DIM BETR(160),BETI(160)                        ' matrix inversion
130 DIM PHIR(160),PHII(160),PHI2(160)             'real, imag, and ^2 wave funct
140 PACKET%=-1:  POT%=-1                           'flags indicating which
150 TIME%=-1                                       ' parameters need to be entered
160 VSAVE%=0:  GSAVE%=0                            'flags to avoid repeated calcs
170 DX=1:  XMIN=-80                                'space step, minimum x value
180 TBEGIN=0                                       'begin at time=0
190 '
200 IF POT% THEN GOSUB 1000                        'input potential
210 IF PACKET% THEN GOSUB 7000                     'input wave packet
220 IF TIME% THEN GOSUB 8000                       'input time step
230 IF TIME% OR POT% THEN GOSUB 9000              'calculate GAMMA(=alpha)
240 '
250 IF FIRST% THEN GOSUB 10000                     'display commands
260 IF GRAPHICS% THEN GOSUB 13000 ELSE GOSUB 12000        'initialize output
270 IF GSAVE% THEN GOTO 300                        'output hasn't changed
280 GOSUB 11000                                    'find prob and <x>
290 IF GRAPHICS% THEN GOSUB 14000 ELSE GOSUB 15000       'output PHI, V(x)
300 GSAVE%=0:  POT%=0:  PACKET%=0:  TIME%=0:      'reset flags
310 '
320 TIME=TBEGIN                                    'set initial time
330 WHILE K$<>"e" AND K$<>"t"
340     TIME=TIME+DT                               'time increment
350     GOSUB 16000                                'calculate BETA, CHI, PHI
360     GOSUB 11000                                'find prob and <x>
370     IF GRAPHICS% THEN GOSUB 14000 ELSE GOSUB 15000     'output
380     K$=INKEY$                                  'time loop continues until
390 WEND                                           ' user types 'e' or 't'
400 '
410 IF K$<>"t" THEN GOTO 460
420     TIME%=-1                                   'flag to enter DT subroutine
430     TBEGIN=TIME                                'keep the same time
440     GSAVE%=-1                                  'don't redraw the graph
450     GOTO 490
460 IF K$<>"e" THEN GOTO 490
470     GOSUB 17000                                'display commands
480     TBEGIN=0:  TIME=0                          'begin at time 0
490 K$=""                                          'reset K$ to enter while loop
500 GOTO 200                                       'begin again
510 '
1000 '********************************************************************
1010 'subroutine to input the 1-dimensional potential
1020 'input variables:   ONCE%, V(I%), VSAVE%, VZERO, VZSAVE, XMIDDLE
1030 'output variables:  IMIDDLE%, ONCE%, V(I%), V%(I%), VSAVE%, VSCALE,XMIDDLE
1040 'global variables:  GRAPHICS%, NPTS%
1050 'local variables:   ANSWER$, E$, F$, H$, I%, RESPONSE$
1060 '********************************************************************
1070 GOSUB 3000                                    'input analytic form,Vzero,etc
1080 '
1090 IF NOT GRAPHICS% THEN RETURN
1100 '
1110 IF NOT VSAVE% OR (VZSAVE=VZERO) THEN GOTO 1160
1120     FOR I%=0 TO NPTS%                         'rescale V if Vzero changed
1130         V(I%)=V(I%)*VZERO/VZSAVE
1140     NEXT I%
1150 '
1160 IF VZERO<>0 THEN GOTO 1380                    'special treatment if Vzero=0
1170     IF NOT ONCE% THEN RETURN                  'print following only once
```

```
1180  ONCE%=0                                          'reset flag
1190  CLS:  LOCATE 6,9
1200  PRINT "Because you have chosen Vzero=0 you cannot alter the potential";
1210  LOCATE 8,11
1220  PRINT "from the screen.  If you wish to begin altering from a zero";
1230  LOCATE 10,10
1240  PRINT "potential, begin with a narrow square well (for example) with";
1250  LOCATE 12,15
1260  PRINT "a value for Vzero near the final value you desire.";
1270  LOCATE 14,16
1280  PRINT "Do you wish to 1)continue with zero potential or";
1290  LOCATE 16,22
1300  INPUT; "2)begin again with the analytic form";RESPONSE$
1310  IF RESPONSE$="1" THEN RETURN
1320  IF RESPONSE$="2" THEN GOTO 1070
1330  LOCATE 18,24:  BEEP
1340  PRINT "Try again, the response is 1 or 2";
1350  LOCATE 16,1:   PRINT SPACE$(80);
1360  GOTO 1290
1370  '
1380  CLS                                        'directions for altering V
1390  LOCATE 2,3,0
1400  PRINT "You may now use the keyboard to alter V(x) at any of the 160";
1410  LOCATE 2,62
1420  PRINT " lattice points:";
1430  LOCATE 6,17
1440  PRINT CHR$(24)+CHR$(25)+" keys increase and decrease V(x) at fixed x.";
1450  LOCATE 9,7
1460  PRINT "PgUp and Home keys increase and decrease x without altering V(x).";
1470  LOCATE 12,12
1480  PRINT CHR$(26)+CHR$(27)+" keys increase and decrease x and change";
1490  LOCATE 12,54
1500  PRINT " V(x+dx) to V(x).";
1510  LOCATE 17,13
1520  PRINT "You will then be able to alter the potential again, so";
1530  LOCATE 18,25
1540  PRINT " don't be afraid to experiment.";
1550  '
1560  LOCATE 24,8:  BEEP
1570  INPUT;"Do you wish to 1)alter the potential or 2)use the analytic form";H$
1580  IF H$="2" THEN RETURN
1590  IF H$="1" THEN GOTO 1640
1600    LOCATE 24,1:  PRINT SPACE$(79);:  BEEP
1610    LOCATE 23,35:  PRINT "Try again";
1620    GOTO 1560
1630  '
1640  VSCALE=170/(2*ABS(VZERO))                  'scale V
1650  FOR I%=0 TO NPTS%
1660    V%(I%)=170-(V(I%)+ABS(VZERO))*VSCALE
1670  NEXT I%
1680  GOSUB 5000                                 'alter V from keyboard
1690  VSAVE%=0                                   'reset flag
1700  '
1710  LOCATE 23,1:  PRINT SPACE$(79);
1720  LOCATE 25,1:  PRINT SPACE$(79);
1730  LOCATE 24,4:  BEEP
1740  PRINT "Are you 1)happy with the potential or 2)do you wish to alter";
1750  LOCATE 24,64
```

```
1760 INPUT; " it further";E$
1770 IF E$="1" THEN GOTO 1820
1780 IF E$="2" GOTO 1950
1790    LOCATE 23,35:  PRINT "Try again,";
1800    GOTO 1730
1810 '
1820 FOR I%=23 TO 25                          'clear bottom of screen
1830    LOCATE I%,1:  PRINT SPACE$(79);
1840 NEXT I%
1850 LOCATE 23,33                             'input new XMIDDLE
1860 PRINT USING "Xmiddle=+##.##";XMIDDLE;
1870 LOCATE 24,21
1880 INPUT; "Do you wish to change this value (y/n)";ANSWER$
1890 IF ANSWER$="n" THEN RETURN
1900 LOCATE 25,32
1910 INPUT; "Enter new value";XMIDDLE
1920 IMIDDLE%=(XMIDDLE-XMIN)
1930 RETURN
1940 '
1950 FOR I%=23 TO 25                          'clear bottom of screen
1960    LOCATE I%,1:  PRINT SPACE$(79);
1970 NEXT I%
1980 LOCATE 24,13:  BEEP
1990 PRINT "Would you like to 1)begin again with the analytic form";
2000 LOCATE 25,18
2010 INPUT; "or 2)alter the potential now on the screen";F$
2020 IF F$="1" THEN GOTO 1070
2030 IF F$="2" THEN GOTO 1680
2040    LOCATE 23,25                          'error trapping
2050    PRINT "Try again,";
2060    GOTO 1980
2070 '
3000 '*******************************************************************
3010 'subroutine to choose the analytic form of the potential
3020 'input  variables:  FIRST%, VSAVE%
3030 'output variables:  A, IMIDDLE%, IMONOMID%, OPT$, VXO, VZERO, VZSAVE,
3040 '                    XMIDDLE
3050 'global variables:  XMIN
3060 'local  variables:  none
3070 '*******************************************************************
3080 CLS
3090 IF NOT VSAVE% THEN GOTO 3130             'if the old potential is to
3100    VZSAVE=VZERO                          ' be saved, only need to input
3110    GOTO 3380                             ' new Vzero
3120 '
3130 LOCATE 1,8,0                             'display options for V(x)
3140 PRINT "In this program you have a choice of one-dimensional potentials:";
3150 LOCATE 4,15
3160 PRINT "1)Square barrier:  V(x)=Vzero for X0-A <= X <= X0+A";
3170 LOCATE 6,11
3180 PRINT "2)Gaussian barrier:  V(x)=Vzero*exp(-(X-X0)^2*log(20)/A^2)";
3190 LOCATE 8,11
3200 PRINT "3)Potential step:  V(x)=Vzero*(atn((X-X0)*6.3/A)*2/PI)+1)/2";
3210 LOCATE 10,22
3220 PRINT USING "4)Parabolic well:  V(x)=Vzero*(X/##)^2";ABS(XMIN);
3230 LOCATE 13,5
3240 PRINT "All of these potentials have walls (V(x)=infinity";
3250 PRINT USING ") at X=+## and X=##,";XMIN,ABS(XMIN);
```

```
3260 LOCATE 14,6
3270 PRINT "giving the boundary conditions that the wave ";
3280 LOCATE 14,51
3290 PRINT "function is zero there.";
3300 '
3310 LOCATE 17,19:  BEEP                              'enter analytic form
3320 INPUT; "Enter the number of the desired potential";OPT$
3330 IF (OPT$="1" OR OPT$="2" OR OPT$="3" OR OPT$="4") GOTO 3380
3340    LOCATE 17,1:  PRINT SPACE$(79);:  LOCATE 16,20
3350    PRINT "Try again, the responses are 1,2,3, or 4";
3360    GOTO 3310
3370 '
3380 IF FIRST% GOTO 3410                              'enter Vzero
3390    LOCATE 19,22
3400    PRINT USING "Your last value of Vzero was +###.###";VZERO;
3410 LOCATE 20,33
3420 INPUT; "Vzero(height)=";VZERO
3430 IF VSAVE% THEN RETURN
3440 IF OPT$="4" THEN GOTO 3580
3450 '
3460    IF FIRST% GOTO 3490                           'enter A (half width)
3470       LOCATE 21,25
3480       PRINT USING "Your last value of A was ##.##";A;
3490    LOCATE 22,18
3500    INPUT; "A(half width at one twentieth full height)=";A
3510 '
3520    IF FIRST% GOTO 3550                           'enter X0
3530       LOCATE 23,24
3540       PRINT USING "Your last value of X0 was +##.##";VX0;
3550    LOCATE 24,35
3560    INPUT; "X0(center)=";VX0
3570 '
3580 FOR I%=19 TO 24                                  'clear bottom of screen
3590    LOCATE I%,1,0:  PRINT SPACE$(79);
3600 NEXT I%
3610 '
3620 IF OPT$="4" THEN LOCATE 19,33:  PRINT USING "Vzero=+###.###";VZERO;
3630 IF OPT$="4" THEN GOTO 3680
3640 '
3650    LOCATE 19,20
3660    PRINT USING "Vzero=+###.###      A=##.##      X0=+##.##";VZERO,A,VX0;
3670 '
3680 LOCATE 21,15                                     'enter XMIDDLE
3690 PRINT "The program will calculate probability and <x> for";
3700 LOCATE 22,23
3710 PRINT "left and right sides of the lattice";
3720 LOCATE 23,13
3730 INPUT; "Enter the value of X which separates left from right";XMIDDLE
3740 IF XMIDDLE<=ABS(XMIN) AND XMIDDLE>=XMIN GOTO 3790
3750    LOCATE 24,19:  BEEP
3760    PRINT USING "Try again, the limits are +## <= X <= ##";XMIN,ABS(XMIN);
3770    LOCATE 23,1:  PRINT SPACE$(79);
3780    GOTO 3720
3790 IMIDDLE%=(XMIDDLE-XMIN)
3800 IMONOMID%=IMIDDLE%/2
3810 '
3820 GOSUB 4000                                       'calculate V(x)
3830 '
```

```
3840 RETURN
3850 '
4000 '******************************************************************
4010 'subroutine to calculate the potential
4020 'input   variables:   A, OPT$, VX0, VZERO
4030 'output variables:   V(I%)
4040 'global variables:   NPTS%, XMIN
4050 'local   variables:   I%
4060 '******************************************************************
4070 IF OPT$<>"1" THEN GOTO 4140              'evaluate the potential
4080    FOR I%=0 TO NPTS%                     'at lattice points
4090       X=XMIN+I%                          'Square Well
4100       IF X>=(VX0-A) AND X<=(VX0+A) THEN V(I%)=VZERO ELSE V(I%)=0
4110    NEXT I%
4120    RETURN
4130 '
4140 IF OPT$<>"2" THEN GOTO 4210             'Gaussian Barrier
4150    FOR I%=0 TO NPTS%
4160       X=XMIN+I%
4170       V(I%)=VZERO*EXP(-(X-VX0)^2*LOG(20)/A^2)
4180    NEXT I%
4190    RETURN
4200 '
4210 IF OPT$<>"3" THEN GOTO 4280             'Potential Step
4220    FOR I%=0 TO NPTS%
4230       X=XMIN+I%
4240       V(I%)=VZERO*((ATN((X-VX0)*6.313/A)*2/3.14159)+1)/2
4250    NEXT I%
4260    RETURN
4270 '
4280 IF OPT$<>"4" THEN RETURN                'Parabolic Well
4290    FOR I%=0 TO NPTS%
4300       X=XMIN+I%
4310       V(I%)=VZERO*(X/XMIN)^2
4320    NEXT I%
4330    RETURN
4340 '
5000 '******************************************************************
5010 'subroutine to alter the analytic potential from the graphics screen
5020 'input   variables:   V(I%), V%(I%), VSCALE, VZERO
5030 'output variables:   V(I%), V%(I%)
5040 'global variables:   MONO%, NPTS%, XMIN
5050 'local   variables:   CURSOR$, I%, IOLD%, K$, V%, VOLD%
5060 '******************************************************************
5070 IF MONO% THEN GOSUB 19000               'switch to graphics screen
5080 GOSUB 6000                              'draw the axes
5090 '
5100 FOR I%=0 TO NPTS%                       'graph the potential
5110    LINE (I%*4,V%(I%))-((I%*4)+2,V%(I%))
5120 NEXT I%
5130 '
5140 LOCATE 24,8: BEEP                       'display instructions
5150 PRINT  CHR$(24)+CHR$(25)+CHR$(26)+CHR$(27)+" Home and PgUp keys move the";
5160 LOCATE 24,40:  PRINT " cursor to alter the potential.";
5170 LOCATE 25,27:  PRINT "Type return when finished.";
5180 '
5190 V%=V%(80):  I%=80                       'begin at the bottom of well
5200 LINE (I%*4+1,V%+1)-(I%*4-1,V%-1)
```

```
5210 LINE (I%*4-1,V%+1)-(I%*4+1,V%-1)
5220 LOCATE 23,27:  PRINT USING "X=+##.###   V(x)=+###.###";(XMIN+DX*I%),V(I%);
5230 '
5240 K$=INKEY$
5250     IF K$=CHR$(13) GOTO 5690              'typing return exits from loop
5260     IF LEN(K$)<2 GOTO 5240               'cursors are special two
5270     CURSOR$=RIGHT$(K$,1)                 'character ASCII codes
5280     VOLD%=V%:  IOLD%=I%
5290     IF (CURSOR$=CHR$(72) OR CURSOR$=CHR$(80)) THEN GOTO 5410
5300     IF (CURSOR$=CHR$(75) OR CURSOR$=CHR$(77)) THEN GOTO 5480
5310 '
5320     IF CURSOR$=CHR$(73) THEN I%=I%+1      'cursor moves along
5330     IF CURSOR$=CHR$(71) THEN I%=I%-1      'the potential curve
5340     IF I%<0 THEN I%=NPTS%                'cursor wraps around x axis
5350     IF I%>NPTS% THEN I%=0
5360     V%=(170-(V(I%)+ABS(VZERO))*VSCALE)   'evaluate new variables
5370     LINE (IOLD%*4,VOLD%)-(IOLD%*4+2,VOLD%) 'redraw line at old cursor pos.
5380     LINE (I%*4,V%)-(I%*4+2,V%),0         'erase line at new cursor pos.
5390     GOTO 5580
5400 '
5410     IF CURSOR$=CHR$(72) THEN V%=V%-1      'cursor up
5420     IF CURSOR$=CHR$(80) THEN V%=V%+1      'cursor down
5430     IF V%>170 THEN V%=170                'keep potential -1<V<1
5440     IF V%<0 THEN V%=0
5450     V(I%)=-ABS(VZERO)-(V%-170)/VSCALE
5460     GOTO 5580
5470 '
5480     IF CURSOR$=CHR$(77) THEN I%=I%+1      'cursor right
5490     IF CURSOR$=CHR$(75) THEN I%=I%-1      'cursor left
5500     IF I%<0 THEN I%=NPTS%                'cursor wraps around x axis
5510     IF I%>NPTS% THEN I%=0
5520     V%=(170-(V(I%)+ABS(VZERO))*VSCALE)
5530     LINE (IOLD%*4,VOLD%)-(IOLD%*4+2,VOLD%)
5540     LINE (I%*4,V%)-(I%*4+2,V%),0
5550     V%=VOLD%                             'V at the new position
5560     V(I%)=V(IOLD%)                       'is the same as V at last pos.
5570 '
5580     LINE (IOLD%*4+1,VOLD%+1)-(IOLD%*4-1,VOLD%-1),0  'erase old cursor
5590     LINE (IOLD%*4-1,VOLD%+1)-(IOLD%*4+1,VOLD%-1),0
5600     LINE (I%*4+1,V%+1)-(I%*4-1,V%-1)           'draw new cursor
5610     LINE (I%*4-1,V%+1)-(I%*4+1,V%-1)
5620     LINE (0,0)-(0,170)                        'redraw axes
5630     LINE (0,85)-(640,85)
5640 '
5650     LOCATE 23,27
5660     PRINT USING "X=+##.###   V(x)=+###.###";(XMIN+DX*I%),V(I%);
5670 GOTO 5240
5680 '
5690 FOR I%=0 TO NPTS%                        'put new values in V% array
5700     V%(I%)=170-(V(I%)+ABS(VZERO))*VSCALE
5710 NEXT I%
5720 '
5730 RETURN
5740 '
6000 '***********************************************************************
6010 'subroutine to draw and label axes for V(x) vs. x
6020 'input variables:  none
6030 'output variables:  none
```

```
6040 'global variables:  XMIN
6050 'local   variables:  COL%, I%, MARK%, ROW%
6060 '******************************************************************
6070 CLS:  SCREEN 2,0,0,0
6080 '
6090 LINE (0,0)-(0,170)                         'draw and label axes
6100 LINE (0,85)-(640,85)
6110 LOCATE 3,2:   PRINT "V(x)";                'V legend
6120 LOCATE 12,78: PRINT "x"                    'X legend
6130 '
6140 FOR I%=0 TO 4
6150    MARK%=I%*(170/4)
6160    LINE (0,MARK%)-(3,MARK%)                'V ticks
6170    ROW%=I%*5.5:  IF I%=0 THEN ROW%=1       'V labels
6180    LOCATE ROW%,2:  PRINT USING "+###.#";(1-I%*.5)*ABS(VZERO);
6190 NEXT I%
6200 '
6210 FOR I%=1 TO 4
6220    MARK%=I%*(640/4)                        'X ticks
6230    LINE (MARK%,83)-(MARK%,87)
6240    COL%=I%*20-2:  IF I%=4 THEN COL%=COL%-2 'X labels
6250    LOCATE 10,COL%:  PRINT USING "+##.#";(XMIN+I%*40);
6260 NEXT I%
6270 '
6280 RETURN
6290 '
7000 '******************************************************************
7010 'subroutine to input wave packet parameters
7020 'input variables:   LPROB, RPROB
7030 'output variables:  K0, NORM, PACKET$, PHII(I%), PHIR(I%), PHISCALE,
7040 '                   PHI2(I%), SIGMA, X0
7050 'global variables:  FIRST%, NPTS%, XMIN
7060 'local  variables:  GAUSS, IX%, LORENTZ, PHI2MAX, SQN
7070 '******************************************************************
7080 CLS
7090 '
7100 LOCATE 1,10                               'input analytic form
7110 PRINT "You have two choices for the analytic form of the wave packet:";
7120 LOCATE 3,12
7130 PRINT "1)Gaussian:     PHI=EXP(i*K0*X)*EXP(-(X-X0)^2*log(2)/SIGMA^2)";
7140 LOCATE 4,12
7150 PRINT "2)Lorentzian:   PHI=EXP(i*K0*X)*SIGMA^2/(SIGMA^2+(X-X0)^2)";
7160 LOCATE 6,18
7170 INPUT; "Enter the number corresponding to your choice";PACKET$
7180 IF PACKET$="1" OR PACKET$="2" GOTO 7240
7190    LOCATE 7,23:   BEEP
7200    PRINT "Try again, the response is 1 or 2";
7210    LOCATE 6,1:   PRINT SPACE$(80);
7220    GOTO 7160
7230 '
7240 LOCATE 9,15                               'display lattice parameters
7250 PRINT "(The lattice parameters are ";
7260 PRINT USING "+## <= X <= ##   and DX=1)";XMIN,ABS(XMIN);
7270 '
7280 IF FIRST% GOTO 7310                       'input K0
7290    LOCATE 11,25
7300    PRINT USING "Your last K0 value was +##.###";K0;
7310 LOCATE 12,15
```

```
7320 PRINT USING "Remember wave packet energy= K0^2 and Vzero=+###.###";VZERO;
7330 LOCATE 13,38
7340 INPUT; "K0=";K0
7350 '
7360 IF FIRST% GOTO 7390                                'input SIGMA
7370   LOCATE 15,24
7380   PRINT USING "Your last SIGMA value was ###.##";SIGMA;
7390 LOCATE 16,23
7400 INPUT; "SIGMA(half width at half height)=";SIGMA
7410 '
7420 IF FIRST% GOTO 7450                                'input SIGMA
7430   LOCATE 18,25
7440   PRINT USING "Your last X0 value was ###.##";X0;
7450 LOCATE 19,35
7460 INPUT; "X0(center)=";X0
7470 '
7480 PHIR(0)=0:      PHII(0)=0                  'Dirichlet boundary conditions
7490 PHIR(NPTS%)=0:  PHIR(NPTS%)=0
7500 '
7510 IF PACKET$="1" THEN GOTO 7620
7520 '
7530   FOR IX%=1 TO NPTS%-1                     'find PHI at time=0
7540     X=XMIN+IX%                             ' for a Lorentzian packet
7550     LORENTZ=SIGMA^2/(SIGMA^2+(X-X0)^2)
7560     PHIR(IX%)=COS(K0*X)*LORENTZ
7570     PHII(IX%)=SIN(K0*X)*LORENTZ
7580     PHI2(IX%)=LORENTZ^2
7590   NEXT IX%
7600   GOTO 7700
7610 '
7620 FOR IX%=1 TO NPTS%-1                       'find PHI at time=0
7630   X=XMIN+IX%                               ' for a Gaussian packet
7640   GAUSS=.9*EXP(-(X-X0)^2*LOG(2)/SIGMA^2)
7650   PHIR(IX%)=COS(K0*X)*GAUSS
7660   PHII(IX%)=SIN(K0*X)*GAUSS
7670   PHI2(IX%)=GAUSS^2
7680 NEXT IX%
7690 '
7700 GOSUB 11000                               'find the normalization
7710 NORM=LPROB+RPROB: SQN=SQR(NORM)
7720 PHI2MAX=0
7730 FOR I%=0 TO NPTS%                         'normalize PHI and find PHI max
7740     PHIR(I%)=PHIR(I%)/SQN
7750     PHII(I%)=PHII(I%)/SQN
7760     PHI2(I%)=PHI2(I%)/NORM
7770     IF PHI2(I%)>PHI2MAX THEN PHI2MAX=PHI2(I%)
7780 NEXT I%
7790 IF GRAPHICS% THEN PHISCALE=58/(2*PHI2MAX) ELSE PHISCALE=19/(2*PHI2MAX)
7800 '
7810 RETURN
7820 '
8000 '**********************************************************************
8010 'subroutine to input DT
8020 'input  variables:  none
8030 'output variables:  DT, XDT, XDT4
8040 'global variables:  FIRST%
8050 'local  variables:  none
8060 '**********************************************************************
```

```
8070 LOCATE 24,1:  PRINT SPACE$(79);                          'clear bottom of screen
8080 LOCATE 25,1:  PRINT SPACE$(79);
8090 IF FIRST% GOTO 8120                                      'input DT
8100    LOCATE 24,25
8110    PRINT USING "Your last time step was ##.###";DT;
8120 LOCATE 25,33:  INPUT; "DT=time step=";DT
8130 '
8140 XDT=1/DT                                                'XDT=dx^2/dt
8150 XDT4=4*XDT                                              'where dx=space step=1
8160 '
8170 RETURN
8180 '
9000 '***********************************************************************
9010 'subroutine to calculate alpha and gamma
9020 'input   variables:  V(I%), XDT
9030 'output variables:  GAMI(I%), GAMR(I%)
9040 'global variables:  NPTS%
9050 'local   variables:  AZI, AZR(I%), DENOM, I%
9060 '***********************************************************************
9070 FOR I%=0 TO NPTS%                                       'evaluate Azero real and imag
9080    AZR(I%)=-2-V(I%)                                     'for the lattice
9090 NEXT I%
9100 AZI=2*XDT
9110 '
9120 DENOM=AZR(NPTS%-1)^2+AZI^2                              'evaluate GAMMA at NPTS%-1 to
9130 GAMR(NPTS%-1)=-AZR(NPTS%-1)/DENOM                       'begin the backward recursion
9140 GAMI(NPTS%-1)=AZI/DENOM
9150 '
9160 FOR I%=NPTS%-1 TO 1 STEP -1                             'backward recursion
9170    DENOM=(GAMR(I%)+AZR(I%-1))^2+(GAMI(I%)+AZI)^2
9180    GAMR(I%-1)=-(GAMR(I%)+AZR(I%-1))/DENOM
9190    GAMI(I%-1)=(GAMI(I%)+AZI)/DENOM
9200 NEXT I%
9210 '
9220 RETURN
9230 '
10000 '***********************************************************************
10010 'subroutine to display commands
10020 'input   variables:  none
10030 'output variables:  FIRST%
10040 'global variables:  none
10050 'local   variables:  none
10060 '***********************************************************************
10070 CLS
10080 '
10090 LOCATE 1,13
10100 PRINT "The program will calculate the time development of the";
10110 LOCATE 3,24
10120 PRINT "wave packet until you type either";
10130 LOCATE 6,14
10140 PRINT "e to end the time loop and proceed to new parameters";
10150 LOCATE 8,10
10160 PRINT "or t to change the time step and then continue the time loop";
10170 '
10180 LOCATE 14,21
10190 PRINT "The scales for graphing are as follows";
10200 LOCATE 17,18
10210 PRINT "PHI^2 scale runs from 0 to 2*PHI^2(X=X0,T=0)";
```

```
10220 LOCATE 19,23
10230 PRINT "V(x) scale runs from Vmin to Vmax";
10240 IF GRAPHICS% THEN GOTO 10290
10250   LOCATE 21,15
10260   PRINT "For clarity, only non-zero values of V are plotted";
10270 '
10280 '
10290 LOCATE 25,31,1
10300 PRINT "Type c to continue";
10310 IF INKEY$<>"c" THEN GOTO 10310
10320 '
10330 LOCATE ,,0                                'turn off cursor
10340 FIRST%=0
10350 '
10360 RETURN
10370 '
11000 '**********************************************************************
11010 'subroutine to find normalization and <x>
11020 'input  variables:   IMIDDLE%, PHI2(I%)
11030 'output variables:   LPROB, LX, RPROB, RX, TPROB, TX
11040 'global variables:   NPTS%, XMIN
11050 'local  variables:   I%
11060 '**********************************************************************
11070 LPROB=0:   LX=0                           'zero sums
11080 FOR I%=1 TO IMIDDLE%-1                    'use trapezoidal rule
11090    LPROB=LPROB+PHI2(I%)                    ' first point was zero
11100    X=XMIN+I%
11110    LX=LX+X*PHI2(I%)
11120 NEXT I%
11130 LPROB=LPROB+.5*PHI2(IMIDDLE%)             'special treatment for last
11140 LX=LX+.5*XMIDDLE*PHI2(IMIDDLE%)           'lattice point
11150 '
11160 RPROB=.5*PHI2(IMIDDLE%)                   'special treatment for
11170 RX=.5*XMIDDLE*PHI2(MIDDLE%)               'first point
11180 FOR I%=IMIDDLE%+1 TO NPTS%-1
11190    RPROB=RPROB+PHI2(I%)
11200    X=XMIN+I%
11210    RX=RX+X*PHI2(I%)
11220 NEXT I%
11230 '
11240 TPROB=LPROB+RPROB                         'find total normalization
11250 IF LPROB=0 THEN LX=-8 ELSE LX=LX/LPROB    'normalize <x>
11260 IF RPROB=0 THEN RX=8 ELSE RX=RX/RPROB
11270 TX=LPROB*LX+RPROB*RX                      'total <x> is a weighted sum
11280 '
11290 RETURN
11300 '
12000 '**********************************************************************
12010 'subroutine to prepare the monochrome screen for output
12020 'input  variables:   POT%, V(I%)
12030 'output variables:   V%(I%)
12040 'global variables:   NPTS%
12050 'local  variables:   I%, VMAX, VMIN, VSCALE
12060 '**********************************************************************
12070 IF NOT POT% THEN RETURN                   'potential hasn't changed
12080 '
12090 VMAX=V(0):   VMIN=V(0)                    'find maximum V(x)
12100 FOR I%=2 TO NPTS% STEP 2
```

```
12110     IF V(I%)>VMAX THEN VMAX=V(I%)
12120     IF V(I%)<VMIN THEN VMIN=V(I%)
12130 NEXT I%
12140 IF (VMAX-VMIN)=0 THEN VSCALE=0 ELSE VSCALE=19/(VMAX-VMIN)
12150 '
12160 FOR I%=0 TO NPTS% STEP 2                'fill graphing array for V(x)
12170     V%(I%)=20-(V(I%)-VMIN)*VSCALE
12180 NEXT I%
12190 '
12200 RETURN
12210 '
13000 '*********************************************************************
13010 'subroutine to prepare the graphics screen for output
13020 'input    variables:  IMIDDLE%, GSAVE%, K0, POT%, VZERO, XMIDDLE
13030 'output variables:   N%, V1%(I%), V2%(I%), V3%(I%)
13040 'global variables:   MONO%, NPTS%
13050 'local    variables:  COL%, I%, ROW%, V%, VMAX, VMIN, VSCALE, X%, XMID%
13060 '*********************************************************************
13070 IF GSAVE% THEN GOTO 13350              'don't redraw graph
13080 '
13090     N%=1                               'start in top box
13100 '
13110     IF MONO% THEN GOSUB 19000          'switch to graphics
13120     CLS:   SCREEN 2,0,0,0
13130 '
13140     LINE (160,0)-(639,58),1,B          'draw top box
13150     FOR I%=0 TO 160                    'draw lines to indicate
13160         X%=160+I%*3                    ' lattice points
13170         LINE (X%,58)-(X%,62)
13180     NEXT I%
13190     LINE (160,62)-(639,120),1,B        'draw middle box
13200     FOR I%=0 TO 160                    'draw lines to indicate
13210         X%=160+I%*3                    ' lattice points
13220         LINE (X%,120)-(X%,124)
13230     NEXT I%
13240     LINE (160,124)-(639,182),1,B       'draw bottom box
13250 '
13260     FOR I%=0 TO 2                      'print titles for 3 boxes
13270         ROW%=1+I%*8
13280         LOCATE ROW%,  1:  PRINT  "time=";
13290         LOCATE ROW%+1,4:  PRINT "probability    <x>";
13300         LOCATE ROW%+2,1:  PRINT "Left";
13310         LOCATE ROW%+3,1:  PRINT "Right";
13320         LOCATE ROW%+4,1:  PRINT "Total";
13330     NEXT I%
13340 '
13350 LOCATE 24,1:  PRINT SPACE$(79);        'clear bottom of screen
13360 LOCATE 25,1:  PRINT SPACE$(79);
13370 FOR I%=0 TO 4                          'x ticks and labels
13380     COL%=20+I%*15:   IF COL%=80 THEN COL%=77
13390     X%=160+I%*120
13400     LINE (X%,180)-(X%,184)
13410     LOCATE 24,COL%:  PRINT USING "+##";(I%*40-80);
13420 NEXT I%
13430 XMID%=160+IMIDDLE%*3                    'XMIDDLE tick
13440 LINE (XMID%,180)-(XMID%,184)
13450 '
13460 LOCATE 25,1
```

```
13470 PRINT USING "E=###.###   Vzero=+###.###";K0^2,VZERO;
13480 LOCATE 25,30:   PRINT USING "Xmiddle=+##.##";XMIDDLE;
13490 LOCATE 25,50:   PRINT "e:end time loop    t:change DT";
13500 IF GSAVE% THEN RETURN
13510 '
13520 IF NOT POT% THEN RETURN
13530 '
13540 VMAX=V(0):   VMIN=V(0)                          'find VMAX and VSCALE
13550 FOR I%=1 TO NPTS%
13560    IF V(I%)>VMAX THEN VMAX=V(I%)
13570    IF V(I%)<VMIN THEN VMIN=V(I%)
13580 NEXT I%
13590 IF (VMAX-VMIN)=0 THEN VSCALE=0 ELSE VSCALE=58/(VMAX-VMIN)
13600 '
13610 FOR I%=0 TO NPTS%                               'fill graphing arrays for
13620    V%=(V(I%)-VMIN)*VSCALE                        ' V(x) in all three boxes
13630    V1%(I%)=58-V%
13640    V2%(I%)=120-V%
13650    V3%(I%)=182-V%
13660 NEXT I%
13670 '
13680 RETURN
13690 '
14000 '**************************************************************************
14010 'subroutine to graph wave function and potential on graphics screen
14020 'input  variables:  LPROB, LX, N%, PHI2(I%), PHISCALE, RPROB, RX, TIME,
14030 '                    TPROB, TX, V1%(I%), V2%(I%), V3%(I%)
14040 'output variables:  N%
14050 'global variables:  NPTS%
14060 'local  variables:  PHI%, PHIOLD%, ROW%, TOP%, VOLD%, X%, XOLD%
14070 '**************************************************************************
14080 TOP%=(N%-1)*62                                  'line number corresponding to
14090 LINE (161,TOP%+1)-(638,TOP%+57),0,BF            ' top of current box;clear box
14100 '
14110 ROW%=N%*8-7                                     'print current values
14120 LOCATE ROW%,6:   PRINT USING "###.##";TIME;
14130 LOCATE ROW%+2,7:   PRINT USING "#.###   +##.###";LPROB,LX
14140 LOCATE ROW%+3,7:   PRINT USING "#.###   +##.###";RPROB,RX
14150 LOCATE ROW%+4,7:   PRINT USING "#.###   +##.###";TPROB,TX
14160 '
14170 XOLD%=160                                       'graph PHI
14180 PHIOLD%=TOP%+58
14190 FOR I%=0 TO NPTS%
14200    PHI%=(TOP%+58)-PHI2(I%)*PHISCALE
14210    IF PHI%<TOP% THEN PHI%=TOP%                   'cutoff PHI if it's too large
14220    X%=160+I%*3
14230    LINE (X%,PHI%)-(XOLD%,PHIOLD%)
14240    XOLD%=X%
14250    PHIOLD%=PHI%
14260 NEXT I%
14270 '
14280 ON N% GOTO 14300,14360,14420                    'V% is different for each box
14290 '
14300 VOLD%=V1%(0)                                    'top box
14310 FOR I%=1 TO NPTS%
14320    LINE ((160+I%*3),V1%(I%))-((157+I%*3),VOLD%)
14330    VOLD%=V1%(I%)
14340 NEXT I%:   GOTO 14480
```

```
14350 '
14360 VOLD%=V2%(0)                                        'middle box
14370 FOR I%=1 TO NPTS%
14380     LINE ((160+I%*3),V2%(I%))-((157+I%*3),VOLD%)
14390     VOLD%=V2%(I%)
14400 NEXT I%:  GOTO 14480
14410 '
14420 VOLD%=V3%(0)                                        'bottom box
14430 FOR I%=1 TO NPTS%
14440     LINE ((160+I%*3),V3%(I%))-((157+I%*3),VOLD%)
14450     VOLD%=V3%(I%)
14460 NEXT I%:  GOTO 14480
14470 '
14480 N%=(N% MOD 3)+1                                     'which box is next?
14490 '
14500 RETURN
14510 '
15000 '*********************************************************************
15010 'subroutine for output to a text-only screen
15020 'input  variables:  IMONOMID%, K0, LPROB, LX, PHI2(I%), PHISCALE, RPROB,
15030 '                   RX, TIME, TPROB, TX, V%(I%), V(I%), VZERO, XMIDDLE
15040 'output variables:  none
15050 'global variables:  NPTS%
15060 'local  variables:  COL%, I%, ROW%
15070 '*********************************************************************
15080 CLS
15090 '
15100 LOCATE 20,1,0:  PRINT STRING$(80,196);      'draw axis
15110 FOR I%=0 TO 4
15120     COL%=I%*20:  IF COL%=0 THEN COL%=1
15130     LOCATE 20,COL%:  PRINT CHR$(197)         'X ticks
15140     IF COL%=80 THEN COL%=78
15150     LOCATE 21,COL%:  PRINT USING "+##";(I%*40-80);   'X labels
15160 NEXT I%
15170 LOCATE 20,IMONOMID%:  PRINT CHR$(197)        'XMIDDLE tick
15180 '
15190 LOCATE 22,1:   PRINT USING "time=###.##";TIME;       'print current value
15200 LOCATE 22,26:  PRINT USING "xmiddle=+##.##";XMIDDLE;
15210 LOCATE 22,53:  PRINT "x:wave packet   o:potential";
15220 '
15230 LOCATE 23,27:  PRINT "Left";
15240 LOCATE 23,47:  PRINT "Right";
15250 LOCATE 23,67:  PRINT "Total";
15260 LOCATE 24,1:   PRINT "probability/<x>";
15270 LOCATE 24,23:  PRINT USING "#.###/+##.###";LPROB,LX;
15280 LOCATE 24,43:  PRINT USING "#.###/+##.###";RPROB,RX;
15290 LOCATE 24,63:  PRINT USING "#.###/+##.###";TPROB,TX;
15300 LOCATE 25,1
15310 PRINT USING "E=###.###   Vzero=+###.###";K0^2,VZERO;
15320 LOCATE 25,51,0:  PRINT "e:end time loop   t:change dt";
15330 '
15340 FOR I%=2 TO NPTS% STEP 2                    'graph non-zero potential
15350     LOCATE V%(I%),I%/2:  IF V(I%)<>0 THEN PRINT "o";
15360 NEXT I%
15370 '
15380 FOR I%=2 TO NPTS% STEP 2                    'graph PHI
15390     ROW%=20-PHI2(I%)*PHISCALE
15400     IF ROW%<1 THEN ROW%=1
```

```
15410     LOCATE ROW%,I%/2:  PRINT "x";
15420 NEXT I%
15430 '
15440 RETURN
15450 '
16000 '*********************************************************************
16010 'subroutine to calculate BETA, CHI, PHI and PHI^2
16020 'input  variables:  GAMI(I%), GAMR(I%), PHII(I%), PHIR(I%), XDT4
16030 'output variables:  PHI2(I%)
16040 'global variables:  NPTS%
16050 'local  variables:  BETI(I%), BETR(I%), CHII, CHIR, IX%, TEMP
16060 '*********************************************************************
16070 BETR(NPTS%-1)=0                              'find initial BETA values for
16080 BETI(NPTS%-1)=0                              'backward recursion
16090 FOR IX%=(NPTS%-1) TO 1 STEP -1
16100     BETR(IX%-1)=GAMR(IX%)*(BETR(IX%)+XDT4*PHII(IX%))
16110     BETR(IX%-1)=BETR(IX%-1)-GAMI(IX%)*(BETI(IX%)-XDT4*PHIR(IX%))
16120     BETI(IX%-1)=GAMR(IX%)*(BETI(IX%)-XDT4*PHIR(IX%))
16130     BETI(IX%-1)=BETI(IX%-1)+GAMI(IX%)*(BETR(IX%)+XDT4*PHII(IX%))
16140 NEXT IX%
16150 '
16160 CHIR=0: CHII=0                               'give initial Chi for
16170 FOR IX%=1 TO NPTS%-1                         'forward recursion
16180     TEMP=CHIR
16190     CHIR=GAMR(IX%)*CHIR-GAMI(IX%)*CHII+BETR(IX%-1)
16200     CHII=GAMR(IX%)*CHII+GAMI(IX%)*TEMP+BETI(IX%-1)
16210     PHIR(IX%)=CHIR-PHIR(IX%)
16220     PHII(IX%)=CHII-PHII(IX%)
16230     PHI2(IX%)=PHIR(IX%)^2+PHII(IX%)^2
16240 NEXT IX%
16250 '
16260 RETURN
16270 '
17000 '*********************************************************************
17010 'subroutine to display commands
17020 'input  variables:  K0, NORM, PACKET$, SIGMA, X0
17030 'output variables:  PACKET%, PHII(I%), PHIR(I%), PHI2(I%), POT%, TIME%,
17040 '                   VSAVE%
17050 'global variables:  GRAPHICS%, NPTS%, XMIN
17060 'local  variables:  A$, ANS$, B$, C$, D$, GAUSS, IX%, LORENTZ
17070 '*********************************************************************
17080 CLS
17090 '
17100 LOCATE 3,22                                  'end command
17110 INPUT; "Do you wish to end the program (y/n)";A$
17120 IF A$="y" THEN END
17130 '
17140 LOCATE 8,11                                  'change wave packet command
17150 PRINT "Do you wish to change the wave packet (y/n) (if you respond";
17160 LOCATE 9,18
17170 INPUT; "'n', PHI(x) will return to its initial value)";B$
17180 IF B$="y" OR B$="n" THEN GOTO 17230
17190     LOCATE 10,36:  BEEP
17200     PRINT "Try again";
17210     LOCATE 9,1:    PRINT SPACE$(80);
17220     GOTO 17160
17230 IF B$="y" THEN PACKET%=-1
17240 '
```

```
17250 LOCATE 13,25                                    'change time command
17260 INPUT; "Do you wish to change DT (y/n)";C$
17270 IF C$="y" OR C$="n" THEN GOTO 17320
17280    LOCATE 14,36:   BEEP
17290    PRINT "Try again";
17300    LOCATE 13,1:    PRINT SPACE$(80);
17310    GOTO 17250
17320 IF C$="y" THEN TIME%=-1
17330 '
17340 LOCATE 18,20                                    'change potential command
17350 INPUT; "Do you wish to change the potential (y/n)";D$
17360 IF D$="y" OR D$="n" THEN GOTO 17410
17370    LOCATE 19,36:   BEEP
17380    PRINT "Try again";
17390    LOCATE 18,1:  PRINT SPACE$(80);
17400    GOTO 17340
17410 IF D$="y" THEN POT%=-1
17420 '
17430 IF NOT POT% OR NOT GRAPHICS% GOTO 17540      'more options for V
17440    LOCATE 21,18
17450    PRINT "Do you wish to 1)begin with the analytic form";
17460    LOCATE 22,19
17470    PRINT "or 2)with the current form of the potential";
17480    LOCATE 23,10
17490    PRINT "(You will be able to alter Vzero while preserving the shape";
17500    LOCATE 24,23
17510    INPUT; "of V(x) if you choose option 2)";ANS$
17520    IF ANS$="2" THEN VSAVE%=-1 ELSE VSAVE%=0
17530 '
17540 IF PACKET% THEN RETURN                          'return PHI to original value
17550 IF PACKET$="1" THEN GOTO 17660                  ' unless  it is to be changed
17560 '
17570    FOR IX%=1 TO NPTS%-1                          'find PHI at time=0
17580       X=XMIN+IX%
17590       LORENTZ=SIGMA^2/(SIGMA^2+(X-X0)^2)/SQR(NORM)
17600       PHIR(IX%)=COS(K0*X)*LORENTZ
17610       PHII(IX%)=SIN(K0*X)*LORENTZ
17620       PHI2(IX%)=LORENTZ^2
17630    NEXT IX%
17640    RETURN
17650 '
17660 FOR IX%=1 TO NPTS%-1                            'find PHI at time=0
17670    X=XMIN+IX%
17680    GAUSS=.9*EXP(-(X-X0)^2*LOG(2)/SIGMA^2)/SQR(NORM)
17690    PHIR(IX%)=COS(K0*X)*GAUSS
17700    PHII(IX%)=SIN(K0*X)*GAUSS
17710    PHI2(IX%)=GAUSS^2
17720 NEXT IX%
17730 RETURN
17740 '
18000 '******************************************************************
18010 'subroutine to display the header screen
18020 'input  variables: none
18030 'output variables: GRAPHICS%,MONO%
18040 'global variables: none
18050 'local  variables: G$,M$,ROW%
18060 '******************************************************************
18070 SCREEN 0: CLS: KEY OFF                          'program begins in text mode
```

```
18071 '
18072 LOCATE 1,30: COLOR 15                          'display book title
18073 PRINT "COMPUTATIONAL PHYSICS";
18074 LOCATE 2,40: COLOR 7
18075 PRINT "by";
18076 LOCATE 3,33: PRINT "Steven E. Koonin";
18077 LOCATE 4,14
18078 PRINT "Copyright 1985, Benjamin/Cummings Publishing Company";
18080 '
18090 LOCATE 5,10                                    'draw the box
18100 PRINT CHR$(201)+STRING$(59,205)+CHR$(187);
18110 FOR ROW%=6 TO 19
18120     LOCATE ROW%,10: PRINT CHR$(186);
18130     LOCATE ROW%,70: PRINT CHR$(186);
18140 NEXT ROW%
18150 LOCATE 20,10
18160 PRINT CHR$(200)+STRING$(59,205)+CHR$(188);
18170 '
18180 COLOR 15                                       'print title, etc.
18190 LOCATE  7,36:  PRINT "EXAMPLE 7";
18200 COLOR 7
18210 LOCATE 10,22:  PRINT "Time dependent Schroedinger equation";
18220 LOCATE 13,36:  PRINT "**********"
18230 LOCATE 15,25:  PRINT "Type 'e' to end while running."
18240 LOCATE 16,22:  PRINT "Type ctrl-break to stop at a prompt."
18250 '
18260 LOCATE 19,16: BEEP                             'get screen configuration
18270 INPUT "Does your computer have graphics capability (y/n)";G$
18280 IF LEFT$(G$,1)="y" OR LEFT$(G$,1)="n" GOTO 18310
18290     LOCATE 19,12:  PRINT SPACE$(58): BEEP
18300     LOCATE 18,35:  PRINT "Try again,":  GOTO 18260
18310 IF LEFT$(G$,1)="y" GOTO 18340
18320     GRAPHICS%=0:  MONO%=-1
18330     GOTO 18430
18340 GRAPHICS%=-1
18350 LOCATE 18,15:  PRINT SPACE$(55)
18360 LOCATE 19,15:  PRINT SPACE$(55)
18370 LOCATE 19,15: BEEP
18380 INPUT "Do you also have a separate screen for text (y/n)";M$
18390 IF LEFT$(M$,1)="y" OR LEFT$(M$,1)="n" GOTO 18410  'error trapping
18400     LOCATE 18,35:  PRINT "Try again,":  GOTO 18360
18410 IF LEFT$(M$,1)="y" THEN MONO%=-1 ELSE MONO%=0
18420 '
18430 LOCATE 21,9
18440 PRINT "This program calculates the time evolution of a one-dimensional";
18450 LOCATE 22,8
18460 PRINT"wave packet impinging on a potential barrier (or well).  The wave";
18470 LOCATE 23,10
18480 PRINT "is confined within a box with edges at -/+80.  The lattice has";
18490 LOCATE 24,24
18500 PRINT "160 points; units are hbar=2π=1.";
18510 LOCATE 25,31,1:  BEEP
18520 PRINT "Type c to continue";
18530 IF INKEY$<>"c" GOTO 18530
18540 LOCATE ,,0
18550 '
18560 IF MONO% AND GRAPHICS% THEN GOSUB 20000        'switch to text screen
18570 '
```

```
18580 RETURN
18590 '
19000 '*********************************************************************
19010 'subroutine to switch from mono to graphics screen
19020 'input   variables: none
19030 'output  variables: none
19040 'global  variables: none
19050 'local   variables: none
19060 '*********************************************************************
19070 DEF SEG=0
19080 POKE &H410,  (PEEK(&H410) AND &HCF) OR &H10
19090 SCREEN 0:  WIDTH 80: LOCATE ,,0
19100 RETURN
19110 '
20000 '*********************************************************************
20010 'subroutine to switch from graphics to mono screen
20020 'input   variables: none
20030 'output  variables: none
20040 'global  variables: none
20050 'local   variables: none
20060 '*********************************************************************
20070 DEF SEG=0
20080 POKE &H410,  (PEEK(&H410) OR &H30)
20090 SCREEN 0:  WIDTH 80:  LOCATE ,,0
20100 RETURN
20110 '
```

B.8 Example 8

This program simulates the two-dimensional Ising model using the algorithm of Metropolis *et al.*. Given the size of the lattice and an initial random configuration of spins (subroutine 1000), the coupling and magnetic field are input (lines 2090-2160), as are the number of thermalization sweeps, the number and size of the groups of observables, and the sampling frequency (lines 2230-2380). After the thermalization sweeps are performed (loop 220-300), data taking for the groups begins (loop 420-640). The two basic subroutines are 3000, which does a Metropolis sweep of the lattice, and 5000, which calculates the energy and magnetization for the configuration of the lattice every FREQ% sweeps. After each sweep (either thermalization or data taking), the lattice is displayed by subroutine 4000 using the ASCII block characters 219, 220, and 223 to display two rows of spins at a time, black indicating $S_\alpha=-1$ and white indicating $S_\alpha=+1$. During the data taking, both group and grand sums for the energy and magnetization are accumulated (lines 580-610 and subroutine 6000) so that the effects of sweep-to-sweep correlations can be monitored, as discussed immediately above Exercise 8.6. The estimated energy, magnetization, susceptibility, and heat capacity (all per spin) are displayed below the lattice as the sweeps proceed.

A representative run of this program can be had by choosing a lattice with vertical and horizontal dimensions of 20, coupling and magnetic field values of

$$J=0.3, \ B=0,$$

20 thermalization sweeps, and 10 groups of 5 samples each with a sampling frequency of 5.

```
10  '*********************************************************************
20  'Example 8: Monte Carlo simulation of the two-dimensional Ising model
21  'COMPUTATIONAL PHYSICS by Steven E. Koonin
22  'Copyright 1985, Benjamin/Cummings Publishing Company
30  '*********************************************************************
40  GOSUB 7000                              'display the header
50  GOSUB 8000                              'is output poke or print
60  '
70  DIM S%(44,79)                           'S%(I%,J%)=+-1 contains the spins
80  DIM R(5,2)                              'R(3+F%/2,(S%+3)/2) is flip prob.
90  '                                       ' when the spin is S% and the sum
100 '                                       ' of the neighboring spins is F%
110 '
120 GOSUB 1000                              'initialize the lattice
130 GOSUB 2000                              'get parameters for this run
140 '
150 TOP%=12-NY%/4                           'top line of display
160 BOT%=13+NY%/4                           'bottom line of display
170 '
180 LOCATE TOP%,1                           'print header information
```

```
190 PRINT USING "Nx=## Ny=## J=+##.#### B=+##.####";NX%,NY%,JJ,B;
200 '
210 IF NTHERM%<=0 THEN GOTO 320                'do thermalization sweeps
220   FOR SWEEP%=1 TO NTHERM%
230       IF INKEY$="e" THEN GOTO 660          'check for interrupt
240       LOCATE TOP%,35                       'print header information
250       PRINT USING "Thermalization sweep ### of ###";SWEEP%,NTHERM%;
260       LOCATE TOP%,73
270       PRINT USING "Acc=.###";ACCEPT%/NSPINS;
280       GOSUB 3000                           'do a sweep
290       GOSUB 4000                           'display the lattice
300   NEXT SWEEP%
310 '
320 MORE%=NGROUP%                              'prepare to start a run
330 SUMM=0: SUMM2=0: SUMSIGM=0                 'zero total sums
340 SUME=0: SUME2=0: SUMSIGE=0
350 SUMCHI=0: SUMCHI2=0
360 SUMCB=0:  SUMCB2=0
370 '
380 CLS                                        'display header information
390 LOCATE TOP%,1
400 PRINT USING "Nx=## Ny=## J=+##.#### B=+##.####";NX%,NY%,JJ,B;
410 '
420 FOR IGROUP%=NGROUP%-MORE%+1 TO NGROUP%     'loop over groups
430    GROUPM=0: GROUPM2=0                      'zero group sums
440    GROUPE=0: GROUPE2=0
450    FOR SWEEP%=1 TO FREQ%*SIZE%             'loop over sweeps in a group
460       IF INKEY$="e" THEN GOTO 660          'check for interrupts
470       GOSUB 3000                           'do a sweep of the lattice
480       LOCATE TOP%,35                       'print header information
490       PRINT USING "Sample ### of ###";SWEEP%/FREQ%,SIZE%;
500       LOCATE TOP%,52
510       PRINT USING " in group ### of ###";IGROUP%,NGROUP%;
520       LOCATE TOP%,73
530       PRINT USING "acc=.###";ACCEPT%/NSPINS;
540       GOSUB 4000                           'display the lattice
550 '                                          'sometimes compute the observables
560       IF SWEEP% MOD FREQ%<>0 THEN GOTO 620
570           GOSUB 5000                       'compute E and M for this lattice
580           GROUPM=GROUPM+MAG%               'update group sums
590           GROUPM2=GROUPM2+MAG%~2
600           GROUPE=GROUPE+ENER
610           GROUPE2=GROUPE2+ENER~2
620    NEXT SWEEP%
630    GOSUB 6000                              'display group and total observables
640 NEXT IGROUP%
650 '
660 BEEP                                       'prompt and wait for response
670 LOCATE TOP%,66
680 COLOR 15: PRINT "Type c         ";: COLOR 7
690 KB$=INKEY$: IF KB$="" THEN GOTO 690
700 '
710 CLS: LOCATE 10,20: BEEP
720 INPUT "How many more groups to run (<=0 to stop)";MORE%
730 IF MORE%>0 THEN GOTO 770                   'start a new run
740    LOCATE 12,20: BEEP
750    INPUT "Do you wish to change the lattice size (y/n)";KB$
760    IF KB$="y" THEN GOTO 120 ELSE GOTO 130
```

```
770 NGROUP%=NGROUP%+MORE%: GOTO 380
780 '
1000 '*******************************************************************
1010 'subroutine to initialize the lattice
1020 'input  variables: none
1030 'output variables: B,JJ,NSPINS,NX%,NY%
1040 'global variables: S%
1050 'local  variables: I%,J%
1060 '*******************************************************************
1070 CLS
1080 '
1090 LOCATE 10,20: BEEP
1100 INPUT "Enter vertical dimension of lattice (<=44,even)";NY%
1110 IF NY%>0 AND NY%<=44 AND NY% MOD 2=0 THEN GOTO 1130 ELSE GOTO 1090
1120 '
1130 LOCATE 13,20: BEEP
1140 INPUT "Enter horizontal dimension of lattice (<=79)";NX%
1150 IF NX%>0 AND NX%<=79 THEN GOTO 1170 ELSE GOTO 1130
1160 '
1170 NSPINS=CSNG(NX%*NY%)                     'total number of spins
1180 '
1190 FOR I%=1 TO NY%                          'initial state is half up, half down
1200    FOR J%=1 TO NX%
1210       IF RND<.5 THEN S%(I%,J%)=1 ELSE S%(I%,J%)=-1
1220    NEXT J%
1230 NEXT I%
1240 '
1250 JJ=0: B=0                                'initial couplings are zero
1260 '
1270 LOCATE 16,20: BEEP
1280 RANDOMIZE                                'seed the random number generator
1290 RETURN
1300 '
2000 '*******************************************************************
2010 'subroutine to input the parameters for a run
2020 'input  variables: none
2030 'output variables: FREQ%,NGROUP%,NTHERM%,SIZE%
2040 'global variables: B,JJ,R
2050 'local  variables: I%
2060 '*******************************************************************
2070 CLS
2080 '                                        'get new couplings
2090 LOCATE 12,20
2100 PRINT USING "The current value of J is ##.#####";JJ
2110 LOCATE 13,20: BEEP
2120 INPUT "Enter the  new value of J";JJ
2130 LOCATE 14,20
2140 PRINT USING "The current value of B is ##.#####";B
2150 LOCATE 15,20: BEEP
2160 INPUT "Enter the  new value of B";B
2170 '
2180 FOR I%=1 TO 5                            'calculate the flip probs.
2190    R(I%,2)=EXP(-2*(JJ*(2*I%-6)+B))
2200    R(I%,1)=1/R(I%,2)
2210 NEXT I%
2220 '
2230 LOCATE 16,20: BEEP                       'get number of thermal sweeps
2240 INPUT "Enter number of thermalization sweeps";NTHERM%
```

```
2250 IF NTHERM%<0 THEN NTHERM%=0
2260 '
2270 LOCATE 17,20: BEEP                              'get group number and size
2280 INPUT "Enter number of groups to run";NGROUP%
2290 LOCATE 18,20
2300 PRINT USING "The current number of samples in each group is ###";SIZE%
2310 LOCATE 19,22: BEEP
2320 INPUT "Enter new number of samples in each group";SIZE%
2330 '
2340 LOCATE 20,20                                   'get sampling frequency
2350 PRINT USING "The current sampling frequency is ###";FREQ%
2360 LOCATE 21,22: BEEP
2370 INPUT "Enter new sampling frequency";FREQ%
2380 IF FREQ%>0 THEN GOTO 2400 ELSE GOTO 2340
2390 '
2400 CLS
2410 RETURN
2420 '
3000 '*********************************************************************
3010 'subroutine to do a sweep of the lattice
3020 'input    variables: none
3030 'output variables: ACCEPT%
3040 'global variables: NX%,NY%,S%,R
3050 'local    variables: F%,I%,IM%,IP%,J%,JM%,JP%,SPIN%
3060 '*********************************************************************
3070 ACCEPT%=0                                      'zero acceptance count
3080 '
3090 FOR I%=1 TO NY%                                'vertical loop
3100     IF I%<NY% THEN IP%=I%+1 ELSE IP%=1         'I% index of upper neighbor
3110     IF I%>1    THEN IM%=I%-1 ELSE IM%=NY%      'I% index of lower neighbor
3120     FOR J%=1 TO NX%                            'horizontal loop
3130         IF J%<NX% THEN JP%=J%+1 ELSE JP%=1     'J% index of right neighbor
3140         IF J%>1    THEN JM%=J%-1 ELSE JM%=NX%  'J% index of left neighbor
3150         SPIN%=S%(I%,J%)                        'spin at this site
3160 '                                              'sum of four neighbor spins
3170         F%=S%(IP%,J%)+S%(IM%,J%)+S%(I%,JP%)+S%(I%,JM%)
3180 '                                              'decide if spin is to flip
3190         IF RND>R(3+F%/2,(3+SPIN%)/2) THEN GOTO 3220
3200             S%(I%,J%)=-SPIN%                   'flip the spin
3210             ACCEPT%=ACCEPT%+1                  'increment acceptance count
3220     NEXT J%
3230 NEXT I%
3240 RETURN
3250 '
4000 '*********************************************************************
4010 'subroutine to display the lattice
4020 'input    variables: S%
4030 'output variables: none
4040 'global variables: NX%,NY%,POKING%,TOP%
4050 'local    variables: I%,J%,LEDGE%,PLACE%,SY%
4060 '*********************************************************************
4070 LEDGE%=160*TOP%+2*CINT(40-NX%/2)    'buffer location of left edge of
4080 '                                   ' current two rows
4090 '
4100 FOR I%=2 TO NY% STEP 2              'loop over pairs of rows
4110     PLACE%=LEDGE%                   'current place in buffer
4120     FOR J%=1 TO NX%                 'loop over columns
4130 '                                   'find ASCII symbol for these two sites
```

```
4140        IF S%(I%,J%)=-1 THEN GOTO 4170
4150         IF S%(I%-1,J%)=1 THEN SY%=219 ELSE SY%=223
4160         GOTO 4190
4170        IF S%(I%-1,J%)=1 THEN SY%=220 ELSE SY%=0
4180    '
4190        IF NOT POKING% THEN GOTO 4230
4200          POKE PLACE%,SY%                    'poke the symbol into the buffer
4210          PLACE%=PLACE%+2                     'increment buffer location
4220          GOTO 4250
4230        LOCATE TOP%+I%/2,39-NX%/2+J% 'use LOCATE/PRINT method
4240        PRINT CHR$(SY%);
4250      NEXT J%
4260      LEDGE%=LEDGE%+160                       'left-edge location of next rows
4270 NEXT I%
4280 RETURN
4290 '
5000 '*****************************************************************
5010 'subroutine to calculate the observables (energy and magnetization)
5020 'input   variables: S%
5030 'output variables: ENER,MAG%
5040 'global variables: B,JJ,NX%,NY%
5050 'local   variables: I%,IM%,J%,JM%,SUMSS%
5060 '*****************************************************************
5070 MAG%=0                               'zero magnetization for this lattice
5080 SUMSS%=0                             'zero sum of interactions with upper
5090 '                                    ' and left neighbors
5100 FOR I%=1 TO NY%                                      'vertical loop
5110    IF I%>1 THEN IM%=I%-1 ELSE IM%=NY%               'I% for upper nghbr
5120    FOR J%=1 TO NX%                                   'horizontal loop
5130       IF J%>1 THEN JM%=J%-1 ELSE JM%=NX%            'J% for left  nghbr
5140       SUMSS%=SUMSS%+S%(I%,J%)*(S%(IM%,J%)+S%(I%,JM%)) 'interaction sum
5150       MAG%=MAG%+S%(I%,J%)                            'magnetization sum
5160    NEXT J%
5170 NEXT I%
5180 '
5190 ENER=(-JJ*SUMSS%-B*MAG%)                             'total energy
5200 RETURN
5210 '
6000 '*****************************************************************
6010 'subroutine to compute and display group and total quantities
6020 'input   variables: none
6030 'output variables: none
6040 'global variables: BOT%,GROUPE,GROUPE2,GROUPM,GROUPM2,IGROUP%,NSPINS,
6050 '                   SIZE%,SUMCB,SUMCB2,SUMCHI,SUMCHI2,SUME,SUME2,SUMSIGE,
6060 '                   SUMM,SUMM,SUMSIGM
6070 'local   variables: CB,CHI,E,M,SIGCB,SIGCHI,SIGE,SIGE1,SIGM,SIGM1
6080 '*****************************************************************
6090 GROUPM=GROUPM/SIZE%:   GROUPM2=GROUPM2/SIZE%         'find group averages
6100 GROUPE=GROUPE/SIZE%:   GROUPE2=GROUPE2/SIZE%
6110 CHI=GROUPM2-GROUPM^2:  CB=GROUPE2-GROUPE^2
6120 SIGM=SQR(CHI/SIZE%):   SIGE=SQR(CB/SIZE%)            'group uncertainties
6130 '
6140 LOCATE BOT%,1                                        'display group values
6150 PRINT USING "Group ###";IGROUP%;
6160 LOCATE BOT%,11
6170 PRINT USING "E=+#.###+-#.###";GROUPE/NSPINS,SIGE/NSPINS;
6180 LOCATE BOT%,27
6190 PRINT USING "M=+#.###+-#.###";GROUPM/NSPINS,SIGM/NSPINS;
```

```
6200 LOCATE BOT%,43
6210 PRINT USING "Chi=##.### Cb=##.###";CHI/NSPINS,CB/NSPINS;
6220 '                                        'update total sums
6230 SUMM=SUMM+GROUPM: SUMM2=SUMM2+GROUPM^2: SUMSIGM=SUMSIGM+SIGM^2
6240 SUME=SUME+GROUPE: SUME2=SUME2+GROUPE^2: SUMSIGE=SUMSIGE+SIGE^2
6250 SUMCHI=SUMCHI+CHI: SUMCHI2=SUMCHI2+CHI^2
6260 SUMCB=SUMCB+CB:     SUMCB2=SUMCB2+CB^2
6270 '                                        'display total values
6280 LOCATE 25,1
6290 PRINT "Tot";
6300 E=SUME/IGROUP%: SIGE=SQR((SUME2/IGROUP%-E^2)/IGROUP%)
6310 SIGE1=SQR(SUMSIGE)/IGROUP%
6320 LOCATE 25,5
6330 PRINT USING "E=+#.###+-.###/.###";E/NSPINS,SIGE/NSPINS,SIGE1/NSPINS;
6340 M=SUMM/IGROUP%: SIGM=SQR((SUMM2/IGROUP%-M^2)/IGROUP%)
6350 SIGM1=SQR(SUMSIGM)/IGROUP%
6360 LOCATE 25,25
6370 PRINT USING "M=+#.###+-.###/.###";M/NSPINS,SIGM/NSPINS,SIGM1/NSPINS;
6380 CHI=SUMCHI/IGROUP%: SIGCHI=SQR((SUMCHI2/IGROUP%-CHI^2)/IGROUP%)
6390 LOCATE 25,45
6400 PRINT USING "chi=##.###+-#.###";CHI/NSPINS,SIGCHI/NSPINS;
6410 CB=SUMCB/IGROUP%: SIGCB=SQR((SUMCB2/IGROUP%-CB^2)/IGROUP%)
6420 LOCATE 25,63
6430 PRINT USING "Cb=##.###+-##.###";CB/NSPINS,SIGCB/NSPINS;
6440 '
6450 RETURN
6460 '
7000 '****************************************************************
7010 'subroutine to display the header screen
7020 'input  variables: none
7030 'output variables: none
7040 'global variables: none
7050 'local  variables: KB$
7060 '****************************************************************
7070 SCREEN 0: CLS: KEY OFF                   'program begins in text mode
7080 '
7090 LOCATE 1,30: COLOR 15                    'display book title
7100 PRINT "COMPUTATIONAL PHYSICS";
7110 LOCATE 2,40: COLOR 7
7120 PRINT "by";
7130 LOCATE 3,33: PRINT "Steven E. Koonin";
7140 LOCATE 4,14
7150 PRINT "Copyright 1985, Benjamin/Cummings Publishing Company";
7160 '
7170 LOCATE 5,10                              'draw the box
7180 PRINT CHR$(201)+STRING$(59,205)+CHR$(187);
7190 FOR ROW%=6 TO 19
7200     LOCATE ROW%,10: PRINT CHR$(186);
7210     LOCATE ROW%,70: PRINT CHR$(186);
7220 NEXT ROW%
7230 LOCATE 20,10
7240 PRINT CHR$(200)+STRING$(59,205)+CHR$(188);
7250 '
7260 COLOR 15
7270 LOCATE  7,36:  PRINT "EXAMPLE 8";
7280 COLOR 7
7290 LOCATE  9,19:  PRINT "Monte Carlo simulation of the 2-D Ising model"
7300 LOCATE 10,27:  PRINT "using the Metropolis algorithm"
```

```
7310 LOCATE 13,36:  PRINT "**********"
7320 LOCATE 15,25:  PRINT "Type 'e' to end while running."
7330 LOCATE 16,22:  PRINT "Type ctrl-break to stop at a prompt."
7340 '
7350 BEEP: LOCATE 19,28
7360 PRINT "Press any key to continue";
7370 KB$=INKEY$: IF KB$="" THEN GOTO 7370
7380 RETURN
7390 '
8000 '***********************************************************************
8010 'subroutine to print cautionary message about POKEing
8020 'input   variables: none
8030 'output variables: POKING%
8040 'global variables: none
8050 'local   variables: ANSWER$,OPT$
8060 '***********************************************************************
8070 CLS
8080 '
8090 LOCATE 1,34,0
8100 PRINT "C A U T I O N"
8110 LOCATE 3,9
8120 PRINT "For efficient screen display this program uses the POKE command";
8130 LOCATE 5,9
8140 PRINT "which places data directly into memory.  The location given in";
8150 LOCATE 7,8
8160 PRINT "POKE (syntax:  POKE location,data) is an offset from the current";
8170 LOCATE 9,8
8180 PRINT "segment as given by the DEF SEG statement.  This program sets the";
8190 LOCATE 11,10
8200 PRINT "segment to the beginning of either the graphics or monochrome";
8210 LOCATE 13,8
8220 PRINT "screen buffer for the IBM PC.  If your memory map isn't the same";
8230 LOCATE 15,8
8240 PRINT "as an IBM, POKE may write over important data in RAM.  Check your";
8250 LOCATE 17,9
8260 PRINT "BASIC manual (DEF SEG statement) for the segment of the screen";
8270 LOCATE 19,11
8280 PRINT "buffer and compare with line 8520 in this program to ensure";
8290 LOCATE 21,33
8300 PRINT "smooth running.";
8310 '
8320 LOCATE 24,1                                  'input method of screen display
8330 PRINT "Do you wish to use 1)the POKE method, or 2)the slower and sure";
8340 INPUT; " LOCATE method";OPT$
8350 IF OPT$="1" OR OPT$="2" THEN GOTO 8400
8360    LOCATE 25,24:  BEEP
8370    PRINT "Try again, the response is 1 or 2";
8380    LOCATE 24,1:  PRINT SPACE$(79);
8390    GOTO 8320
8400 IF OPT$="1" THEN POKING%=-1 ELSE POKING%=0
8410 IF NOT POKING% THEN RETURN
8420 '
8430 LOCATE 24,1:  PRINT SPACE$(79);              'input monitor type
8440 LOCATE 25,1:  PRINT SPACE$(79);
8450 LOCATE 24,13
8460 INPUT; "Are you currently on a 1)graphics or 2)text only screen";ANSWER$
8470 IF ANSWER$="1" OR ANSWER$="2" THEN GOTO 8520
8480    LOCATE 25,24:  BEEP
```

```
8490    PRINT "Try again, the response is 1 or 2";
8500    LOCATE 24,1:  PRINT SPACE$(79);
8510    GOTO 8450
8520 IF ANSWER$="1" THEN DEF SEG=&HB800 ELSE DEF SEG=&HB000        'set memory
8530 LOCATE ,,0
8540 '
8550 RETURN
8560 '
```

Appendix C

Programs for the Projects

This appendix contains the source code listing for the programs described in each of the projects in the text. The format of these listings and the properties of the programs are identical to those described for the examples at the beginning of Appendix B.

C.1 Project I

This program calculates and graphs, for a given energy, the deflection function for scattering from the Lennard-Jones potential at NB impact parameters covering the range from BMIN to BMAX. The basic task is to evaluate Eq. (I.8) for each impact parameter required (loop 230-330). This is done in subroutine 2000, where a simple search is used to locate RMIN, the turning point, and the integrals are evaluated by Simpson's rule, with a change of variable to regulate the singularities. Subroutine 4000 graphs the deflection function found.

Representative output from this program can be obtained by entering

$$E=1$$

and

$$BMIN,BMAX,NB=0.1,2.4,20 .$$

```
10  '••••••••••••••••••••••••••••••••••••••••••••••••••••••••••••••••••••••••••
20  'Project 1:  Scattering by a central 6-12 potential
21  'COMPUTATIONAL PHYSICS by STEVEN E. KOONIN
22  'Copyright 1985, Benjamin/Cummings Publishing Company
30  '••••••••••••••••••••••••••••••••••••••••••••••••••••••••••••••••••••••••••
40  GOSUB 5000                                      'display header screen
50  '
60  PI=3.141593                                     'define constants, functions
70  NBMAX%=300                                      'max number of B values
80  DIM THETA(300)                                  'array for scattering angles
90  'redimension THETA if you increase NBMAX%
100 RMAX=2.5                                        'max R for which V<>0
110 NPTS%=40                                        'number of quadrature points
120 TOLR=.0001                                      'tolerance for t.p. search
130 DEF FNV(R)=4*(R^(-12)-R^(-6))                   '6-12 central potential
```

```
140 DEF FNI1(R)=1/R^2/SQR(1-(B/R)^2)              'first integrand
150 DEF FNI2(R)=1/R^2/SQR(1-(B/R)^2-FNV(R)/E)    'second integrand
160 '
170 GOSUB 1000                                   'input E and B parameters
180 '
190 CLS: LOCATE 2,29                             'print title of output
200 PRINT USING "Incident Energy=###.##";E:
210 PRINT ""
220 '
230 FOR IB%=0 TO NB%-1                           'calculate THETA for each B
240   B=BMIN+IB%*DB                              'current value of B
250   IF INKEY$="e" THEN GOTO 170                'typing e will end program
260   GOSUB 2000                                 'integrate to find THETA(B)
270 '
280   THETA(IB%)=2*B*(INT1-INT2)*180/PI 'store angle in degrees
290 '
300   IF ((IB%+1) MOD 18)=0 THEN GOSUB 3000    'clear screen if full
310   LOCATE ,21:  PRINT USING "B=#.###     ";B;
320   PRINT USING "Rmin=#.###    Theta=####.##";RMIN,THETA(IB%)
330 NEXT IB%
340 '
350 IF GRAPHICS% THEN GOSUB 4000                 'graph the deflection function
360 GOTO 170                                     'get new parameters and begin again
370 '
1000 '*************************************************************
1010 'subroutine to input E and B parameters
1020 'input  variables: none
1030 'output variables: E,BMAX,BMIN,DB,NB%
1040 'global variables: GRAPHICS%,MONO%
1050 'local  variables: none
1060 '*************************************************************
1070 LOCATE 22,12,1: BEEP                              'prompt for E
1080 INPUT "Enter the incident energy E in units of Vzero or 0 to end";E
1090 IF E=>0 GOTO 1140
1100   LOCATE 23,25:  BEEP
1110   PRINT "E must be greater than zero";
1120   LOCATE 22,10: PRINT SPACE$(59)
1130   GOTO 1070                                  'prompt for E again
1140 IF E=0 THEN END
1150 '
1160 LOCATE 22,1: PRINT SPACE$(80);               'prompt for B parameters
1170 LOCATE 23,1: PRINT SPACE$(80);
1180 LOCATE 22,1
1190 PRINT "B is the impact parameter, and NB is the number of steps ";
1200 PRINT "used to find THETA(B).";
1210 LOCATE 23,26
1220 BEEP: INPUT; "Enter BMIN, BMAX, and NB";BMIN,BMAX,NB%
1230 DB=(BMAX-BMIN)/(NB%-1)                       'step in B
1240 '
1250 IF BMAX<=0 OR BMIN<=0 OR NB%<=0 THEN GOTO 1320   'check for errors in B
1260 IF BMAX<=BMIN OR BMAX>=RMAX THEN GOTO 1350    ' parameters
1270 IF NB%>NBMAX% THEN GOTO 1390
1280 IF GRAPHICS% AND MONO% THEN GOSUB 7000        'switch to mono
1290 LOCATE ,,0
1300 RETURN
1310 '
1320 LOCATE 22,1: PRINT SPACE$(80): BEEP            'error handling
1330    LOCATE 22,26:  PRINT "All values must be positive.";
```

```
1340     GOTO 1430
1350 LOCATE 22,1: PRINT SPACE$(80):   BEEP
1360     LOCATE 22,20
1370     PRINT USING "The limits on BMAX are BMIN<BMAX<RMAX=#.##";RMAX
1380     GOTO 1430
1390 LOCATE 22,1: PRINT SPACE$(80):   BEEP
1400     LOCATE 22,8
1410     PRINT "The number of B values=(BMAX-BMIN)/DB must be less than";
1420     PRINT USING " NBMAX=###";NBMAX%
1430 LOCATE 23,20: PRINT SPACE$(59)                'prompt again for
1440 GOTO 1210                                      ' B parameters
1450 '
2000 '****************************************************************
2010 'subroutine to find THETA(B)
2020 'input  variables: B, FNI1(R), FNI2(R), FNV(R)
2030 'output variables: INT1, INT2
2040 'global variables: E, NPTS%, RMAX
2050 'local  variables: DR, H, I%, R, RMIN, SUM, U, UMAX
2060 '****************************************************************
2070 RMIN=RMAX: DR=.2                      'inward search for turning point
2080 WHILE DR>TOLR
2090     RMIN=RMIN-DR
2100     IF (1-(B/RMIN)^2-FNV(RMIN)/E)>0 GOTO 2120
2110     RMIN=RMIN+DR: DR=DR/2
2120 WEND
2130 '
2140 SUM=0                                 'first integral by rectangle rule
2150 UMAX=SQR(RMAX-B)                      'change variable to U=SQR(R-B)
2160 H=UMAX/NPTS%                          ' to handle singularity
2170 FOR I%=1 TO NPTS%
2180     U=H*(I%-.5)
2190     R=U^2+B
2200     SUM=SUM+U*FNI1(R)
2210 NEXT I%
2220 INT1=2*H*SUM
2230 '
2240 SUM=0                                 'second integral by rectangle rule
2250 UMAX=SQR(RMAX-RMIN)                   'change variable to U=SQR(R-RMIN)
2260 H=UMAX/NPTS%                          ' to handle singularity
2270 FOR I%=1 TO NPTS%
2280     U=H*(I%-.5)
2290     R=U^2+RMIN
2300     SUM=SUM+U*FNI2(R)
2310 NEXT I%
2320 INT2=2*H*SUM
2330 '
2340 RETURN
2350 '
3000 '****************************************************************
3010 'subroutine to clear screen when full
3020 'input  variables: none
3030 'output variables: none
3040 'global variables: E
3050 'local  variables: none
3060 '****************************************************************
3070 LOCATE 24,31,1:   BEEP
3080 PRINT "Type c to continue ";
3090 IF INKEY$<>"c" THEN GOTO 3090
```

```
3100 CLS:   LOCATE 2,29,0
3110 PRINT USING "Incident Energy=###.##";E: PRINT ""
3120 RETURN
3130 '
4000 '**********************************************************************
4010 'subroutine to graph the deflection function (THETA vs. B)
4020 'input  variables: B,THETA
4030 'output variables: none
4040 'global variables: BMIN,BMAX,DB,E,MONO%,NB%,RMAX
4050 'local  variables: I%,MARK%,X%,XOLD%,XSCALE,Y%,YOLD%,YSCALE
4060 '**********************************************************************
4070 LOCATE 24,28,1:   BEEP                      'prompt for command and wait
4080 PRINT "Type g to begin graphing ";
4090 IF INKEY$<>"g" GOTO 4090
4100 LOCATE ,,0
4110 '
4120 IF MONO% THEN GOSUB 6000                    'switch to graphics screen
4130 SCREEN 2,0,0,0                              'switch to hi res graphics
4140 CLS: KEY OFF
4150 '
4160 LINE (60,1)-(60,199)                        'THETA axis
4170 LINE (60,100)-(640,100)                     'B   axis
4180 '
4190 FOR MARK%=1 TO 5                            'B ticks
4200     X%=60+MARK%*579/5
4210     LINE (X%,98)-(X%,102)
4220 NEXT MARK%
4230 '
4240 FOR I%=1 TO 5                               'B labels
4250     LOCATE 14, (6+I%*14)
4260     PRINT USING "#.###";BMIN+(I%*(BMAX-BMIN)/5)
4270 NEXT I%
4280 LOCATE 14,9: PRINT USING "#.###";BMIN
4290 '
4300 LOCATE 12,79:  PRINT "b"                    'B legend
4310 '
4320 FOR MARK%=0 TO 8                            'THETA ticks
4330     Y%=MARK%*199/8
4340     LINE (58,Y%)-(62,Y%)
4350 NEXT MARK%
4360 '
4370 FOR I%=4 TO -3 STEP -1                      'THETA labels
4380     LOCATE (22-(I%+3)*3),4
4390     PRINT USING "+###";I%*45;
4400 NEXT I%
4410 '
4420 LOCATE 11,1:   PRINT "T";                   'THETA legend
4430 LOCATE 12,1:   PRINT "H"
4440 LOCATE 13,1:   PRINT "E"
4450 LOCATE 14,1:   PRINT "T"
4460 LOCATE 15,1:   PRINT "A"
4470 '
4480 LOCATE 1,61:   PRINT "Deflection function"  'print plot title
4490 LOCATE 2,63:   PRINT USING "Energy=###.###";E
4500 '
4510 XSCALE=580/(BMAX-BMIN):  YSCALE=199/360     'scales for the screen
4520 '
4530 FOR IB%=0 TO NB%-1                          'plot THETA for each B
```

```
4540    X%=60+XSCALE*IB%*DB                      'location of this point
4550    Y%=199-YSCALE*(THETA(IB%)+180)
4560    IF Y%>199 THEN Y%=199                    'keep above screen bottom
4570    LINE (X%-2,Y%)-(X%+2,Y%)                 'mark point with a +
4580    LINE (X%,Y%-2)-(X%,Y%+2)
4590    IF IB%=0 GOTO 4610                       'don't connect 1'st point
4600       LINE (XOLD%,YOLD%)-(X%,Y%)            'connect to previous point
4610    XOLD%=X%: YOLD%=Y%                       'update previous point
4620 NEXT IB%
4630 '
4640 LOCATE 24,41:  BEEP                         'prompt for command and wait
4650 PRINT "Type c to continue with new parameters";
4660 IF INKEY$<>"c" GOTO 4660
4670 '
4680 IF MONO% THEN GOSUB 7000                    'switch back to mono screen
4690 SCREEN 0: WIDTH 80: LOCATE ,,1,12,13        'switch back to text mode
4700 RETURN
4710 '
5000 '*******************************************************************
5010 'subroutine to display the header screen
5020 'input   variables: none
5030 'output  variables: GRAPHICS%,MONO%
5040 'global  variables: none
5050 'local   variables: G$,M$,ROW%
5060 '*******************************************************************
5070 SCREEN 0: CLS: KEY OFF                      'program begins in text mode
5080 '
5090 LOCATE 1,30: COLOR 15                       'display book title
5100 PRINT "COMPUTATIONAL PHYSICS";
5110 LOCATE 2,40: COLOR 7
5120 PRINT "by";
5130 LOCATE 3,33: PRINT "Steven E. Koonin";
5140 LOCATE 4,14
5150 PRINT "Copyright 1985, Benjamin/Cummings Publishing Company";
5160 '
5170 LOCATE 5,10                                 'draw the box
5180 PRINT CHR$(201)+STRING$(59,205)+CHR$(187);
5190 FOR ROW%=6 TO 19
5200    LOCATE ROW%,10: PRINT CHR$(186);
5210    LOCATE ROW%,70: PRINT CHR$(186);
5220 NEXT ROW%
5230 LOCATE 20,10
5240 PRINT CHR$(200)+STRING$(59,205)+CHR$(188);
5250 '
5260 COLOR 15                                    'print title, etc.
5270 LOCATE  7,36:  PRINT "PROJECT 1";
5280 COLOR 7
5290 LOCATE 10,23:  PRINT "Scattering by a central 6-12 potential"
5300 LOCATE 13,36:  PRINT "**********"
5310 LOCATE 15,26:  PRINT "Type 'e' to end while running."
5320 LOCATE 16,23:  PRINT "Type ctrl-break to stop at a prompt."
5330 '
5340 LOCATE 19,16: BEEP                          'get screen configuration
5350 INPUT "Does your computer have graphics capability (y/n)";G$
5360 IF LEFT$(G$,1)="y" OR LEFT$(G$,1)="n" GOTO 5390
5370    LOCATE 19,12:  PRINT SPACES$(58):  BEEP
5380    LOCATE 18,35:  PRINT "Try again,":  GOTO 5340
5390 IF LEFT$(G$,1)="y" GOTO 5420
```

```
5400    GRAPHICS%=0:   MONO%=-1
5410    RETURN
5420 GRAPHICS%=-1
5430 LOCATE 18,15:   PRINT SPACE$(55)
5440 LOCATE 19,15:   PRINT SPACE$(55)
5450 LOCATE 19,15: BEEP
5460 INPUT "Do you also have a separate screen for text (y/n)";M$
5470 IF LEFT$(M$,1)="y" OR LEFT$(M$,1)="n" GOTO 5490   'error trapping
5480    LOCATE 18,35: PRINT "Try again,":   GOTO 5440
5490 IF LEFT$(M$,1)="y" THEN MONO%=-1 ELSE MONO%=0
5500 RETURN
5510 '
6000 '******************************************************************
6010 'subroutine to switch from mono to graphics screen
6020 'input  variables: none
6030 'output variables: none
6040 'global variables: none
6050 'local  variables: none
6060 '******************************************************************
6070 DEF SEG=0
6080 POKE &H410, (PEEK(&H410) AND &HCF) OR &H10
6090 SCREEN 0:  WIDTH 40: LOCATE ,,1,6,7
6100 RETURN
6110 '
7000 '******************************************************************
7010 'subroutine to switch from graphics to mono screen
7020 'input  variables: none
7030 'output variables: none
7040 'global variables: none
7050 'local  variables: none
7060 '******************************************************************
7070 DEF SEG=0
7080 POKE &H410, (PEEK(&H410) OR &H30)
7090 SCREEN 0:  WIDTH 80:  LOCATE ,,1,12,13
7100 RETURN
7110 '
```

C.2 Project II

This program constructs a series of white dwarf models for a specified electron fraction YE with central densities ranging in equal logarithmic steps between the values of RHO1 and RHO2 specified. For each model (loop 190-290), the dimensionless equations (II.18) for the mass and density are integrated by the 4'th-order Runge-Kutta algorithm. An empirically scaled radial step (line 230) is used for each model and initial conditions for the integration are determined by a Taylor series expansion of the differential equations about $r=0$ (lines 240-270). Integration proceeds until the density has fallen below 10^3 gm cm^{-3} (loop 2100-2150, subroutines 3000,4000), whereupon the internal and gravitational energies are calculated (lines 5080-5160) and the model displayed.

Representative output can be obtained by entering

Ye=1,

an initial central density of 1E5, and a final central density of 1E11, and requesting 4 models.

```
10  '********************************************************************
20  'Project 2: The structure of white dwarf stars
21  'COMPUTATIONAL PHYSICS by Steven E. Koonin
22  'Copyright 1985 by Benjamin/Cummings Publishing Company
30  '********************************************************************
40  GOSUB 7000                                   'display header screen
50  '                                            'define constants, functions
60  MEOVERMP=1/1836                              'electron/nucleon mass ratio
70  DEF FNGAMMA(X)=X*X/3/SQR(X*X+1)              'function for the pressure
80  DEF FNPART1(X)=X*(1+2*X^2)*SQR(1+X^2)        'functions for electron energy
90  DEF FNPART2(X)=LOG(X+SQR(1+X^2))
100 DEF FNEPSRHO(X)=.375*(FNPART1(X)-FNPART2(X))
110 '
120 JMAX%=300                                    'maximum number of radial steps
130 DIM RHOSTOR(300)                             'array for dimensionless rho
140 DIM MSTOR(300)                               'array for dimensionless mass
150 '
160 GOSUB 1000                                   'input models to calculate
170 RHOCUT=1000!/RHO0                            'dimensionless cutoff dens.
180 '
190 FOR MODEL%=1 TO NMODELS%                     'loop over models
200     IF INKEY$="e" THEN GOTO 350
210     TEMP=(RHO2/RHO1)^((MODEL%-1)/(NMODELS%-1))'central density for this one
220     RHOCENT=(RHO1/RHO0)*TEMP
230     DR=(3*.001/RHOCENT)^.33333/3             'radial step
240     RSTART=DR/10                             'first point is at dr/10
250     MSTOR(0)=RSTART^3*RHOCENT/3              'initial mass
260     TEMP=RSTART^2/6/FNGAMMA(RHOCENT^.333333) 'initial density
270     RHOSTOR(0)=RHOCENT*(1-TEMP*RHOCENT)
280     GOSUB 2000                               'integrate the model
290     GOSUB 5000                               'output the results
300 NEXT MODEL%
```

```
310 IF GRAPHICS% THEN LOCATE 16,58 ELSE LOCATE ,31
320 PRINT "Type c to continue";: BEEP
330 IF INKEY$="c" THEN GOTO 160 ELSE GOTO 330
340 '
350 IF GRAPHICS% THEN LOCATE 16,46 ELSE LOCATE ,22
360 BEEP: INPUT "Do you really want to end (y/n)";KB$
370 IF KB$="y" THEN STOP ELSE GOTO 160
380 '
1000 '*****************************************************************
1010 'subroutine to input the parameters of the models to be calculated
1020 'input   variables: none
1030 'output variables: M0,NMODELS%,RHO0,RHO1,RHO2,YE
1040 'global variables: none
1050 'local   variables: none
1060 '*****************************************************************
1070 CLS
1080 LOCATE 9,20,1
1090 BEEP: INPUT "Enter electron fraction, Ye (>0,<=1)";YE
1100 IF YE>0 AND YE<=1 THEN GOTO 1120 ELSE GOTO 1080
1110 '
1120 LOCATE 11,14
1130 PRINT "A series of models will be calculated in which the"
1140 LOCATE 12,14
1150 PRINT "central density is increased from a starting value"
1160 LOCATE 13,17
1170 PRINT "to a final value in equal logarithmic steps"
1180 '
1190 LOCATE 15,20
1200 BEEP: INPUT "Enter initial central density (>=1e5,<=1e15)";RHO1
1210 IF RHO1>=100000 AND RHO1<=1E+15 THEN GOTO 1220 ELSE GOTO 1190
1220 LOCATE 16,20
1230 BEEP: INPUT "Enter  final  central density (>=1e5,<=1e15)";RHO2
1240 IF RHO2>=100000 AND RHO2<=1E+15 THEN GOTO 1250 ELSE GOTO 1220
1250 LOCATE 17,20
1260 BEEP: INPUT "Enter number of models to calculate (>=2)";NMODELS%
1270 IF NMODELS%>=2 THEN GOTO 1290 ELSE GOTO 1250
1280 '
1290 R0=7.72E+08*YE                         'scaling radius
1300 M0=5.67E+33*YE^2                       'scaling mass
1310 RHO0=979000!/YE                        'scaling density
1320 '
1330 CLS
1340 RETURN
1350 '
2000 '*****************************************************************
2010 'subroutine to integrate the model for a given central density
2020 'input   variables: none
2030 'output variables: RHOSTOR,MSTOR
2040 'global variables: J%,JMAX%,M,R,RHO,RHOCUT
2050 'local   variables: none
2060 '*****************************************************************
2070 R=RSTART: J%=0                         'initial conditions
2080 RHO=RHOSTOR(0): M=MSTOR(0)
2090 '
2100 WHILE RHO>RHOCUT AND J%<JMAX%          'outward integration
2110    J%=J%+1
2120    GOSUB 3000                          'take a Runge-Kutta step
2130    IF RHO<=RHOCUT THEN RHO=RHOCUT      'keep rho above the cutoff
```

```
2140    RHOSTOR(J%)=RHO: MSTOR(J%)=M                    'store the new values
2150 WEND
2160 '
2170 RETURN
2180 '
3000 '******************************************************************
3010 'subroutine to take a 4'th order Runge-Kutta step of DR in the diff. eqs.
3020 'input   variables: none
3030 'output  variables: none
3040 'global  variables: DMDR,DR,DRHODR,M,R,RHO
3050 'local   variables: K1M,K1RHO,K2M,K2RHO,K3M,K3RHO,K4M,K4RHO
3060 '******************************************************************
3070 GOSUB 4000                                        'compute k1
3080 K1M=DR*DMDR: K1RHO=DR*DRHODR
3090 '
3100 R=R+DR/2                                           'compute k2
3110 M=M+.5*K1M
3120 RHO=RHO+.5*K1RHO
3130 GOSUB 4000
3140 K2M=DR*DMDR: K2RHO=DR*DRHODR
3150 '
3160 M=M+.5*(K2M-K1M)                                   'compute k3
3170 RHO=RHO+.5*(K2RHO-K1RHO)
3180 GOSUB 4000
3190 K3M=DR*DMDR: K3RHO=DR*DRHODR
3200 '
3210 R=R+DR/2                                           'compute k4
3220 M=M+K3M-.5*K2M
3230 RHO=RHO+K3RHO-.5*K2RHO
3240 GOSUB 4000
3250 K4M=DR*DMDR: K4RHO=DR*DRHODR
3260 '
3270 M=M-K3M+(K1M+2*K2M+2*K3M+K4M)/6                    'new values of M and RHO
3280 RHO=RHO-K3RHO+(K1RHO+2*K2RHO+2*K3RHO+K4RHO)/6
3290 '
3300 RETURN
3310 '
4000 '******************************************************************
4010 'subroutine to calculate the derivatives of the mass and density
4020 'input   variables: M,R,RHO
4030 'output  variables: DMDR,DRHODR
4040 'global  variables: FNGAMMA
4050 'local   variables: none
4060 '******************************************************************
4070 IF RHO<=0 THEN RETURN
4080 '
4090 DRHODR=-M*RHO/R^2/FNGAMMA(RHO^.333333)
4100 DMDR=R^2*RHO
4110 RETURN
4120 '
5000 '******************************************************************
5010 'subroutine to output the results of the model calculated
5020 'input   variables: MSTOR,RHOSTOR
5030 'output  variables: none
5040 'global  variables: DR,FNEPSRHO,GRAPHICS%,J%,MEOVERMP,MMAX,MODEL%,MO,
5050 '                   RHOCENT,RHOMIN,RHOSCALE,RHOO,RMAX,RO,YE
5060 'local   variables: EGRAV,EINT,EO,I%,MOLD%,R,ROLD%,X%,Y%,XOLD%,YOLD%
5070 '******************************************************************
```

```
5080 EGRAV=0: EINT=0                                        'compute energies
5090 FOR I%=1 TO J%-1
5100     R=I%*DR
5110     EGRAV=EGRAV+MSTOR(I%)*RHOSTOR(I%)*R
5120     EINT=EINT+R^2*FNEPSRHO(RHOSTOR(I%)^.333333)
5130 NEXT I%
5140 E0=YE*MEOVERMP*9*(M0/1E+31)
5150 EGRAV=-EGRAV*DR*E0
5160 EINT=EINT*DR*E0
5170 '
5180 IF GRAPHICS% THEN GOTO 5370
5190 '                                                      'monochrome output
5200     IF MODEL%>1 THEN GOTO 5240
5210         LOCATE 25,24
5220         PRINT USING "White dwarf stars with Ye=#.###";YE
5230         LOCATE 1,1
5240     LOCATE , 1: PRINT USING "Model ##";MODEL%;
5250     LOCATE ,10: PRINT USING "Central density=+#.####^^^^";RHOCENT*RHO0;
5260     LOCATE ,38: PRINT USING "Mass=+#.####^^^^";MSTOR(J%)*M0;
5270     LOCATE ,55: PRINT USING "Radius=+#.####^^^^";J%*DR*R0
5280     LOCATE , 3: PRINT "Energies in 10^51 erg:";
5290     LOCATE ,26: PRINT USING "Egrav=+#.####^^^^";EGRAV;
5300     LOCATE ,44: PRINT USING "Eint=+#.####^^^^";EINT;
5310     LOCATE ,61: PRINT USING "Etot=+#.####^^^^";EGRAV+EINT
5320     LOCATE ,10: PRINT USING "Radial step=+#.####^^^^";DR*R0;
5330     LOCATE ,40: PRINT USING "Number of steps=###";J%
5340     PRINT
5350     RETURN
5360 '                                                      'graphics output
5370 IF MODEL%=1 THEN GOSUB 6000                            'prepare graphics screen
5380 '
5390 LOCATE  5,55: PRINT USING "Model ##";MODEL%;
5400 LOCATE  5,69: PRINT USING "Ye=#.####";YE;
5410 LOCATE  6,53: PRINT USING "Radial step      +#.####^^^^";DR*R0
5420 LOCATE  7,53: PRINT USING "Number of steps       ###";J%
5430 LOCATE  9,53: PRINT USING "Central density +#.####^^^^";RHOCENT*RHO0;
5440 LOCATE 10,53: PRINT USING "Mass            +#.####^^^^";MSTOR(J%)*M0;
5450 LOCATE 11,53: PRINT USING "Radius          +#.####^^^^";J%*DR*R0;
5460 LOCATE 12,53: PRINT USING "Egrav (10^51)   +#.####^^^^";EGRAV;
5470 LOCATE 13,53: PRINT USING "Eint  (10^51)   +#.####^^^^";EINT;
5480 LOCATE 14,53: PRINT USING "Etot  (10^51)   +#.####^^^^";EGRAV+EINT;
5490 '
5500 XOLD%=30                                               'plot the density
5510 YOLD%=183-LOG(RHOSTOR(0)/RHOMIN)*RHOSCALE
5520 FOR I%=1 TO J%
5530     R=DR*I%*R0/1E+08
5540     X%=30+R*300/RMAX
5550     Y%=183-LOG(RHOSTOR(I%)/RHOMIN)*RHOSCALE
5560     LINE (XOLD%,YOLD%)-(X%,Y%)
5570     XOLD%=X%: YOLD%=Y%
5580 NEXT I%
5590 '
5600 X%=30+300*DR*J%*R0/1E+08/RMAX                          'put point on mass-radius
5610 Y%=85-85*MSTOR(J%)*M0/1E+33/MMAX                       ' plot
5620 IF MODEL%>1 THEN GOTO 5640
5630     ROLD%=X%: MOLD%=Y%
5640 LINE (ROLD%,MOLD%)-(X%,Y%)
5650 LINE (X%,Y%-2)-(X%,Y%+2)
```

```
5660 ROLD%=X%: MOLD%=Y%
5670 '
5680 RETURN
5690 '
6000 '****************************************************************
6010 'subroutine to prepare the graphics display
6020 'input   variables: J%
6030 'output  variables: MMAX,RHOMIN,RHOSCALE,RMAX
6040 'global  variables: DR,RHOCUT,RHO0,RHO2,R0,YE
6050 'local   variables: LABEL,LRHOMAX%,LRHOMIN%,MARK%,RHOMAX,X%,Y%
6060 '****************************************************************
6070 RMAX=J%*DR*R0/1E+08                          'max radius (10^8 cm) on plot
6080 MMAX=15*YE^2                                 'max mass   (10^33 g) on plot
6090 '
6100 FOR MARK%=0 TO 5
6110    X%=3+MARK%*7                              'radius labels
6120    IF MARK%>=3 THEN X%=X%+1
6130    LABEL=RMAX*MARK%/5
6140    LOCATE 24,X%
6150    PRINT USING "##.#";LABEL;
6160    LOCATE 12,X%
6170    PRINT USING "##.#";LABEL;
6180 '
6190    Y%=11-MARK%*2                             'mass labels
6200    LOCATE Y%,1
6210    LABEL=MARK%*MMAX/5
6220    IF LABEL <10 THEN PRINT USING "#.#";LABEL;
6230    IF LABEL>=10 THEN PRINT USING  "##";LABEL;
6240 '
6250    Y%=85-85*MARK%/5                          'mass ticks
6260    LINE (30,Y%)-(33,Y%)
6270 '
6280    X%=30+MARK%*60                            'radius ticks
6290    LINE (X%,183)-(X%,180)
6300    LINE (X%,85)-(X%,82)
6310 NEXT MARK%
6320 '
6330 LOCATE 12,23: PRINT "R";                    'radius legends
6340 LOCATE 24,23: PRINT "r";
6350 LOCATE 12,43: PRINT "x10^8 cm";
6360 LOCATE 24,43: PRINT "x10^8 cm";
6370 '
6380 LOCATE 17,1: PRINT "log";                   'rho legend
6390 LOCATE 18,1: PRINT "rho";
6400 '
6410 LOCATE 1,6: PRINT "M (10^33 g)";            'mass legend
6420 '
6430 LINE (30,183)-(330,183)                      'draw the axes
6440 LINE (30,183)-(30,98)
6450 LINE (30,85)-(330,85)
6460 LINE (30,85)-(30,0)
6470 '
6480 LRHOMAX%=CINT(LOG(RHO2)/LOG(10) +1)          'density is on a log scale
6490 RHOMAX=10^LRHOMAX%
6500 LRHOMIN%=CINT(LOG(RHO0*RHOCUT)/LOG(10))
6510 RHOMIN=10^LRHOMIN%
6520 RHOSCALE=85/LOG(RHOMAX/RHOMIN)
6530 '
```

```
6540 LOCATE 23,2                                  'rho labels
6550 PRINT USING "##";LRHOMIN%;
6560 LOCATE 13,2
6570 PRINT USING "##";LRHOMAX%;
6580 '
6590 FOR MARK%=LRHOMIN%+1 TO LRHOMAX%            'rho ticks
6600    Y%=183-(MARK%-LRHOMIN%)*85/(LRHOMAX%-LRHOMIN%)
6610    LINE (30,Y%)-(33,Y%)
6620 NEXT MARK%
6630 '
6640 RHOMIN=RHOMIN/RHO0                          'make RHOMIN dimensionless
6650 '
6660 RETURN
6670 '
7000 '**************************************************************************
7010 'subroutine to display the header screen
7020 'input  variables: none
7030 'output variables: GRAPHICS%,MONO%
7040 'global variables: none
7050 'local  variables: G$,M$,ROW%
7060 '**************************************************************************
7070 SCREEN 0: CLS: KEY OFF                      'program begins in text mode
7080 '
7090 LOCATE 1,30: COLOR 15                       'display book title
7100 PRINT "COMPUTATIONAL PHYSICS";
7110 LOCATE 2,40: COLOR 7
7120 PRINT "by";
7130 LOCATE 3,33: PRINT "Steven E. Koonin";
7140 LOCATE 4,14
7150 PRINT "Copyright 1985, Benjamin/Cummings Publishing Company";
7160 '
7170 LOCATE 5,10                                 'draw the box
7180 PRINT CHR$(201)+STRING$(59,205)+CHR$(187);
7190 FOR ROW%=6 TO 19
7200    LOCATE ROW%,10: PRINT CHR$(186);
7210    LOCATE ROW%,70: PRINT CHR$(186);
7220 NEXT ROW%
7230 LOCATE 20,10
7240 PRINT CHR$(200)+STRING$(59,205)+CHR$(188);
7250 '
7260 COLOR 15                                    'print title, etc.
7270 LOCATE  7,36:  PRINT "PROJECT 2";
7280 COLOR 7
7290 LOCATE 10,23:
7300 PRINT "The structure of white dwarf stars";
7310 LOCATE 13,36:  PRINT "**********"
7320 LOCATE 15,25:  PRINT "Type 'e' to end while running."
7330 LOCATE 16,22:  PRINT "Type ctrl-break to stop at a prompt."
7340 '
7350 LOCATE 19,16: BEEP                          'get screen configuration
7360 INPUT "Does your computer have graphics capability (y/n)";G$
7370 IF LEFT$(G$,1)="y" OR LEFT$(G$,1)="n" GOTO 7400
7380    LOCATE 19,12:  PRINT SPACE$(58):  BEEP
7390    LOCATE 18,35:  PRINT "Try again,":  GOTO 7350
7400 IF LEFT$(G$,1)="y" GOTO 7430
7410    GRAPHICS%=0:  MONO%=-1
7420    GOTO 7520
7430 GRAPHICS%=-1
```

```
7440 LOCATE 18,15:  PRINT SPACE$(55)
7450 LOCATE 19,15:  PRINT SPACE$(55)
7460 LOCATE 19,15: BEEP
7470 INPUT "Do you also have a separate screen for text (y/n)";M$
7480 IF LEFT$(M$,1)="y" OR LEFT$(M$,1)="n" GOTO 7500   'error trapping
7490    LOCATE 18,35:  PRINT "Try again,":  GOTO 7450
7500 IF LEFT$(M$,1)="y" THEN MONO%=-1 ELSE MONO%=0
7510 '
7520 LOCATE 21,13
7530 PRINT "This program calculates a series of white dwarf models";
7540 LOCATE 22,18
7550 PRINT "with fixed Ye and increasing central density";
7560 LOCATE 23,12
7570 PRINT "All input and output quantities are expressed in cgs units";
7580 LOCATE 24,31
7590 PRINT "Type c to continue";
7600 IF INKEY$<>"c" GOTO 7600
7610 LOCATE ,,0
7620 '
7630 IF GRAPHICS% THEN GOSUB 8000              'switch to graphics screen
7640 '
7650 RETURN
7660 '
8000 '*************************************************************
8010 'subroutine to switch from mono to graphics screen
8020 'input  variables: none
8030 'output variables: none
8040 'global variables: none
8050 'local  variables: none
8060 '*************************************************************
8070 DEF SEG=0
8080 POKE &H410,  (PEEK(&H410) AND &HCF) OR &H10
8090 SCREEN 2,0,0,0:  WIDTH 80: LOCATE ,,0
8100 RETURN
8110 '
```

C.3 Project III

This program solves the Hartree-Fock equations in the filling approximation for atomic systems with electrons in the $1s$, $2s$, and $2p$ shells. The nuclear charge, occupations of the shells, and lattice parameters are defined in subroutine 1000. The heart of the program is the subroutines 4000, 5000, 6000, and 9000, which calculate the total energy (III.26,27) and single particle energies (III.30) for a given set of radial wavefunctions. Note that in doing so, the density is taken to be a weighted average of the current and previous density (lines 5080-5100, 5140-5170). The integrals required are evaluated by the trapezoidal rule. These subroutines are also used by subroutine 2000 to find the optimal scaled hydrogenic wavefunctions assumed to start the iterations. The single particle eigenvalue equations (III.28) are solved as inhomogeneous boundary value problems (subroutine 7000) using the single particle energies calculated from the previous iteration. Care is taken to ensure that the $2s$ orbital is orthogonal to that of the $1s$ (lines 7540-7620). At each iteration, the various energies are displayed, as are graphs of the partial and total electron densities.

For a representative calculation, one can input a nuclear charge of

$$Z=6,$$

an occupation of 2 electrons in each of the $1s$, $2s$, and $2p$ orbitals, a radial step size of 0.01 Å, and an outer lattice radius of 2 Å and get a fair description of the neutral Carbon atom.

```
10  '*****************************************************************
20  'Project 3: Hartree-Fock solutions of small atomic systems in the filling
30  '              approximation
31  'COMPUTATIONAL PHYSICS by Steven E. Koonin
32  'Copyright 1985, Benjamin/Cummings Publishing Company
40  '*****************************************************************
50  GOSUB 10000                          'display header screen
60  IF GRAPHICS% THEN GOSUB 11000        'if graphics,switch to that display
70  '                                    'set up arrays and define constants
80  NRMAX%=700                           'maximum number of lattice points
90  NSTATE%=3                            'maximum number of states
100 DIM PSTOR(700,3)                     'array for the radial wavefunctions
110 DIM PSIIN(700),PSIOUT(700)           'array for homogeneous solutions
120 DIM FOCK(700,3)                      'array for Fock terms
130 DIM PHI(700),RHO(700)                'direct potential and density
140 DIM NOCC(3)                          'occupation numbers of the states
150 DIM EXH(3)                           's.p. energies of the states
160 DIM L%(3),ID$(3)                     'angular momenta and labels of states
170 DIM FACTORIAL(10)                    'array for factorials
180 HBM=7.6359                           'hbar^2/m for the electron
190 E2=14.409                            'square of electron charge
200 ABOHR=HBM/E2                         'Bohr radius
210 '
```

```
220 GOSUB 1000                               'input parameters of system
230 GOSUB 2000                               'find optimal hydrogenic wavefunction
240 MIX=.5                                    'mixing of old and new densities
250 '
260 FOR NIT%=1 TO 40                         'main iteration loop
270    FOR STATE%=1 TO NSTATE%              'loop over states to find new w.f.
280       IF INKEY$="e" THEN GOTO 420        'check if time to end
290       IF NOCC(STATE%)=0 THEN GOTO 330   'don't consider empty orbitals
300       E=EXH(STATE%)                       's.p. energy from previous w.f.
310       IF E>0 THEN E=-10                   'keep the energy negative
320       GOSUB 7000                          'solve the s.p. wave equation
330    NEXT STATE%
340    GOSUB 4000                            'calculate new energies and output
350 NEXT NIT%
360 '
370 LOCATE 10,25: BEEP                         'prompt and wait for command
380 PRINT "Press any key to continue"
390 IF INKEY$="" THEN GOTO 390
400 GOTO 220                                   'start again with a new system
410 '
420 LOCATE 10,25: BEEP:                        'prompt and wait for command
430 INPUT "Do you really want to end (y/n)";KB$
440 IF KB$="y" THEN END
450 IF KB$<>"n" THEN GOTO 420 ELSE GOTO 220  'start again with a new system
460 '
1000 '*********************************************************************
1010 'subroutine to input the parameters of the ion to be calculated
1020 'input   variables: none
1030 'output  variables: DR,ID$,L%,NE,NOCC,NR%,RMAX,Z,ZE2
1040 'global  variables: NRMAX%
1050 'local   variables: none
1060 '*********************************************************************
1070 CLS
1080 '
1090 LOCATE 8,15
1100 INPUT "Enter nuclear charge, Z";Z          'get the nuclear charge
1110 ZE2=Z*E2
1120 '
1130 L%(1)=0: ID$(1)="1s"                       'L values and labels
1140 L%(2)=0: ID$(2)="2s"
1150 L%(3)=1: ID$(3)="2p"
1160 '                                           'get the state occupations
1170 LOCATE 9,15
1180 INPUT "Enter number of electrons in the 1s state (>=0,<=2)";NOCC(1)
1190 IF NOCC(1)>=0 AND NOCC(1)<=2 THEN GOTO 1210
1200    BEEP: GOTO 1170
1210 LOCATE 10,15
1220 INPUT "Enter number of electrons in the 2s state (>=0,<=2)";NOCC(2)
1230 IF NOCC(2)>=0 AND NOCC(2)<=2 THEN GOTO 1250
1240    BEEP: GOTO 1210
1250 LOCATE 11,15
1260 INPUT "Enter number of electrons in the 2p state (>=0,<=6)";NOCC(3)
1270 IF NOCC(3)>=0 AND NOCC(3)<=6 THEN GOTO 1300
1280    BEEP: GOTO 1250
1290 '                                           'get the lattice parameters
1300 LOCATE 12,15
1310 INPUT "Enter radial step size (in Angstroms)"; DR
1320 LOCATE 13,15
```

```
1330 INPUT "Enter outer radius of the lattice (in Angstroms)";RMAX
1340 NR%=CINT(RMAX/DR)                            'number of lattice points
1350 IF NR%<=NRMAX% THEN GOTO 1430                'must be < max. allowed
1360    BEEP: LOCATE 14,15
1370    PRINT USING "The resulting lattice has ##### points.";NR%
1380    LOCATE 15,15
1390    PRINT USING "The maximum number of points allowed is ####.";NRMAX%
1400    LOCATE 16,15
1410    PRINT "Try again or exit and redimension arrays in the code."
1420    GOTO 1300
1430 NE=0                                         'count number of electrons
1440 FOR STATE%=1 TO NSTATE%
1450    NE=NE+NOCC(STATE%)
1460 NEXT STATE%
1470 '
1480 LOCATE 19,25
1490 PRINT "HANG ON, I'M WORKING ON IT!"          'message for the impatient
1500 RETURN
1510 '
2000 '********************************************************************
2010 'subroutine to find the optimal hydrogenic variational wavefunction
2020 'input  variables: none
2030 'output variables: none
2040 'global variables: DR,ELAST,ETOT,KTOT,MIX,NE,NIT%,PFLAG%,VEETOT,VENTOT,
2050 '                  VEXTOT,Z,ZSTAR
2060 'local  variables: none
2070 '********************************************************************
2080 ZSTAR=Z
2090 GOSUB 3000                                   'build Z*=Z wavefunctions
2100 MIX=1                                        'don't average with old density
2110 PFLAG%=0                                     'turn off energy printout
2120 GOSUB 4000                                   'calculate the energy
2130 ZSTAR=-Z*(VENTOT+VEETOT+VEXTOT)/(2*KTOT)     'optimal ZSTAR using virial
2140 GOSUB 3000                                   'build best wavefunctions
2150 '
2160 CLS                                          'print the header lines
2170 LOCATE 1, 5: PRINT USING "Z=##.##";Z;
2180 LOCATE 1,13: PRINT USING "Number of electrons=##.##";NE;
2190 LOCATE 1,39: PRINT USING "Radial step=#.###";DR;
2200 LOCATE 1,57: PRINT USING "Maximum radius=#.###";DR*NR%
2210 LOCATE 2,13
2220 PRINT USING "Hydrogenic orbitals give Z*=##.###";ZSTAR;
2230 '
2240 PFLAG%=-1                                    'turn on energy printout
2250 NIT%=0: ELAST=0                              '0'th iteration
2260 GOSUB 4000                                   'calculate energy and output
2270 LOCATE 2,47                                  'print energy on header line
2280 PRINT USING ", Total energy=+####.##";ETOT
2290 RETURN
2300 '
3000 '********************************************************************
3010 'subroutine to construct normalized hydrogenic wavefunctions
3020 'input  variables: ZSTAR
3030 'output variables: PSTOR
3040 'global variables: DR,NOCC,NR%,NSTATE%
3050 'local  variables: ERSTAR2,J%,NORM,RSTAR,STATE%
3060 '********************************************************************
3070 FOR J%=0 TO NR%                              'loop over radial points
```

```
3080      RSTAR=J%*DR*ZSTAR/ABOHR            'scaled radius
3090      ERSTAR2=EXP(-RSTAR/2)              'useful exponential
3100      IF NOCC(1)=0 THEN GOTO 3120
3110         PSTOR(J%,1)=RSTAR*ERSTAR2^2     '1s wavefunction
3120      IF NOCC(2)=0 THEN GOTO 3140
3130         PSTOR(J%,2)=(2-RSTAR)*RSTAR*ERSTAR2    '2s wavefunction
3140      IF NOCC(3)=0 THEN GOTO 3160
3150         PSTOR(J%,3)=RSTAR^2*ERSTAR2    '2p wavefunction
3160 NEXT J%
3170 '                                      'normalization of orbitals
3180 FOR STATE%=1 TO NSTATE%
3190      IF NOCC(STATE%)=0 GOTO 3280        'only treat occupied ones
3200      NORM=0
3210      FOR J%=1 TO NR%                    'find integral of psi^2
3220         NORM=NORM+PSTOR(J%,STATE%)^2
3230      NEXT J%
3240      NORM=1/SQR(NORM*DR)
3250      FOR J%=1 TO NR%                    'normalize to unity
3260         PSTOR(J%,STATE%)=PSTOR(J%,STATE%)*NORM
3270      NEXT J%
3280 NEXT STATE%
3290 RETURN
3300 '
4000 '****************************************************************
4010 'subroutine to calculate and print the energies of a normalized
4020 '          set of single-particle wavefunctions
4030 'input   variables: PFLAG%,PSTOR
4040 'output  variables: ETOT,EXH,KTOT,VEETOT,VENTOT,VEXTOT
4050 'global  variables: DR,ELAST,GRAPHICS%,ID$,HBM,L%,NOCC,NR%,NSTATE%,PHI,ZE2
4060 'local   variables: KEN,J%,LL1,PM,PZ,PZ2,R,STATE%,VCENT,VEE,VEN,VEX
4070 '****************************************************************
4080 GOSUB 5000                             'calculate the density and Fock terms
4090 GOSUB 6000                             'find direct potential
4100 '
4110 KTOT=0                                 'zero total kinetic energy
4120 VENTOT=0                               'zero total electron-nucleus attraction
4130 VEETOT=0                               'zero total electron-electron repulsion
4140 VEXTOT=0                               'zero total exchange energy
4150 '
4160 IF NOT PFLAG% THEN GOTO 4230           'no output if print flag is off
4170    LOCATE 3,26: PRINT USING "------- Iteration ## ------";NIT%
4180    LOCATE 4, 7: PRINT "State L Nocc";
4190    LOCATE 4,21: PRINT "Kinetic    Ven";
4200    LOCATE 4,40: PRINT "Vee       Vex";
4210    LOCATE 4,59: PRINT "V          e"
4220 '
4230 FOR STATE%=1 TO NSTATE%                'loop over states
4240    IF NOCC(STATE%)=0 THEN GOTO 4620    'consider only occupied states
4250       KEN=0: VEN=0: VEE=0: VEX=0       'zero energies for this state
4260       VCENT=0                          'zero centrifugal energy
4270       LL1=L%(STATE%)*(L%(STATE%)+1)    'useful constant
4280       PM=0
4290       FOR J%=1 TO NR%                  'loop over the lattice
4300          R=J%*DR                       'radius at this point
4310          PZ=PSTOR(J%,STATE%)           'wavefunction at this point
4320          PZ2=PZ^2                      'square of the wavefunction
4330          KEN=KEN+(PZ-PM)^2             'contribution to kinetic
4340          VCENT=VCENT+PZ2*LL1/R^2       'contribution to centrifugal
```

```
4350          VEN=VEN-PZ2/R                              'contribution to e-nucleus
4360          VEE=VEE+PHI(J%)*PZ2                        'contribution to e-e repulsion
4370          VEX=VEX+FOCK(J%,STATE%)*PZ                 'contribution to exchange
4380          PM=PZ                                      'roll the wavefunction
4390       NEXT J%
4400       VCENT=VCENT*DR*HBM/2                          'multiply integrals by constants
4410       KEN=KEN*HBM/(2*DR)
4420       VEE=VEE*DR
4430       VEN=VEN*ZE2*DR
4440       VEX=VEX*DR
4450       KEN=KEN+VCENT                                 'kinetic includes centrifugal
4460       EXH(STATE%)=KEN+VEN+VEE+VEX                   'store the s.p. energy
4470       KTOT=KTOT+KEN*NOCC(STATE%)                    'add contributions to total E
4480       VENTOT=VENTOT+VEN*NOCC(STATE%)
4490       VEETOT=VEETOT+VEE*NOCC(STATE%)
4500       VEXTOT=VEXTOT+VEX*NOCC(STATE%)
4510       IF NOT PFLAG% GOTO 4620                       'no output if print flag off
4520         LOCATE ,9
4530         PRINT ID$(STATE%);
4540         LOCATE ,13
4550         PRINT USING "# #.##";L%(STATE%),NOCC(STATE%);
4560         LOCATE ,19
4570         PRINT USING " ####.## +####.## ####.##";KEN,VEN,VEE;
4580         LOCATE ,44
4590         PRINT USING " +####.##";VEX;
4600         LOCATE ,53
4610         PRINT USING " +####.## +####.##";VEN+VEE+VEX,EXH(STATE%)
4620    NEXT STATE%
4630    '
4640    VEETOT=VEETOT/2                                  'don't double-count
4650    VEXTOT=VEXTOT/2
4660    ETOT=KTOT+VENTOT+VEETOT+VEXTOT                   'total energy
4670    '
4680    IF NOT PFLAG% THEN GOTO 4800                     'no output if print flag off
4690      LOCATE , 7: PRINT "TOTALS";
4700      LOCATE ,14: PRINT USING "##.##";NE;
4710      LOCATE ,20: PRINT USING "####.##";KTOT;
4720      LOCATE ,28: PRINT USING "+####.##";VENTOT;
4730      LOCATE ,37: PRINT USING "####.##";VEETOT;
4740      LOCATE ,45: PRINT USING "+####.##";VEXTOT;
4750      LOCATE ,54: PRINT USING "+####.##";VENTOT+VEETOT+VEXTOT
4760      LOCATE ,15: PRINT USING "Total energy=+####.##";ETOT;
4770      LOCATE ,38: PRINT USING "Previous total energy=+####.##";ELAST
4780      ELAST=ETOT                                     'current E is next last E
4790      IF GRAPHICS% THEN GOSUB 9000                   'plot the density
4800    RETURN
4810    '
5000    '*************************************************************************
5010    'subroutine to compute the density and the Fock terms
5020    'input    variables: MIX,PSTOR
5030    'output   variables: FOCK,RHO
5040    'global   variables: DR,E2,L%,LAM%,LAMSTART%,LAMSTOP%,L1%,L2%,NR%,NSTATE%,
5050    '                    THREEJ
5060    'local    variables: DF,FAC,I%,J%,R,RLAM,RLAM1,STATE%,STATE2%,SUM,TERM
5070    '*************************************************************************
5080    FOR J%=1 TO NR%                                  'a fraction of old density
5090       RHO(J%)=(1-MIX)*RHO(J%)
5100    NEXT J%
```

```
5110 '
5120 FOR STATE%=1 TO NSTATE%                              'loop over states
5130    IF NOCC(STATE%)=0 GOTO 5470                       'treat only occupied states
5140    FOR J%=1 TO NR%                                   'contribution to density
5150       RHO(J%)=RHO(J%)+MIX*NOCC(STATE%)*PSTOR(J%,STATE%)^2
5160       FOCK(J%,STATE%)=0
5170    NEXT J%
5180    FOR STATE2%=1 TO NSTATE%                          'loop over second states
5190       IF NOCC(STATE2%)=0 GOTO 5460                   'treat only occupied ones
5200       L1%=L%(STATE%)
5210       L2%=L%(STATE2%)
5220       LAMSTART%=ABS(L1%-L2%)                         'limits on lambda sum
5230       LAMSTOP%=L1%+L2%
5240       FOR LAM%=LAMSTART% TO LAMSTOP% STEP 2'loop for lambda sum
5250          GOSUB 8000                                  'calculate square of 3-j
5260          FAC=-E2/2*NOCC(STATE2%)*THREEJ              'factor in Fock term
5270          SUM=0
5280          FOR I%=1 TO NR%                             'outward integral for Fock
5290             R=I%*DR: RLAM=R^LAM%
5300             TERM=PSTOR(I%,STATE2%)*PSTOR(I%,STATE%)*RLAM/2
5310             SUM=SUM+TERM
5320             DF=PSTOR(I%,STATE2%)*FAC*SUM*DR/(RLAM*R)
5330             FOCK(I%,STATE%)=FOCK(I%,STATE%)+DF
5340             SUM=SUM+TERM
5350          NEXT I%
5360          SUM=0
5370          FOR I%=NR% TO 1 STEP -1                     'inward integral for Fock
5380             R=I%*DR: RLAM1=R^(LAM%+1)
5390             TERM=PSTOR(I%,STATE2%)*PSTOR(I%,STATE%)/RLAM1/2
5400             SUM=SUM+TERM
5410             DF=PSTOR(I%,STATE2%)*FAC*SUM*DR*RLAM1/R
5420             FOCK(I%,STATE%)=FOCK(I%,STATE%)+DF
5430             SUM=SUM+TERM
5440          NEXT I%
5450       NEXT LAM%
5460    NEXT STATE2%
5470 NEXT STATE%
5480 RETURN
5490 '
6000 '**********************************************************************
6010 'subroutine to solve Poisson's equation for the direct potential
6020 'input   variables: RHO
6030 'output  variables: PHI
6040 'global  variables: DR,E2,NR%
6050 'local   variables: CON,J%,M,R,SM,SP,SUM,SZ
6060 '**********************************************************************
6070 SUM=0                                                'quadrature for r=0 value
6080 FOR J%=1 TO NR%
6090    SUM=SUM+RHO(J%)/CSNG(J%)
6100 NEXT J%
6110 '                                                    'outward integration
6120 CON=DR^2/12                                          'useful constant
6130 SM=0: SZ=-E2*RHO(1)/DR                               'first rolling values of S
6140 PHI(0)=0: PHI(1)=SUM*E2*DR                           'initial conditions for PHI
6150 FOR J%=1 TO NR%-1                                    'Numerov integration outward
6160    SP=-E2*RHO(J%+1)/((J%+1)*DR)
6170    PHI(J%+1)=2*PHI(J%)-PHI(J%-1)+CON*(10*SZ+SP+SM)
6180    SM=SZ: SZ=SP
```

```
6190 NEXT J%
6200 '                                                    'subtract linear behavior
6210 M=(PHI(NR%)-PHI(NR%-10))/(10*DR)
6220 FOR J%=1 TO NR%
6230     R=J%*DR
6240     PHI(J%)=PHI(J%)/R-M                              'factor of 1/r for true potl
6250 NEXT J%
6260 RETURN
6270 '
7000 '************************************************************************
7010 'subroutine to solve the single-particle wavefunction as an
7020 '            inhomogeneous boundary-value problem
7030 'input   variables: E,FOCK,PHI,STATE%
7040 'output  variables: PSTOR
7050 'global  variables: DR,HBM,NR%,ZE2
7060 'local   variables: DRHBM,I%,J%,K2M,K2P,K2Z,LL1,NORM,NR2%,PSIIN,PSIOUT,R,
7070 '                   SUM,TERM,WRON
7080 '************************************************************************
7090 DRHBM=DR^2/HBM/6                                     'useful constant
7100 LL1=L%(STATE%)*(L%(STATE%)+1)*HBM/2                  'constant for centrifugal potl.
7110 '                                                    'integrate outward homogeneous soln
7120 K2M=0
7130 K2Z=DRHBM*(E-PHI(1)+(ZE2-LL1/DR)/DR)
7140 PSIOUT(0)=0: PSIOUT(1)=1E-10
7150 FOR J%=2 TO NR%
7160     R=DR*J%
7170     K2P=DRHBM*(E-PHI(J%)+(ZE2-LL1/R)/R)
7180     PSIOUT(J%)=(PSIOUT(J%-1)*(2-10*K2Z)-PSIOUT(J%-2)*(1+K2M))/(1+K2P)
7190     K2M=K2Z: K2Z=K2P
7200 NEXT J%
7210 '                                                    'integrate inward homogeneous soln
7220 K2P=0
7230 R=(NR%-1)*DR
7240 K2Z=DRHBM*(E-PHI(NR%-1)+(ZE2-LL1/R)/R)
7250 PSIIN(NR%)=0
7260 PSIIN(NR%-1)=1E-10
7270 FOR J%=NR%-2 TO 1 STEP -1
7280     R=DR*J%
7290     K2M=DRHBM*(E-PHI(J%)+(ZE2-LL1/R)/R)
7300     PSIIN(J%)=(PSIIN(J%+1)*(2-10*K2Z)-PSIIN(J%+2)*(1+K2P))/(1+K2M)
7310     K2P=K2Z: K2Z=K2M
7320 NEXT J%
7330 '                                                    'Wronskian at middle of mesh
7340 NR2%=NR%/2
7350 WRON=(PSIIN(NR2%+1)-PSIIN(NR2%-1))/(2*DR)*PSIOUT(NR2%)
7360 WRON=WRON-(PSIOUT(NR2%+1)-PSIOUT(NR2%-1))/(2*DR)*PSIIN(NR2%)
7370 '                                                    'outward integral in Green's soln
7380 SUM=0:
7390 FOR I%=1 TO NR%
7400     TERM=-PSIOUT(I%)*FOCK(I%,STATE%)/2
7410     SUM=SUM+TERM
7420     PSTOR(I%,STATE%)=PSIIN(I%)*SUM*DR
7430     SUM=SUM+TERM
7440 NEXT I%
7450 '                                                    'inward integral in Green's soln
7460 SUM=0
7470 FOR I%=NR% TO 1 STEP -1
7480     TERM=-PSIIN(I%)*FOCK(I%,STATE%)/2
```

```
7490     SUM=SUM+TERM
7500     PSTOR(I%,STATE%)=(PSTOR(I%,STATE%)+PSIOUT(I%)*SUM*DR)/WRON
7510     SUM=SUM+TERM
7520 NEXT I%
7530 '
7540 IF STATE%<>2 THEN GOTO 7640              'keep 2s state orthogonal to 1s
7550     SUM=0
7560     FOR I%=1 TO NR%
7570        SUM=SUM+PSTOR(I%,1)*PSTOR(I%,2)
7580     NEXT I%
7590     SUM=SUM*DR
7600     FOR I%=1 TO NR%
7610        PSTOR(I%,2)=PSTOR(I%,2)-SUM*PSTOR(I%,1)
7620     NEXT I%
7630 '                                        'normalize the solution
7640 SUM=0
7650 FOR I%=1 TO NR%
7660     SUM=SUM+PSTOR(I%,STATE%)^2
7670 NEXT I%
7680 NORM=1/SQR(SUM*DR)
7690 FOR I%=1 TO NR%
7700     PSTOR(I%,STATE%)=PSTOR(I%,STATE%)*NORM
7710 NEXT I%
7720 RETURN
7730 '
8000 '********************************************************************
8010 'subroutine to calculate square of the 3-j coefficient appearing    '
8020 '                in the exchange energy
8030 'input   variables: LAM%,L1%,L2%
8040 'output  variables: THREEJ
8050 'global  variables: none
8060 'local   variables: DELTA,FACTORIAL,IMAX%,P%
8070 '********************************************************************
8080 IMAX%=L1%+L2%+LAM%+1                      'calculate array of factorials
8090 FACTORIAL(0)=1
8100 FOR I%=1 TO IMAX%
8110     FACTORIAL(I%)=I%*FACTORIAL(I%-1)
8120 NEXT I%
8130 '                                         'calculate the (3-j)^2
8140 DELTA=FACTORIAL(L1%+L2%-LAM%)*FACTORIAL(-L1%+L2%+LAM%)
8150 DELTA=DELTA*FACTORIAL(L1%-L2%+LAM%)/FACTORIAL(IMAX%)
8160 P%=(L1%+L2%+LAM%)/2
8170 THREEJ=DELTA*FACTORIAL(P%)^2
8180 THREEJ=THREEJ/FACTORIAL(P%-L1%)^2
8190 THREEJ=THREEJ/FACTORIAL(P%-L2%)^2
8200 THREEJ=THREEJ/FACTORIAL(P%-LAM%)^2
8210 RETURN
8220 '
9000 '********************************************************************
9010 'subroutine to graph the densities
9020 'input   variables: PSTOR,RHO
9030 'output  variables: none
9040 'global  variables: DR,NR%,NSTATE%
9050 'local   variables: HSCALE,I%,IMAX%,MARK%,RHOMAX,STATE%,VSCALE,X%,XOLD%,
9060 '                   Y%,YOLD%
9070 '********************************************************************
9080 LINE (0,80)-(639,199),0,BF               'clear lower half of display
9090 '
```

```
9100 RHOMAX=0                                    'find maximum density
9110 FOR I%=1 TO NR%
9120   IF RHO(I%)>RHOMAX THEN RHOMAX=RHO(I%)
9130 NEXT I%
9140 '
9150 VSCALE=98/RHOMAX                            'scale factors
9160 HSCALE=600/(NR%*DR)
9170 LINE (39,180)-(639,180)                     'r axis
9180 LINE (39,180)-(39,80)                       'rho axis
9190 LOCATE 12,50
9200 PRINT "Plot of 4*pi*r^2*rho (in A^-1)"      'plot title
9210 FOR MARK%=0 TO 5                            'r ticks, labels
9220   X%=MARK%*120+39
9230   LINE (X%,180)-(X%,177)
9240   X%=4+MARK%*15
9250   IF MARK%=5 THEN X%=76
9260   LOCATE 24,X%: PRINT USING "##.#";MARK%*RMAX/5;
9270 NEXT MARK%
9280 LOCATE 24,42: PRINT "r (A)";                'r legend
9290 FOR MARK%=0 TO 5                            'rho ticks, labels
9300   Y%=180-MARK%*20
9310   LINE (39,Y%)-(36,Y%)
9320   Y%=23-MARK%*2
9330   IF MARK%>=2 THEN Y%=Y%-1
9340   IF MARK%>=4 THEN Y%=Y%-1
9350   LOCATE Y%,1: PRINT USING "##.#";MARK%*RHOMAX/.9799999/5;
9360 NEXT MARK%
9370 '
9380 XOLD%=39: YOLD%=180                         'plot the total density
9390 FOR I%=1 TO NR%
9400   X%=39+I%*DR*HSCALE
9410   Y%=180-VSCALE*RHO(I%)
9420   LINE (XOLD%,YOLD%)-(X%,Y%)
9430   XOLD%=X%: YOLD%=Y%
9440 NEXT I%
9450 LINE (480,100)-(520,100)                    'total legend
9460 LOCATE 13,67: PRINT "Total"
9470 '
9480 FOR STATE%=1 TO NSTATE%                     'plot densities of each state
9490   IF NOCC(STATE%)=0 THEN GOTO 9810
9500   FOR I%=1 TO NR%                           'loop over lattice
9510     X%=39+I%*DR*HSCALE
9520     Y%=180-VSCALE*PSTOR(I%,STATE%)^2*NOCC(STATE%)
9530     IF STATE%<> 1 THEN GOTO 9550
9540       PSET(X%,Y%): GOTO 9600                '1s symbol
9550     IF STATE%<>2 THEN GOTO 9590
9560       LINE (X%-2,Y%)-(X%,Y%)                '2s symbol
9570       LINE (X%-1,Y%-1)-(X%-1,Y%+1)
9580       GOTO 9600
9590     LINE (X%-2,Y%-1)-(X%,Y%+1),,B           '2p symbol
9600   NEXT I%
9610   IMAX%=40/(DR*HSCALE)
9620   IF STATE%<>1 THEN GOTO 9680
9630     FOR I%=0 TO IMAX%                       '1s legend
9640       PSET (480+40*I%/IMAX%,108)
9650     NEXT I%
9660     LOCATE 14,67: PRINT "1s"
9670     GOTO 9810
```

```
9680      IF STATE%<>2 THEN GOTO 9760
9690        FOR I%=0 TO IMAX%                        '2s legend
9700          X%=480+40*I%/IMAX%: Y%=116
9710          LINE (X%-2,Y%)-(X%,Y%)
9720          LINE (X%-1,Y%-1)-(X%-1,Y%+1)
9730        NEXT I%
9740        LOCATE 15,67: PRINT "2s"
9750        GOTO 9810
9760        FOR I%=0 TO IMAX%                        '2p legend
9770          X%=480+40*I%/IMAX%: Y%=124
9780          LINE (X%-2,Y%-1)-(X%,Y%+1),,B
9790        NEXT I%
9800        LOCATE 16,67: PRINT "2p"
9810 NEXT STATE%
9820 RETURN
9830 '
10000 '***********************************************************
10010 'subroutine to display the header screen
10020 'input   variables: none
10030 'output  variables: GRAPHICS%,MONO%
10040 'global  variables: none
10050 'local   variables: G$,M$,ROW%
10060 '***********************************************************
10070 SCREEN 0: CLS: KEY OFF                      'program begins in text mode
10080 '
10090 LOCATE 1,30: COLOR 15                       'display book title
10100 PRINT "COMPUTATIONAL PHYSICS";
10110 LOCATE 2,40: COLOR 7
10120 PRINT "by";
10130 LOCATE 3,33: PRINT "Steven E. Koonin";
10140 LOCATE 4,14
10150 PRINT "Copyright 1985, Benjamin/Cummings Publishing Company";
10160 '
10170 LOCATE 5,10                                 'draw the box
10180 PRINT CHR$(201)+STRING$(59,205)+CHR$(187);
10190 FOR ROW%=6 TO 19
10200    LOCATE ROW%,10: PRINT CHR$(186);
10210    LOCATE ROW%,70: PRINT CHR$(186);
10220 NEXT ROW%
10230 LOCATE 20,10
10240 PRINT CHR$(200)+STRING$(59,205)+CHR$(188);
10250 '
10260 COLOR 15
10270 LOCATE  7,36:  PRINT "PROJECT 3";
10280 COLOR 7
10290 LOCATE  9,16:  PRINT "Hartree-Fock solutions for small atomic systems"
10300 LOCATE 10,26:  PRINT "in the filling approximation"
10310 LOCATE 13,36:  PRINT "**********"
10320 LOCATE 15,25:  PRINT "Type 'e' to end while running."
10330 LOCATE 16,22:  PRINT "Type ctrl-break to stop at a prompt."
10340 LOCATE 22,19:  PRINT "Energies are given in electron volts (eV)"
10350 LOCATE 23,17:  PRINT "Lengths are given in Angstroms (A, 10^-8 cm)"
10360 '
10370 LOCATE 19,16: BEEP                          'get screen configuration
10380 INPUT "Does your computer have graphics capability (y/n)";G$
10390 IF LEFT$(G$,1)="y" OR LEFT$(G$,1)="n" GOTO 10420
10400    LOCATE 19,12:  PRINT SPACE$(58):  BEEP
10410    LOCATE 18,35:  PRINT "Try again,":  GOTO 10370
```

```
10420 IF LEFT$(G$,1)="y" GOTO 10450
10430   GRAPHICS%=0:   MONO%=-1
10440   RETURN
10450 GRAPHICS%=-1
10460 LOCATE 18,15:   PRINT SPACE$(55)
10470 LOCATE 19,15:   PRINT SPACE$(55)
10480 LOCATE 19,15:  BEEP
10490 INPUT "Do you also have a separate screen for text (y/n)";M$
10500 IF LEFT$(M$,1)="y" OR LEFT$(M$,1)="n" GOTO 10520   'error trapping
10510   LOCATE 18,35:  PRINT "Try again,":  GOTO 10470
10520 IF LEFT$(M$,1)="y" THEN MONO%=-1 ELSE MONO%=0
10530 RETURN
10540 '
11000 '*****************************************************************
11010 'subroutine to switch from mono to graphics screen
11020 'input  variables: none
11030 'output variables: none
11040 'global variables: none
11050 'local  variables: none
11060 '*****************************************************************
11070 DEF SEG=0
11080 POKE &H410,  (PEEK(&H410) AND &HCF) OR &H10
11090 SCREEN 2,0,0,0: LOCATE ,,1,6,7
11100 RETURN
11110 '
```

C.4 Project IV

This program finds the partial wave scattering solution for electrons incident on a square-well (radius 1.5 Å), a gaussian well, or the Lenz-Jensen potential, as defined by the loops 3280-3330, 3220-3250, and 3360-3410, respectively. The scattering amplitude is calculated from (IV.4), with the Legendre polynomials evaluated by forward recursion in subroutine 11000. In each partial wave requested from LSTART% to LFINISH% (loop 7130-7020), the radial Schroedinger equation (IV.2) is integrated outward for both the free and scattering wavefunction (lines 7230-7440) and then the phase shifts are found using Eq. (IV.8) at lines 7460-7520; the required spherical Bessel functions are calculated in subroutine 4000. Note that the radial wavefunctions are assumed to vanish on the first four points of the lattice (within 0.02 Å) to avoid numerical problems associated with potentials becoming singular near $r=0$. This boundary condition, which is equivalent to assuming a hard sphere at this radius, has a negligible effect on the phase shifts for the energies at which the integration is accurate. Multiple-of-π ambiguities in the phase shifts are resolved in subroutine 8000. The radial wavefunctions, phase shifts, and partial cross sections are displayed by subroutine 9000 as each partial wave is calculated and, when all of the requested partial waves have been completed, the differential cross section is calculated by subroutine 10000 and graphed by subroutine 12000.

This program will give a representative output when run with a square-well of depth

$$Vzero=50,$$

with an electron energy

$$E=20,$$

and partial waves calculated from

$$Lstart, Lfinish=0,8.$$

```
10  '**********************************************************************
20  'Project 4:  Partial-wave solution of quantum scattering
21  'COMPUTATIONAL PHYSICS by Steven E, Koonin
22  'Copyright 1985, Benjamin/Cummings Publishing Company
30  '**********************************************************************
40  GOSUB 13000                                'display header screen
50  '
60  PI=3.14159                                 'define constants, functions
70  HBARM=7.6359                               'hbar^2/(mass of electron)
80  E2=14.409                                  'electron charge squared
90  RMAX=2:   REXT=.3                          'r is in Angstroms
100 DR=.005                                    'integration step
110 NR%=RMAX/DR:   NREXT%=(RMAX+REXT)/DR       'number of integration points
```

```
120 FIRST%=-1
130 NANG%=36                                       'PL and DSIGMA are dimensioned
140 DIM PL(100,36):   DIM DSIGMA(36)               ' to NANG%=number of angles
150 DIM JL(201):      DIM NL(201)                  'max L=100
160 DIM DELTA(100):   DIM SIG(100):   DIM DELDEG(100)
170 DIM V(500):       DIM VEFF(500)                'dim V and PSI to NREXT%
180 DIM PSI(500):     DIM R%(500)
190 DIM VFREE(500):   DIM PSIF(500)
200 '
210 GOSUB  1000                                    'input V, E, etc
220 '
230 IF FIRST% THEN GOSUB 2000                      'calculate and store P0 and P1
240 '
250 GOSUB 3000                                     'find energy dep variables
260 '
270 GOSUB 5000                                     'prepare for output
280 '
290 GOSUB 7000                                     'integrate Schrodinger Eq
300 '                                              ' and find DELTA(L%)
310 GOSUB 10000                                    'find partial sigma
320 '
330 GOTO 210                                       'begin again with next case
340 '
1000 '*****************************************************************************
1010 'subroutine to input Z, Energy, number of partial waves
1020 'input  variables:  FIRST%
1030 'output variables:  E, FIRST%, K, K2, LFINISH%, LMAX%, LSTART%, OPT$,
1040 '                   VZERO, Z
1050 'global variables:  none
1060 'local  variables:  ESAVE, VZSAVE, ZSAVE
1070 '*****************************************************************************
1080 CLS
1090 '
1100 LOCATE 1,26                                   'input potential
1110 PRINT "The possible potentials are";
1120 LOCATE 2,11
1130 PRINT "1)Lenz-Jensen potential for an electron and a neutral atom";
1140 LOCATE 3,21
1150 PRINT "2)Square well        3)Gaussian well";
1160 LOCATE 4,11
1170 INPUT; "Enter the number corresponding to your choice or 0 to end";OPT$
1180 IF OPT$="0" OR OPT$="1" OR OPT$="2" OR OPT$="3" THEN GOTO 1230
1190    LOCATE 5,20:   BEEP
1200    PRINT "Try again, the responses are 1, 2, or 3";
1210    LOCATE 4,1:    PRINT SPACE$(80);
1220    GOTO 1160
1230 IF OPT$="0" THEN END
1240 IF OPT$="1" GOTO 1360
1250 '
1260    IF FIRST% GOTO 1290                        'input well depth
1270       LOCATE 8,26
1280       PRINT USING "Your last Vzero was=+###.##";VZSAVE
1290    LOCATE 9,27
1300    INPUT; "Vzero=depth of the well=";VZERO
1310    VZSAVE=VZERO
1320    IF OPT$="2" THEN V$="Square Well" ELSE V$="Gaussian Well"
1330    IF OPT$="2" THEN V%=11 ELSE V%=13
1340    GOTO 1490
```

```
1350 '
1360 IF FIRST% THEN GOTO 1390                          'input nuclear charge
1370    LOCATE 8,30
1380    PRINT USING "Your last Z was ###";ZSAVE;
1390 LOCATE 9,23
1400 INPUT; "Z=charge of the atomic nucleus=";Z
1410 IF Z>0 THEN GOTO 1460
1420    LOCATE 10,25:  BEEP
1430    PRINT "Try again, Z must be positive";
1440    LOCATE 9,1:   PRINT SPACE$(79);
1450    GOTO 1390
1460 ZSAVE=Z
1470 V$="Lenz-Jensen":  V%=11
1480 '
1490 IF FIRST% GOTO 1520                               'input electron energy
1500    LOCATE 13,25
1510    PRINT USING "Your last Energy was #####.###";ESAVE;
1520 LOCATE 14,22
1530 INPUT; "E=energy of the electron in eV=";E
1540 IF E>0 AND E<35186! THEN GOTO 1590
1550    LOCATE 15,13:  BEEP
1560    PRINT "For a scattering state with LMAX < 100, 0 < E < 35186";
1570    LOCATE 14,1:   PRINT SPACE$(79)'
1580    GOTO 1520
1590 ESAVE=E
1600 '
1610 K2=2*E/HBARM                                      'square of wave number
1620 K=SQR(K2)                                         'wave number
1630 LMAX%=K*RMAX/2+4                                  'max L value computed
1640 '
1650 LOCATE 18,36                                      'input L range
1660 PRINT USING "Lmax=###";LMAX%;
1670 LOCATE 19,7
1680 PRINT "Enter Lstart, Lfinish; the program will calculate cross";
1690 PRINT " sections for";
1700 LOCATE 20,10
1710 PRINT "for all L values in this range (they may be the same value,";
1720 LOCATE 21,23
1730 INPUT; "and they may be larger than Lmax)";LSTART%,LFINISH%
1740 IF LSTART%>=0 AND LFINISH%>=0 GOTO 1790
1750    LOCATE 22,20:  BEEP
1760    PRINT "Try again, these values must be positive"
1770    LOCATE 21,1:   PRINT SPACE$(80);
1780    GOTO 1700
1790 IF LSTART%<=LFINISH% AND LFINISH%<=100 THEN RETURN
1800    LOCATE 22,20:  BEEP
1810    PRINT "  Try again, Lstart <= Lfinish <= 100   ";
1820    LOCATE 21,1:   PRINT SPACE$(80);
1830    GOTO 1700
1840 '
2000 '**************************************************************************
2010 'subroutine to calculate and store P0 and P1
2020 'input  variables:   none
2030 'output variables:   FIRST%, LTABLE%, PL(0,I%), PL(1,I%)
2040 'global variables:   GRAPHICS%, NANG%
2050 'local  variables:   I%, THETA, X
2060 '**************************************************************************
2070 FOR I%=0 TO NANG%                      'find Legendre polynom for NANG% angles
```

```
2080    THETA=I%*PI/NANG%
2090    X=COS(THETA)
2100    PL(0,I%)=1                              'P0
2110    PL(1,I%)=X                              'P1
2120 NEXT I%
2130 '
2140 LTABLE%=1                                  'indicates that PL up to L=1
2150 '                                          ' are calculated
2160 IF NOT GRAPHICS% THEN FIRST%=0             'reset flag for text only
2170 '
2180 RETURN
2190 '
3000 '********************************************************************
3010 'subroutine to calculate energy-dependent variables and V(r)
3020 'input  variables:   none
3030 'output variables:   EMAX, EMIN, ESCALE, OFFSET%, RSCALE, V(I%), X
3040 'global variables:   DR, K, LMAX%, NR%, NREXT%, REXT, RMAX, VZERO, Z
3050 'local  variables:   I%, R, U, Z6
3060 '********************************************************************
3070 X=K*RMAX                                   'argument of J11, N11
3080 OFFSET%=0                                  'counter runs 0-LMAX%
3090 GOSUB 4000                                 'calculate J1, N1
3100 '
3110 X=K*(RMAX+REXT)                            'argument of J12, N12
3120 OFFSET%=101                                'counter begins 101
3130 GOSUB 4000                                 'calculate J1, N1
3140 '
3150 FOR I%=NR%+1 TO NREXT%                     'V is nearly zero for
3160    V(I%)=0                                 'R>RMAX
3170 NEXT I%
3180 '
3190 IF OPT$="1" THEN GOTO 3360                 'calculate V(I%) at
3200 IF OPT$="2" THEN GOTO 3280                 'each mesh point
3210 '
3220 FOR I%=1 TO NR%                            'Gaussian Well
3230    R=I%*DR
3240    V(I%)=-VZERO*EXP(-R*R)
3250 NEXT I%
3260 GOTO 3430
3270 '
3280 FOR I%=1 TO NR%*3/4                        'Square Well
3290    V(I%)=-VZERO
3300 NEXT I%
3310 FOR I%=NR%*3/4 TO NR%                      'edge is at 3/4 max
3320    V(I%)=0
3330 NEXT I%
3340 GOTO 3430
3350 '
3360 Z6=Z^.1666667
3370 FOR I%=1 TO NR%                            'Lenz-Jensen
3380    R=I%*DR
3390    U=4.5397*Z6*SQR(R)
3400    V(I%)=-((Z*E2)/R)*EXP(-U)*(1+U+U*U*(.3344+U*(.0485+.002647*U)))
3410 NEXT I%
3420 '
3430 EMAX=4*E:  EMIN=-2*E                       'compute quantities
3440 ESCALE=172/(6*E)                           ' for graphing
3450 '
```

```
3460 RSCALE=600/(RMAX+REXT)
3470 '
3480 RETURN
3490 '
4000 '**********************************************************************
4010 'subroutine to calculate spherical Bessel functions
4020 'input    variables:  OFFSET%, X
4030 'output variables:    JL(I%), NL(I%)
4040 'global variables:    LMAX%
4050 'local    variables:  J, JM1, JP1, L%, LTRUE%, LUPPER%, NORM
4060 '**********************************************************************
4070 NL(0+OFFSET%)=-COS(X)/X                            'analytic form for N0
4080 NL(1+OFFSET%)=-(COS(X)/(X*X))-SIN(X)/X             'analytic form for N1
4090 '
4100 FOR L%=(1+OFFSET%) TO (LFINISH%-1+OFFSET%)         'forward recursion
4110     LTRUE%=L%-OFFSET%
4120     NL(L%+1)=(2*LTRUE%+1)*NL(L%)/X-NL(L%-1)
4130 NEXT L%
4140 '
4150 LUPPER%=X+10                                       'L value at which Jl
4160 JP1=0                                              ' is negligible
4170 J=9.999999E-21
4180 '
4190 FOR L%=(LUPPER%+OFFSET%) TO (1+OFFSET%) STEP -1    'backward recursion
4200     LTRUE%=L%-OFFSET%
4210     JM1=(2*LTRUE%+1)*J/X-JP1
4220     IF (L%-1)<=(LFINISH%+OFFSET%) THEN JL(L%-1)=JM1 'save the value if
4230     JP1=J:   J=JM1                                  ' L% is in the region
4240 NEXT L%                                             ' of interest
4250 '
4260 NORM=SIN(X)/X/JL(0+OFFSET%)                        'normalize Jl by using
4270 FOR L%=(0+OFFSET%) TO (LFINISH%+OFFSET%)           ' analytic form of J0
4280     JL(L%)=NORM*JL(L%)
4290 NEXT L%
4300 '
4310 RETURN
4320 '
5000 '**********************************************************************
5010 'subroutine to prepare screen for output
5020 'input    variables:  none
5030 'output variables:    none
5040 'global variables:    E, GRAPHICS%, LMAX%, MONO%, OPT$, V$, V%, VZERO, Z
5050 'local    variables:  COL%
5060 '**********************************************************************
5070 IF GRAPHICS% GOTO 5230
5080 '
5090 CLS
5100 COL%=(80-V%)/2                                     'display name of pot
5110 LOCATE 1,COL%
5120 PRINT V$;
5130 '
5140 IF OPT$="1" GOTO 5180                              'display E, etc
5150 LOCATE 2,17
5160 PRINT USING "Vzero=+##.##      Energy=#####.##      Lmax=##";VZERO,E,LMAX%;
5170 GOTO 5200
5180 LOCATE 2,21
5190 PRINT USING "Z=###      Energy=#####.##      Lmax=##";Z,E,LMAX%
5200 LOCATE 3,1:   PRINT ""                             'scroll text
```

```
5210 RETURN
5220 '
5230 IF FIRST% THEN GOSUB 6000                    'display information
5240 IF MONO% THEN GOSUB 14000                    'switch to graphics
5250 CLS:   SCREEN 2,0,0,0
5260 '
5270 LOCATE 24,15                                 'display V,E, etc
5280 PRINT V$+SPACE$(3);
5290 IF OPT$="1" GOTO 5320
5300    PRINT USING "Vzero=+###.##    E=#####.##    Lmax=##";VZERO,E,LMAX%;
5310    GOTO 5340
5320 PRINT USING "    Z=###       E=#####.##       Lmax=##";Z,E,LMAX%;
5330 '
5340 RETURN
5350 '
6000 '****************************************************************
6010 'subroutine to display information about the graph
6020 'input   variables:   none
6030 'output variables:    FIRST%
6040 'global variables:    none
6050 'local   variables:   none
6060 '****************************************************************
6070 CLS
6080 '
6090 LOCATE 5,11,0
6100 PRINT "The effective potential, the wavefunction of the scattered
6110 LOCATE 7,13
6120 PRINT "particle, and the wavefunction of the free particle are
6130 LOCATE 9,27
6140 PRINT "plotted on the same graph.
6150 LOCATE 13,12
6160 PRINT "The scale is in units 1E=energy of the incident particle."
6170 LOCATE 17,13
6180 PRINT "The wavefunctions are centered about the energy=1 axis."
6190 LOCATE 24,31,1:  BEEP
6200 PRINT "Type c to continue";
6210 IF INKEY$<>"c" THEN GOTO 6210
6220 LOCATE ,,0
6230 '
6240 FIRST%=0
6250 '
6260 RETURN
6270 '
7000 '****************************************************************
7010 'subroutine to integrate Schroedinger equation using Numerov method
7020 'input   variables:   none
7030 'output variables:    DELDEG(I%), DELTA(I%), L%, PSI(I%), PSIF(I%), SIG(I%),
7040 '                     SIGTOT, VEFF(I%)
7050 'global variables:    DR, GRAPHICS%, HBARM, LFINISH%, LMAX%, LSTART%, NR%
7060 '                     NREXT%, PI, REXT, RMAX
7070 'local   variables:   CONST, DENOM, G, I%, J%, KI, KIF, KIM1, KIM1F, KIP1,
7080 '                     KIP1F, LL, NUMER, VFREE(I%)
7090 '****************************************************************
7100 SIGTOT=0                                     'zero total cross sect
7110 CONST=DR*DR/(HBARM*6)
7120 '
7130 FOR L%=LSTART% TO LFINISH%                    'loop over L range
7140     IF INKEY$="e" THEN RETURN 210
```

```
7150  '
7160     LL=LZ%*(LZ%+1)*HBARM/2
7170     FOR JZ%=1 TO NREXT%                              'calculate effective
7180        R=JZ%*DR                                      ' potentials with ang
7190        VFREE(JZ%)=LL/(R*R)                           ' momentum barrier
7200        VEFF(JZ%)=V(JZ%)+VFREE(JZ%)
7210     NEXT JZ%
7220  '
7230     PSI(4)=0:  PSI(5)=1E-25                          'boundary conditions
7240     KIM1=CONST*(E-VEFF(4)):  KI=CONST*(E-VEFF(5))    'K of Numerov method
7250     PSIF(4)=0: PSIF(5)=1E-25                         'same quantities for the
7260     KIM1F=CONST*(E-VFREE(4)): KIF=CONST*(E-VFREE(5))' free wave function
7270  '
7280     FOR IZ%=5 TO NREXT%-1
7290        KIP1=CONST*(E-VEFF(IZ%+1))                    'integrate with V(r)
7300        PSI(IZ%+1)=((2-10*KI)*PSI(IZ%)-(1+KIM1)*PSI(IZ%-1))/(1+KIP1)
7310        IF PSI(IZ%+1)<=1E+15 GOTO 7350                'prevent overflows
7320           FOR MZ%=1 TO IZ%+1
7330              PSI(MZ%)=PSI(MZ%)*.00001
7340           NEXT MZ%
7350        KIM1=KI:  KI=KIP1
7360  '
7370        KIP1F=CONST*(E-VFREE(IZ%+1))                  'integrate without V(r)
7380        PSIF(IZ%+1)=((2-10*KIF)*PSIF(IZ%)-(1+KIM1F)*PSIF(IZ%-1))/(1+KIP1F)
7390        IF PSIF(IZ%+1)<=1E+15 GOTO 7430               'prevent overflows
7400           FOR MZ%=1 TO IZ%+1
7410              PSIF(MZ%)=PSIF(MZ%)*.00001
7420           NEXT MZ%
7430        KIM1F=KIF:  KIF=KIP1F
7440     NEXT IZ%
7450  '
7460     G=RMAX*PSI(NREXT%)/((RMAX+REXT)*PSI(NR%))        'find DELTA
7470     NUMER=G*JL(LZ%)-JL(LZ%+101)
7480     DENOM=G*NL(LZ%)-NL(LZ%+101)
7490  '
7500     DELTA(LZ%)=ATN(NUMER/DENOM)
7510     GOSUB 8000                                       'get rid of ambiguity
7520     DELDEG(LZ%)=180*DELTA(LZ%)/PI                    ' in DELTA
7530  '
7540     SIG(LZ%)=4*PI/K2*(2*LZ%+1)*SIN(DELTA(LZ%))^2     'partial sigma
7550     SIGTOT=SIGTOT+SIG(LZ%)                           'total    sigma
7560  '
7570     IF GRAPHICS% THEN GOSUB 9000:  GOTO 7620         'graphics output
7580  '
7590        LOCATE ,13                                    'mono     output
7600        PRINT USING "L=##        Delta(L)=+###.##";LZ%,DELDEG(LZ%);
7610        PRINT USING "        Sigma(L)=#.###^^^^";SIG(LZ%)
7620 NEXT LZ%
7630  '
7640 IF GRAPHICS% THEN GOTO 7740                          'display sigma total
7650     PRINT "":  LOCATE ,20
7660     PRINT USING "Sigma total=+#.###^^^^";SIGTOT;
7670     PRINT USING "    L range= ## to ###";LSTART%,LFINISH%;
7680     PRINT "": PRINT""
7690     LOCATE ,31:  BEEP:  PRINT "Type c to continue";
7700     IF INKEY$<>"c" GOTO 7700
7710     LOCATE ,1:  PRINT SPACE$(80)
7720     RETURN
```

```
7730 '
7740 LOCATE 23,1:  PRINT SPACE$(80);
7750 LOCATE 23,20
7760 PRINT USING "Sigma total=+#.###^^^^";SIGTOT;
7770 PRINT USING "     L range= ## to ###";LSTART%,LFINISH%;
7780 LOCATE 1,31,1: BEEP:  PRINT "Type c to continue";
7790 IF INKEY$<>"c" THEN GOTO 7790
7800 LOCATE ,,0
7810 '
7820 RETURN
7830 '
8000 '**********************************************************************
8010 'subroutine to resolve the ambiguity in DELTA(L) and find PSIMAX,PSIFMAX
8020 'input  variables:  DELTA(L%), PSI(I%), PSIF(I%)
8030 'output variables:  DELTA(L%), PSIMAX, PSIFMAX
8040 'global variables:  NR%, NREXT%, PI
8050 'local  variables:  Bf%, BUMP%, DPOS%, F%, N%, NODES%, NODESF%
8060 '**********************************************************************
8070 IF DELTA(L%)>0 THEN DPOS%=-1 ELSE DPOS%=0   'positive delta means
8080 '                                           ' an attractive potential
8090 '
8100 NODES%=0:  PSIMAX=0:  BUMP%=0               'count nodes and antinodes
8110 '                                           ' for scattered wave
8120 FOR I%=5 TO NREXT%-1
8130    IF SGN(PSI(I%))<>SGN(PSI(I%-1)) THEN NODES%=NODES%+1:  INODE%=I%
8140    IF ABS(PSI(I%))>PSIMAX THEN PSIMAX=ABS(PSI(I%))
8150    IF SGN(PSI(I%+1)-PSI(I%))<>SGN(PSI(I%)-PSI(I%-1)) THEN BUMP%=BUMP%+1
8160 NEXT I%
8170 '
8180 NODESF%=0:  PSIFMAX=0:  BF%=0               'count nodes and antinodes
8190 '                                           ' for free wave
8200 FOR I%=5 TO NREXT%-1
8210    IF SGN(PSIF(I%))<>SGN(PSIF(I%-1)) THEN NODESF%=NODESF%+1: IFNODE%=I%
8220    IF ABS(PSIF(I%))>PSIFMAX THEN PSIFMAX=ABS(PSIF(I%))
8230    IF SGN(PSIF(I%+1)-PSIF(I%))<>SGN(PSIF(I%)-PSIF(I%-1)) THEN BF%=BF%+1
8240 NEXT I%
8250 '
8260 IF NODES%=NODESF% GOTO 8350
8270 '
8280    N%=BUMP%-BF%
8290    IF ABS(NODES%-NODESF%)>ABS(N%) THEN N%=NODES%-NODESF%
8300    IF N%>0 AND NOT DPOS% THEN DELTA(L%)=DELTA(L%)+PI      'force the sign
8310    IF N%<0 AND DPOS% THEN DELTA(L%)=DELTA(L%)-PI          ' to be correct
8320    DELTA(L%)=DELTA(L%)+(ABS(N%)-1)*PI*SGN(N%)
8330    RETURN
8340 '
8350 IF NODES%=0 GOTO 8400
8360    IF INODE%>IFNODE% AND DPOS% THEN DELTA(L%)=DELTA(L%)-PI
8370    IF INODE%<IFNODE% AND NOT DPOS% THEN DELTA(L%)=DELTA(L%)+PI
8380    RETURN
8390 '
8400 IF ABS(PSI(NR%/2)/PSIMAX)>ABS(PSIF(NR%/2)/PSIFMAX) THEN F%=-1 ELSE F%=0
8410 IF F% AND NOT DPOS% THEN DELTA(L%)=DELTA(L%)+PI
8420 IF NOT F% AND DPOS% THEN DELTA(L%)=DELTA(L%)-PI
8430 '
8440 RETURN
8450 '
9000 '**********************************************************************
```

```
9010 'subroutine to graph the wave function and Veff(r)
9020 'input   variables:   DELDEG(I%), PSI(I%), PSIF(I%), PSIMAX, PSIFMAX,
9030 '                     SIG(I%), VEFF(I%)
9040 'output variables:    none
9050 'global variables:    DR, EMAX, EMIN, ESCALE, NREXT%, REXT, RMAX, RSCALE
9060 'local  variables:    COL%, E%, I%, MARK%, PSI%, PSIOLD%, PSISCALE,
9070 '                     PSIFSCALE, R%, ROW%, VEFF%
9080 '*******************************************************************
9090 LINE (0,0)-(640,174),0,BF                      'clear top of screen
9100 '
9110 LOCATE 25,11                                   'display calculations
9120 PRINT USING "L=##       Delta(L)=+###.###";L%,DELDEG(L%);
9130 PRINT USING "         Sigma(L)=#.###^^^^";SIG(L%);
9140 '
9150 PSISCALE=172/(2*PSIMAX)                        'scale for PSI
9160 PSIFSCALE=172/(2*PSIFMAX)
9170 '
9180 LINE (40,0)-(40,172)                           'E axis
9190 LINE (40,115)-(640,115)                        'R axis
9200 FOR I%=0 TO 636 STEP 8                         'dotted line at
9210    LINE (40+I%,86)-(44+I%,86)                  'energy=E
9220 NEXT I%
9230 '
9240 FOR MARK%=0 TO 6                               'E ticks
9250    E%=MARK%*172/6
9260    LINE (38,E%)-(42,E%)
9270    ROW%=MARK%*21.5/6:  IF ROW%=0 THEN ROW%=1
9280    IF MARK%=2 OR MARK%=4 THEN ROW%=ROW%+1
9290    LOCATE ROW%,2
9300    PRINT USING "+#E";(4-MARK%);                'E labels
9310 NEXT MARK%
9320 '
9330 FOR MARK%=1 TO 4                               'R ticks
9340    R%=40+MARK%*(RMAX/4)*RSCALE
9350    LINE (R%,113)-(R%,117)
9360    COL%=5+16.3*MARK%
9370    LOCATE 16,COL%
9380    PRINT USING "#.#";RMAX*MARK%/4;             'R labels
9390 NEXT MARK%
9400 LINE (638,113)-(638,117)
9410 LOCATE 16,78
9420 PRINT USING "#.#";(RMAX+REXT);
9430 LOCATE 14,76:  PRINT "R";                      'R legend
9440 '
9450 LOCATE 13,1:  PRINT "Veff";                    'V legend
9460 FOR I%=1 TO NREXT% STEP 2                      'graph Veff(r)
9470    IF VEFF(I%)>EMAX THEN VEFF=EMAX:  GOTO 9490
9480    IF VEFF(I%)<EMIN THEN VEFF=EMIN ELSE VEFF=VEFF(I%)
9490    VEFF%=172-(VEFF-EMIN)*ESCALE
9500    R%(I%)=40+(I%*DR)*RSCALE
9510    IF I%=1 GOTO 9530
9520       LINE (R%(I%-2),VOLD%)-(R%(I%),VEFF%)
9530    VOLD%=VEFF%
9540 NEXT I%
9550 '
9560 LINE (56,179)-(76,179)                         'display key for PSI
9570 LOCATE 23,11:  PRINT "Scattered wave function";
9580 FOR I%=1 TO 16 STEP 4
```

```
9590    CIRCLE (384+I%,179),2
9600 NEXT I%
9610 LOCATE 23,52:  PRINT "Free wave function";
9620 '
9630 LOCATE 6,1:  PRINT "PSI";                        'PSI legend
9640 FOR I%=1 TO NREXT% STEP 2                        'graph scattering PSI
9650    PSI%=172-(PSI(I%)+PSIMAX)*PSISCALE
9660    IF I%<>1 THEN LINE (R%(I%-2),PSIOLD%)-(R%(I%),PSI%)
9670    PSIOLD%=PSI%
9680 NEXT I%
9690 '
9700 FOR I%=1 TO NREXT% STEP 2                        'graph free PSI
9710    PSI%=172-(PSIF(I%)+PSIFMAX)*PSIFSCALE
9720    CIRCLE (R%(I%),PSI%),2
9730    IF I%<>1 THEN LINE (R%(I%-2),PSIOLD%)-(R%(I%),PSI%)
9740    PSIOLD%=PSI%
9750 NEXT I%
9760 '
9770 RETURN
9780 '
10000 '************************************************************
10010 'subroutine to calculate dsigma/domega
10020 'input   variables:  DELTA(L%)
10030 'output variables:  DSIGMA(L%)
10040 'global variables:  GRAPHICS%, LFINISH%, LMAX%, LSTART%, NANG%
10050 'local   variables:  CD, FREAL, FIMAG, I%, L%, PL(L%,I%), SD, THETA, X
10060 '************************************************************
10070 IF LFINISH%>LTABLE% THEN GOSUB 11000             'calculate PL's
10080 '
10090 FOR I%=0 TO NANG%                                'find dsigma/domega
10100    FREAL=0:  FIMAG=0                             ' at each of 36 angles
10110    FOR L%=LSTART% TO LFINISH%                    'sum over L values
10120       SD=SIN(DELTA(L%))
10130       CD=COS(DELTA(L%))
10140       FREAL=FREAL+(2*L%+1)*SD*CD*PL(L%,I%)       'real and imaginary
10150       FIMAG=FIMAG+(2*L%+1)*SD*SD*PL(L%,I%)       ' parts of scatt. ampl
10160    NEXT L%
10170    FREAL=FREAL/K
10180    FIMAG=FIMAG/K
10190    DSIGMA(I%)=FREAL*FREAL+FIMAG*FIMAG
10200 '
10210    IF GRAPHICS% GOTO 10260                       'display results
10220       LOCATE ,1
10230       PRINT USING "Theta=###    Real amp=+#.###^^^^";(I%*180/NANG%),FREAL;
10240       PRINT USING "  Imag amp=+#.###^^^^";FIMAG;
10250       PRINT USING "  dSigma/dOmega=+#.###^^^^";DSIGMA(I%)
10260 NEXT I%
10270 IF GRAPHICS% THEN GOTO 10320
10280    LOCATE ,31
10290    BEEP: PRINT "Type c to continue";
10300    IF INKEY$="c" THEN GOTO 10340 ELSE GOTO 10300
10310 '
10320 GOSUB 12000                                      'graph dsigma/domega
10330 '
10340 RETURN
10350 '
11000 '************************************************************
11010 'subroutine to calculate the Legendre polynomials
```

```
11020 'input    variables:   LFINISH%, LTABLE%
11030 'output variables:   LTABLE%, PL(I%,J%)
11040 'global variables:   NANG%, PI
11050 'local    variables:   I%, L%, THETA, X
11060 '***********************************************************
11070 FOR I%=0 TO NANG%                              'find Legendre polynom
11080     THETA=I%*PI/NANG%                          ' for NANG% angles
11090     X=COS(THETA)
11100     FOR L%=LTABLE% TO LFINISH%-1               'forward recursion
11110        PL(L%+1,I%)=((2*L%+1)*X*PL(L%,I%)-L%*PL(L%-1,I%))/(L%+1)
11120     NEXT L%
11130 NEXT I%
11140 '
11150 LTABLE%=LFINISH%                               'now have PL's up to L=LFINISH%
11160 '
11170 RETURN
11180 '
12000 '***********************************************************
12010 'subroutine to graph the differential cross section vs theta
12020 'input    variables:   DSIGMA(L%), SIGTOT
12030 'output variables:   none
12040 'global variables:   E, MONO%, OPT$, V$, V%, VZERO, Z
12050 'local    variables:   COL%, EXPMAX%, EXPMIN%, I%, LSIG, MARK%, NORD%,
12060 '                      ROW%, S%,SIGMAX, SIGMIN, SIGSCALE, T%, TSCALE
12070 '***********************************************************
12080 CLS:   SCREEN 2,0,0,0
12090 '
12100 SIGMAX=0                                       'find largest sigma
12110 FOR I%=0 TO NANG%
12120     IF DSIGMA(I%)>SIGMAX THEN SIGMAX=DSIGMA(I%)
12130 NEXT I%
12140 '
12150 SIGMIN=SIGMAX                                  'find smallest sigma
12160 FOR I%=0 TO NANG%
12170     IF DSIGMA(I%)<SIGMIN THEN SIGMIN=DSIGMA(I%)
12180 NEXT I%
12190 '
12200 EXPMAX%=INT(LOG(SIGMAX)/LOG(10)+1)             'find max and min
12210 EXPMIN%=INT(LOG(SIGMIN)/LOG(10))               ' exponents for sigma
12220 NORD%=EXPMAX%-EXPMIN%                          'orders of magnitude
12230 '
12240 TSCALE=584/NANG%                               'theta scale
12250 SIGSCALE=175/NORD%                             'sigma scale
12260 '
12270 LINE (56,0)-(56,175)                           'sigma axis
12280 LINE (56,175)-(640,175)                        'theta axis
12290 LOCATE 3,1:   PRINT "Sigma";                   'sigma legend
12300 LOCATE 23,68:   PRINT "Theta";                 'theta legend
12310 '
12320 LOCATE 1,(80-V%)                               'display name of pot
12330 PRINT V$;
12340 '
12350 IF OPT$="1" THEN GOTO 12390                    'display E, Z, Vzero
12360     LOCATE 2,54
12370     PRINT USING "Vzero=+###.##    E=#####.##";VZERO,E;
12380     GOTO 12420
12390 LOCATE 2,62
12400 PRINT USING "Z=###    E=#####.##";Z,E;
```

```
12410 '
12420 LOCATE 24,20                                          'display sigma
12430 PRINT USING "Sigma total=+#.###^^^^";SIGTOT;
12440 PRINT USING "     L range= ## to ###";LSTART%,LFINISH%;
12450 '
12460 FOR MARK%=0 TO 4                                      'theta ticks
12470    T%=56+MARK%*584/4
12480    LINE (T%,173)-(T%,177)
12490    COL%=6+MARK%*18
12500    IF MARK%=4 THEN COL%=77                            'theta labels
12510    LOCATE 23,COL%
12520    PRINT USING "###";MARK%*45;
12530 NEXT MARK%
12540 '
12550 FOR MARK%=0 TO NORD%                                  'sigma ticks
12560    S%=MARK%*175/NORD%
12570    LINE (54,S%)-(58,S%)
12580    ROW%=MARK%*22/NORD%:  IF ROW%=0 THEN ROW%=1        'sigma labels
12590    LOCATE ROW%,1
12600    PRINT USING "1E+##";(EXPMAX%-MARK%)
12610 NEXT MARK%
12620 '
12630 '
12640 FOR I%=0 TO NANG%                                     'graph partial sigma
12650    T%=56+I%*TSCALE                                    'on semilog plot
12660    LSIG=LOG(DSIGMA(I%))/LOG(10)
12670    S%=175-(LSIG-EXPMIN%)*SIGSCALE
12680    LINE (T%+2,S%+2)-(T%-2,S%-2)                       'mark points with
12690    LINE (T%+2,S%-2)-(T%-2,S%+2)                       ' an X
12700    IF I%=0 GOTO 12720
12710      LINE (TOLD%,SOLD%)-(T%,S%)
12720    TOLD%=T%
12730    SOLD%=S%
12740 NEXT I%
12750 '
12760 LOCATE 25,31,1:  BEEP:  PRINT "Type c to continue";
12770 IF INKEY$<>"c" THEN GOTO 12770
12780 LOCATE ,,0
12790 '
12800 IF MONO% THEN GOSUB 15000                            'return to text screen
12810 CLS
12820 '
12830 RETURN
12840 '
13000 '***********************************************************************
13010 'subroutine to display the header screen
13020 'input  variables: none
13030 'output variables: GRAPHICS%,MONO%
13040 'global variables: none
13050 'local  variables: G$,M$,ROW%
13060 '***********************************************************************
13070 SCREEN 0: CLS: KEY OFF                                'program begins in text mode
13080 '
13090 LOCATE 1,30: COLOR 15                                 'display book title
13100 PRINT "COMPUTATIONAL PHYSICS";
13110 LOCATE 2,40: COLOR 7
13120 PRINT "by";
13130 LOCATE 3,33: PRINT "Steven E. Koonin";
```

```
13140 LOCATE 4,14
13150 PRINT "Copyright 1985, Benjamin/Cummings Publishing Company";
13160 '
13170 LOCATE 5,10                                    'draw the box
13180 PRINT CHR$(201)+STRING$(59,205)+CHR$(187);
13190 FOR ROW%=6 TO 19
13200    LOCATE ROW%,10: PRINT CHR$(186);
13210    LOCATE ROW%,70: PRINT CHR$(186);
13220 NEXT ROW%
13230 LOCATE 20,10
13240 PRINT CHR$(200)+STRING$(59,205)+CHR$(188);
13250 '
13260 COLOR 15                                       'print title, etc.
13270 LOCATE  7,36:  PRINT "PROJECT 4";
13280 COLOR 7
13290 LOCATE 10,18:  PRINT "Partial-wave solution of quantum scattering";
13300 LOCATE 13,36:  PRINT "**********"
13310 LOCATE 15,25:  PRINT "Type 'e' to end while running."
13320 LOCATE 16,22:  PRINT "Type ctrl-break to stop at a prompt."
13330 '
13340 LOCATE 19,16: BEEP                             'get screen configuration
13350 INPUT "Does your computer have graphics capability (y/n)";G$
13360 IF LEFT$(G$,1)="y" OR LEFT$(G$,1)="n" GOTO 13390
13370    LOCATE 19,12:  PRINT SPACE$(58):  BEEP
13380    LOCATE 18,35:  PRINT "Try again,":  GOTO 13340
13390 IF LEFT$(G$,1)="y" GOTO 13420
13400    GRAPHICS%=0:  MONO%=-1
13410    GOTO 13510
13420 GRAPHICS%=-1
13430 LOCATE 18,15:  PRINT SPACE$(55)
13440 LOCATE 19,15:  PRINT SPACE$(55)
13450 LOCATE 19,15: BEEP
13460 INPUT "Do you also have a separate screen for text (y/n)";M$
13470 IF LEFT$(M$,1)="y" OR LEFT$(M$,1)="n" GOTO 13490  'error trapping
13480    LOCATE 18,35:  PRINT "Try again,":  GOTO 13440
13490 IF LEFT$(M$,1)="y" THEN MONO%=-1 ELSE MONO%=0
13500 '
13510 LOCATE 21,8
13520 PRINT"This program calculates the total and differential cross sections";
13530 LOCATE 22,8
13540 PRINT"and phase shifts for scattering of an electron from a spherically";
13550 LOCATE 23,11
13560 PRINT "symmetric potential using a partial wave decomposition.  All";
13570 LOCATE 24,9
13580 PRINT "energies are in eV, angles in degrees, and lengths in Angstroms.";
13590 LOCATE 25,31,1:  BEEP
13600 PRINT "Type c to continue";
13610 IF INKEY$<>"c" GOTO 13610
13620 LOCATE ,,0
13630 '
13640 IF MONO% AND GRAPHICS% THEN GOSUB 15000        'switch to text screen
13650 '
13660 RETURN
13670 '
14000 '**************************************************************************
14010 'subroutine to switch from mono to graphics screen
14020 'input  variables: none
14030 'output variables: none
```

```
14040 'global variables: none
14050 'local  variables: none
14060 '*****************************************************************
14070 DEF SEG=0
14080 POKE &H410,  (PEEK(&H410) AND &HCF) OR &H10
14090 SCREEN 0:  WIDTH 80: LOCATE ,,0
14100 RETURN
14110 '
15000 '*****************************************************************
15010 'subroutine to switch from graphics to mono screen
15020 'input  variables: none
15030 'output variables: none
15040 'global variables: none
15050 'local  variables: none
15060 '*****************************************************************
15070 DEF SEG=0
15080 POKE &H410,  (PEEK(&H410) OR &H30)
15090 SCREEN 0:  WIDTH 80:  LOCATE ,,0
15100 RETURN
15110 '
```

C.5 Project V

This program solves the schematic shell model described in the text for a specified number of particles and coupling strength. The bases of states with odd and even m are treated separately. For the set of parameters input (subroutine 1000), the tri-diagonal hamiltonian matrices of Eq. (V.9) are constructed (loops 2070-2160 and 2320-2400) and all of the negative eigenvalues found (loop 4240-4430) by searching for the zeros of the determinant of $(H-\lambda I)$, computed by the recursion formula (5.10) in subroutine 5000; the Gerschgorin bounds on the eigenvalues, Eqs. (5.11), are used to guide the search (lines 4070-4220). After an eigenvalue has been found, the associated eigenvector is found by two inverse vector iterations (subroutines 3000 and 7000), if requested. After all of the negative eigenvalues have been calculated, they are displayed (subroutine 6000), together with the expectation values of J_z (if the eigenvectors have been computed).

A demonstration run of this program can be had by entering 14 particles, a coupling constant chi of 1, and by requesting that the eigenvectors be calculated.

```
10  '*********************************************************************
20  'Project 5: Solution of a schematic shell model
21  'COMPUTATIONAL PHYSICS by Steven E. Koonin
22  'Copyright 1985, Benjamin/Cummings Publishing Company
30  '*********************************************************************
40  DIM DIAG(30)                              'diagonal of Hamiltonian
50  DIM LDIAG(30)                             'off-diagonal of Hamiltonian
60  DIM EVAL(30)                              'temp storage for eigenvalues
70  DIM A#(30,30)                             'array for inverse of (E-H)
80  DIM P#(30),Q#(30),R%(30)                  'work space for inversion
90  DIM EVEVEN(30),EVODD(30)                  'even and odd eigenvalues
100 DIM CEVEN(30,30),CODD(30,30)              'even and odd eigenvectors
110 DIM TVECTOR(30),NVECTOR(30)              'temp arrays for eigenvectors
120 '
130 GOSUB 8000                                'display header screen
140 '
150 GOSUB 1000                                'get parameters for a case
160 GOSUB 2000                                'find eigenvalues and vectors
170 GOSUB 6000                                'display results
180 GOTO 150                                  'do the next case
190 '
1000 '*******************************************************************
1010 'subroutine to read in the parameters for a case
1020 'input   variables: none
1030 'output  variables: CHI,JJ%,JJ1%,NPART%,V,VECTORS%
1040 'global  variables: none
1050 'local   variables: KB$
1060 '*******************************************************************
1070 CLS: KEY OFF
1080 '                                            'get # of particles
1090 LOCATE 5,10: BEEP
1100 INPUT "Enter number of particles (even, <=58, <=0 to end)";NPART%
```

```
1110 IF NPART%<=0 THEN END
1120 IF NPART% MOD 2 <>0 OR NPART%>58 THEN GOTO 1090
1130 '
1140 LOCATE 7,10: BEEP                                    'get value of chi
1150 INPUT "Enter the coupling parameter chi"; CHI
1160 '
1170 JJ%=NPART%/2: JJ1%=JJ%*(JJ%+1)                      'value of j and j(j+1)
1180 V=CHI/NPART%                                         'interaction strength
1190 '
1200 LOCATE 9,15                                          'show derived params.
1210 PRINT USING "For these parameters, the quasi-spin is ##";JJ%
1220 LOCATE 10,21
1230 PRINT USING "and the interaction strength V is ##.####";V
1240 '
1250 LOCATE 14,10: BEEP                                   'calculate vectors?
1260 INPUT "Do you want the eigenvectors and <Jz> calculated (y/n)";KB$
1270 IF KB$="y" THEN VECTORS%=-1 ELSE VECTORS%=0
1280 RETURN
1290 '
2000 '***********************************************************************
2010 'subroutine to set up and diagonalize the hamiltonian matrix
2020 'input    variables: none
2030 'output variables: CEVEN,CODD,DIAG,EVEVEN,EVODD,LDIAG
2040 'global variables: EVAL,JJ%,JJ1%,LAMBDA,N%,NFIND%,TVECTOR,V,VECTORS%
2050 'local    variables: I%,ILAM%,M%,TEMP
2060 '***********************************************************************
2070 N%=JJ%+1                                             'number of even-m states
2080 '                                                    'set-up even-m matrix
2090 FOR I%=1 TO N%                                       'loop over even-m states
2100    M%=-JJ%+2*(I%-1)                                  'm for this state
2110    DIAG(I%)=M%                                       'diagonal part of H
2120    IF I%=1 THEN GOTO 2160                            'off-diagonal part of H
2130      TEMP=JJ1%-M%*(M%-1)
2140      TEMP=TEMP*(JJ1%-(M%-1)*(M%-2))
2150      LDIAG(I%)=-V/2*SQR(ABS(TEMP))
2160 NEXT I%
2170 '                                                    'number of evals to find
2180 IF JJ% MOD 2 =0 THEN NFIND%=1+JJ%/2 ELSE NFIND%=(JJ%+1)/2
2190 '
2200 GOSUB 4000                                           'find the evals
2210 '
2220 FOR ILAM%=1 TO NFIND%                                'for each eval found
2230    EVEVEN(ILAM%)=EVAL(ILAM%)                         'store it
2240    IF NOT VECTORS% THEN GOTO 2300                    'sometimes find evector
2250      LAMBDA=EVAL(ILAM%)
2260      GOSUB 3000                                      'find the evector
2270      FOR I%=1 TO N%                                  'store it
2280        CEVEN(I%,ILAM%)=TVECTOR(I%)
2290      NEXT I%
2300 NEXT ILAM%
2310 '                                                    'now do odd-m states
2320 N%=JJ%                                               'number of odd-m states
2330 FOR I%=1 TO N%                                       'loop over odd-m states
2340    M%=-JJ%-1+2*I%                                    'm for each state
2350    DIAG(I%)=M%                                       'diagonal part of H
2360    IF I%=1 THEN GOTO 2400                            'off-diagonal part of H
2370      TEMP=JJ1%-M%*(M%-1)
2380      TEMP=TEMP*(JJ1%-(M%-1)*(M%-2))
```

```
2390      LDIAG(I%)=-V/2*SQR(ABS(TEMP))
2400 NEXT I%
2410 '                                             'number of evals to find
2420 IF JJ% MOD 2 =0 THEN NFIND%=JJ%/2 ELSE NFIND%=(JJ%+1)/2
2430 '
2440 GOSUB 4000                                    'find the evals
2450 '
2460 FOR ILAM%=1 TO NFIND%                          'for each eval found
2470   EVODD(ILAM%)=EVAL(ILAM%)                      'store it
2480   IF NOT VECTORS% THEN GOTO 2540               'sometimes find evector
2490   LAMBDA=EVAL(ILAM%)
2500   GOSUB 3000                                   'find the evector
2510   FOR I%=1 TO N%                                'store it
2520      CODD(I%,ILAM%)=TVECTOR(I%)
2530   NEXT I%
2540 NEXT ILAM%
2550 '
2560 RETURN
2570 '
3000 '***********************************************************************
3010 'subroutine to return an eigenvector of a tri-diagonal matrix
3020 'input    variables: DIAG,LAMBDA,LDIAG,N%,NFIND%
3030 'output variables: TVECTOR
3040 'global variables: A#
3050 'local    variables: I%,J%,NORM,SUM
3060 '***********************************************************************
3070 FOR I%=1 TO N%                                 'load (E-H) into A#
3080   FOR J%=1 TO N%                               'most elements are 0
3090      A#(I%,J%)=0
3100   NEXT J%
3110   A#(I%,I%)=CDBL(LAMBDA-DIAG(I%))              'diagonal element
3120   IF I%<N% THEN A#(I%,I%+1)=-CDBL(LDIAG(I%+1)) 'off-diagonal elements
3130   IF I%>1   THEN A#(I%,I%-1)=-CDBL(LDIAG(I%))
3140   TVECTOR(I%)=RND                              'random initial evector
3150 NEXT I%
3160 '
3170 GOSUB 7000                                     'invert (E-H)
3180 '                                              'first inverse iteration
3190 FOR I%=1 TO N%                                 'apply inverse to TVECTOR
3200   SUM=0                                        ' to produce NVECTOR
3210   FOR J%=1 TO N%
3220      SUM=SUM+A#(I%,J%)*TVECTOR(J%)
3230   NEXT J%
3240   NVECTOR(I%)=SUM
3250 NEXT I%
3260 '                                              'second inverse iteration
3270 NORM=0
3280 FOR I%=1 TO N%                                 'apply inverse to NVECTOR
3290   SUM=0                                        ' to produce TVECTOR
3300   FOR J%=1 TO N%
3310      SUM=SUM+A#(I%,J%)*NVECTOR(J%)
3320   NEXT J%
3330   TVECTOR(I%)=SUM
3340   NORM=NORM+SUM^2
3350 NEXT I%
3360 '
3370 NORM=1/SQR(NORM)                              'normalize final evector
3380 FOR I%=1 TO N%
```

```
3390     TVECTOR(I%)=TVECTOR(I%)*NORM
3400 NEXT I%
3410 RETURN
3420 '
4000 '*********************************************************************
4010 'subroutine to find the eigenvalues of a symmetric tri-diagonal matrix
4020 'input  variables: DIAG,LDIAG,N%,NFIND%
4030 'output variables: EVAL
4040 'global variables: COUNT%,LAMBDA
4050 'local  variables: DLAM,GER,I%,L%,LBOUND,RAD,SPACING,UBOUND
4060 '*********************************************************************
4070 LBOUND=DIAG(1)-ABS(LDIAG(2))                 'find gerschgorin bounds on evals
4080 UBOUND=DIAG(1)+ABS(LDIAG(2))
4090 FOR I%=2 TO N%-1
4100     RAD=ABS(LDIAG(I%+1))+ABS(LDIAG(I%))
4110     GER=DIAG(I%)-RAD
4120     IF GER<LBOUND THEN LBOUND=GER
4130     GER=DIAG(I%)+RAD
4140     IF GER>UBOUND THEN UBOUND=GER
4150 NEXT I%
4160 GER=DIAG(N%)-ABS(LDIAG(N%))
4170 IF GER<LBOUND THEN LBOUND=GER
4180 GER=DIAG(N%)+ABS(LDIAG(N%))
4190 IF GER>UBOUND THEN UBOUND=GER
4200 '
4210 LAMBDA=LBOUND                               'first guess for lowest eval
4220 SPACING=(UBOUND-LBOUND)/N%                  'estimate spacing of evals
4230 '
4240 FOR L%=1 TO NFIND%                          'loop over evals to find
4250     DLAM=SPACING                            'estimate step size
4260     COUNT%=L%-1                             'coarse search to find a LAMBDA
4270     WHILE COUNT%<L%                         ' between L% and L%-1 evals
4280         LAMBDA=LAMBDA+DLAM
4290         GOSUB 5000                          'find number of evals <LAMBDA
4300     WEND
4310 '
4320     LAMBDA=LAMBDA-DLAM
4330     WHILE DLAM>.00001                       'now search to within 1e-5
4340         GOSUB 5000                          'find number of evals <LAMBDA
4350         DLAM=DLAM/2                          'next step is half as big
4360         IF COUNT%>L%-1 THEN GOTO 4390        'if still below eval
4370             LAMBDA=LAMBDA+DLAM               ' take the step
4380             GOTO 4400
4390         LAMBDA=LAMBDA-DLAM                   'step was too far, back-up halfway
4400     WEND
4410 '
4420     EVAL(L%)=LAMBDA                         'store the value found
4430 NEXT L%
4440 RETURN
4450 '
5000 '*********************************************************************
5010 'subroutine to count the number of eigenvalues less than LAMBDA
5020 'input  variables: DIAG,LAMBDA,LDIAG,N%
5030 'output variables: COUNT%,DET
5040 'global variables: none
5050 'local  variables: I%
5060 '*********************************************************************
5070 TEMP1=1                                     '1st term in det recursion
```

```
5080 DET=DIAG(1)-LAMBDA                              '2nd term in det recursion
5090 IF DET<0 THEN COUNT%=1 ELSE COUNT%=0            'count of evals < LAMBDA
5100 FOR I%=2 TO N%                                  'recursion for the det
5110     TEMP=(DIAG(I%)-LAMBDA)*DET-LDIAG(I%)^2*TEMP1
5120     TEMP1=DET: DET=TEMP                         'roll the terms
5130     IF DET*TEMP1<0 THEN COUNT%=COUNT%+1         'count 1 more if sign change
5140     IF ABS(DET)<100000! GOTO 5160              'renormalize det if too big
5150        DET=DET/100000!: TEMP1=TEMP1/100000!
5160 NEXT I%
5170 RETURN
5180 '
6000 '********************************************************************
6010 'subroutine to output the results
6020 'input    variables: none
6030 'output variables: none
6040 'global variables: CEVEN,CHI,CODD,EVEVEN,EVODD,JJ%,NPART%,VECTORS%
6050 'local    variables: COL%,E,I%,ILAM%,JZ,K%,M%,ROW%
6060 '********************************************************************
6070 CLS
6080 LOCATE 1,14                                     'print heading info
6090 PRINT USING "Number of particles=## ";NPART%
6100 LOCATE 1,37
6110 PRINT USING "Coupling constant chi=+###.###";CHI
6120 LOCATE 2,22
6130 PRINT "The states with E<=0 are as follows:";
6140 '
6150 FOR ILAM%=1 TO JJ%+1                            'loop over states found
6160     IF ILAM% MOD 2 =0 THEN GOTO 6270           'odd or even state
6170        K%=(ILAM%+1)/2                           'id# of the even state
6180        E=EVEVEN(K%)                             'energy of this state
6190        IF NOT VECTORS% THEN GOTO 6360           'sometimes find <Jz>
6200           JZ=0
6210           FOR I%=1 TO JJ%+1                     'loop over even basis
6220              M%=-JJ%+2*(I%-1)                   'M for this basis state
6230              JZ=JZ+CEVEN(I%,K%)^2*M%            'contribution to <Jz>
6240           NEXT I%
6250           GOTO 6360
6260 '
6270        K%=ILAM%/2                               'id# for an odd state
6280        E=EVODD(K%)                              'energy of this state
6290        IF NOT VECTORS% THEN GOTO 6360           'sometimes compute <Jz>
6300           JZ=0
6310           FOR I%=1 TO JJ%                       'loop over odd basis
6320              M%=-JJ%-1+2*I%                     'M for this basis state
6330              JZ=JZ+CODD(I%,K%)^2*M%             'contribution to <Jz>
6340           NEXT I%
6350 '
6360        IF ILAM%>22 THEN COL%=40 ELSE COL%=1     'column and row of output
6370        IF ILAM%>22 THEN ROW%=ILAM%-19 ELSE ROW%=ILAM%+2
6380        LOCATE ROW%,COL%
6390        PRINT USING "State=## E=+###.####";ILAM%,E;   'output energy
6400        IF NOT VECTORS% THEN GOTO 6430           'sometimes output <Jz>
6410           LOCATE ROW%,COL%+21
6420           PRINT USING "<Jz>=+##.####";JZ;
6430 NEXT ILAM%
6440 '                                               'prompt and wait
6450 LOCATE 25,28: PRINT "Press any key to continue";: BEEP
6460 IF INKEY$<> "" THEN GOTO 6470 ELSE GOTO 6460
```

```
6470 RETURN
6480 '
7000 '*********************************************************************
7010 'subroutine to invert a symmetric matrix A#(N%,N%) by Gauss-Jordan elim.
7020 ' On output, A# contains the inverse of the input matrix.
7030 ' Main code must dimension A# and auxiliary arrays R%(N%),P#(N%),Q#(N%).
7040 'input   variables: A#,N%
7050 'output variables: A#
7060 'global variables: P#,Q#,R%
7070 'local   variables: BIG,I%,J%,K%,PIVOT1#,TEST
7080 '*********************************************************************
7090 FOR I%=1 TO N%                              'R%=0 tells which rows
7100    R%(I%)=1                                 ' already zeroed
7110 NEXT I%
7120 '
7130 FOR I%=1 TO N%                              'loop over rows
7140    K%=0: BIG=0                              'pivot is largest diag
7150    FOR J%=1 TO N%                           'in rows not yet zeroed
7160       IF R%(J%)=0   THEN GOTO 7200
7170        TEST=ABS(A#(J%,J%))
7180         IF TEST<=BIG THEN GOTO 7200
7190          BIG=TEST: K%=J%
7200    NEXT J%
7210 '
7220    IF K%=0 THEN GOTO 7580                   'all diags=0;A# singulr
7230    IF I%=1 THEN PIVOT1#=A#(K%,K%)           'largest diag of A#
7240    IF ABS(A#(K%,K%)/PIVOT1#)<1E-14 THEN GOTO 7580 'ill-conditioned A#
7250 '
7260    R%(K%)=0: Q#(K%)=1/A#(K%,K%)             'begin zeroing this row
7270    P#(K%)=1: A#(K%,K%)=0
7280    IF K%=1 THEN GOTO 7360
7290      FOR J%=1 TO K%-1                       'elements above diag
7300         P#(J%)=A#(J%,K%)
7310         Q#(J%)=A#(J%,K%)*Q#(K%)
7320         IF R%(J%)=1 THEN Q#(J%)=-Q#(J%)
7330         A#(J%,K%)=0
7340      NEXT J%
7350 '
7360    IF K%=N% THEN GOTO 7440
7370      FOR J%=K%+1 TO N%                      'elements right of diag
7380         P#(J%)=A#(K%,J%)
7390         Q#(J%)=-A#(K%,J%)*Q#(K%)
7400         IF R%(J%)=0 THEN P#(J%)=-P#(J%)
7410         A#(K%,J%)=0
7420      NEXT J%
7430 '
7440    FOR J%=1 TO N%                           'transform all of A#
7450       FOR K%=J% TO N%
7460          A#(J%,K%)=A#(J%,K%)+P#(J%)*Q#(K%)
7470       NEXT K%
7480    NEXT J%
7490 NEXT I%
7500 '
7510 FOR J%=2 TO N%                              'symmetrize A#^-1
7520    FOR K%=1 TO J%-1
7530       A#(J%,K%)=A#(K%,J%)
7540    NEXT K%
7550 NEXT J%
```

```
7560 '
7570 RETURN
7580 BEEP: PRINT "FAILURE IN MATRIX INVERSION";
7590 END
7600 '
8000 '*******************************************************************
8010 'subroutine to display the header screen
8020 'input   variables: none
8030 'output variables: none
8040 'global variables: none
8050 'local   variables: ROW%
8060 '*******************************************************************
8070 SCREEN 0: CLS: KEY OFF                        'program begins in text mode
8080 '
8090 LOCATE 1,30: COLOR 15                         'display book title
8100 PRINT "COMPUTATIONAL PHYSICS";
8110 LOCATE 2,40: COLOR 7
8120 PRINT "by";
8130 LOCATE 3,33: PRINT "Steven E. Koonin";
8140 LOCATE 4,14
8150 PRINT "Copyright 1985, Benjamin/Cummings Publishing Company";
8160 '
8170 LOCATE 5,10                                   'draw the box
8180 PRINT CHR$(201)+STRING$(59,205)+CHR$(187);
8190 FOR ROW%=6 TO 19
8200     LOCATE ROW%,10: PRINT CHR$(186);
8210     LOCATE ROW%,70: PRINT CHR$(186);
8220 NEXT ROW%
8230 LOCATE 20,10
8240 PRINT CHR$(200)+STRING$(59,205)+CHR$(188);
8250 '
8260 COLOR 15                                      'print title, etc.
8270 LOCATE  7,36:  PRINT "PROJECT 5";
8280 COLOR 7
8290 LOCATE 10,23:  PRINT "Solution of a schematic shell model";
8300 LOCATE 13,36:  PRINT "*********"
8310 LOCATE 15,25:  PRINT "Type 'e' to end while running."
8320 LOCATE 16,22:  PRINT "Type ctrl-break to stop at a prompt."
8330 '
8340 LOCATE 21,19: PRINT "This program finds all negative eigenvalues";
8350 LOCATE 22,18: PRINT "of the schematic shell model of Lipkin et al.";
8360 LOCATE 23,18: PRINT "If requested, the eigenvectors and expectation";
8370 LOCATE 24,18: PRINT "values of Jz in each state are also calculated";
8380 LOCATE 25,31,1:  BEEP
8390 PRINT "Type c to continue";
8400 IF INKEY$<>"c" GOTO 8400
8410 LOCATE ,,0
8420 '
8430 RETURN
8440 '
```

C.6 Project VI

This program uses a relaxation method to solve for the stationary, incompressible, viscous flow around a plate in two dimensions; the geometry is as described in the text. The lattice size, plate size and location, lattice Reynolds number, and relaxation parameters are defined in subroutine 1000; a sub-lattice to be relaxed also can be specified. Initial values for the vorticity and stream function are defined in subroutine 2000, either from a previously calculated solution stored in a file (lines 2230-2360) or as those for the free-streaming solution (loop 2380-2430). The main relaxation loop then occurs (210-270). Subroutines 3000 and 4000 relax the stream function and vorticity, respectively, by solving Eqs. (VI.9) and (VI.10). Apart from the special points on the lattice or plate boundaries, the relaxation steps are embodied in lines 3220-3230 and 4270-4300. After each sweep of the sub-lattice, the dimensionless pressure and viscous forces are calculated by integrating the appropriate quantities over the boundaries of the plate (subroutine 6000), and the solution is displayed (subroutine 5000) as an array of characters on both the graphics and text monitors. When the relaxation iterations are interrupted, subroutine 7000 processes the various options for continuing.

To obtain a representative output, choose horizontal and vertical lattice sizes of

$$Nx=40, \ Ny=20,$$

a plate half-width of 6, a plate thickness of 6, and a front-face location of 16. Then enter a lattice Reynolds number of 1.0 and 0.3,0.3 as the relaxation parameters for the stream function and vorticity. Finally, specify the sub-lattice to be relaxed as the entire lattice by entering

$$Nxmax=40, \ Nymax=20,$$

and do not read the starting conditions from a file. Iterations will then converge to a very laminar solution.

```
10  '*****************************************************************
20  'Project 6: 2-D viscous incompressible flow about a rectangular block
21  'COMPUTATIONAL PHYSICS by Steven E. Koonin
22  'Copyright 1985, Benjamin/Cummings Publishing Company
30  '*****************************************************************
40  DIM P(24,79)                    'array for the stream function
50  DIM XSI(24,79)                  'array for the vorticity
60  DIM PLACE%(24,79)               'monitor buffer location of each point
70  DIM IS%(24,79)                  'IS%=1 indicates a boundary point
80  '                                ' IS%=0 for all other points
90  DIM ASKII%(35)                  'array for conversion of P and XSI
100 DIM NEGASKII%(26)               ' to ASCII characters
110 DIM INTERIOR%(10)               'array for the color of interior pts
```

```
120 DIM BORDER%(10)                               'array for the color of border pts
130 FIRST%=-1                                     'flag for printing information once
140 '
150 GOSUB 8000                                    'display header screen
160 GOSUB 9000                                    'display cautionary message
170 GOSUB 1000                                    'input parameters of the calculation
180 GOSUB 2000                                    'initialize the lattice
190 IF FIRST% THEN GOSUB 10000                    'print information about display once
200 IT%=0                                         'zero iteration count
210 WHILE INKEY$<>"e"                             'typing "e" ends the iteration loop
220     GOSUB 5000                                'display the functions
230     GOSUB 6000                                'calculate and display the forces
240     IT%=IT%+1                                 'increment iteration count
250     GOSUB 3000                                'relax P
260     GOSUB 4000                                'relax XSI
270 WEND
280 '
290 LOCATE 25,53,1: BEEP
300 COLOR 7: PRINT "Press any key to continue";
310 IF INKEY$="" GOTO 310
320 GOSUB 7000                                    'display options for continuing
330 IF AGAIN% THEN GOTO 170 ELSE GOTO 200
340 '
1000 '**********************************************************************
1010 'subroutine to input the parameters of the calculation
1020 'input   variables: none
1030 'output variables: JL%, JR%, NX%, NXMAX%, NY%, NYMAX%, RE, RE4, WID%,
1040 '                   WP, WP1, WX, WX1
1050 'global variables: GRAPHICS%, MONO%
1060 'local   variables: THICK%
1070 '**********************************************************************
1080 IF MONO%     THEN GOSUB 12000: CLS                      'clear screens
1090 IF GRAPHICS% THEN GOSUB 11000: CLS
1100 '
1110 LOCATE 10,20,0                                          'input parameters
1120 INPUT "Enter horizontal lattice size, Nx(>0,<=79)";NX%
1130 IF NX%>0 AND NX%<=79 THEN GOTO 1150
1140    BEEP: GOTO 1110
1150 LOCATE 11,20,0
1160 INPUT "Enter vertical    lattice size, Ny(>0,<=24)";NY%
1170 IF NY%>0 AND NY%<=24 THEN GOTO 1190
1180    BEEP: GOTO 1150
1190 LOCATE 12,20,0
1200 INPUT "Enter half-width of plate (in mesh spacings)";WID%
1210 LOCATE 13,20,0
1220 INPUT "Enter thickness  of plate (in mesh spacings)";THICK%
1230 LOCATE 14,20,0
1240 INPUT "Enter lattice location of front face";JL%
1250 JR%=JL%+THICK%-1
1260 LOCATE 15,20
1270 INPUT "Enter lattice Reynolds number";RE
1280 RE4=RE/4
1290 LOCATE 16,20
1300 INPUT "Enter relaxation parameters for PSI,XSI";WP,WX
1310 WP1=1-WP: WX1=1-WX
1320 LOCATE 17,20
1330 INPUT "Enter size of the sublattice to relax, Nxmax, Nymax";NXMAX%,NYMAX%
1340 CLS
```

```
1350 '
1360 RETURN
1370 '
2000 '********************************************************************
2010 'subroutine to initialize the calculation
2020 'input  variables: none
2030 'output variables: IS(I%,J%), P(I%,J%), PLACE%(I%,J%), XSI(I%,J%)
2040 'global variables: NX%, NY%, WID%
2050 'local  variables: BOTTOM%, I%, IC%, IR%, J%, LEDGE%, NXOLD%, NYOLD%, S$
2060 '********************************************************************
2070 LEDGE%=39-NX%/2:   IF NX% MOD 2=1 THEN LEDGE%=LEDGE%+2      'center lattice
2080 BOTTOM%=12+NY%/2:  IF NY% MOD 2=1 THEN BOTTOM%=BOTTOM%-1
2090 '
2100 FOR I%=1 TO NY%                               'establish array for buffer pos
2110    IR%=BOTTOM%-I%+1
2120    FOR J%=1 TO NX%
2130       IC%=LEDGE%+J%-1
2140       PLACE%(I%,J%)=160*(IR%-1)+2*(IC%-1)
2150       IS%(I%,J%)=0
2160    NEXT J%
2170 NEXT I%
2180 '
2190 CLS                                           'read in data file
2200 BEEP: LOCATE 10,20,0
2210 INPUT "Read starting conditions from a file (y/n)";S$
2220 IF S$<>"y" THEN GOTO 2380
2230    BEEP: LOCATE 11,20
2240    INPUT "Enter name of file to read";S$
2250    OPEN S$ FOR INPUT AS #1
2260    INPUT #1,NXOLD%,NYOLD%
2270    IF NXOLD%=NX% AND NYOLD%=NY% THEN GOTO 2300
2280      PRINT "Mismatch in lattice sizes, input file not used"
2290      CLOSE #1: GOTO 2380
2300    FOR I%=1 TO NY%
2310       FOR J%=1 TO NX%
2320          INPUT #1,P(I%,J%),XSI(I%,J%)
2330       NEXT J%
2340    NEXT I%
2350    CLOSE #1
2360    GOTO 2450
2370 '
2380 FOR I%=1 TO NY%
2390    FOR J%=1 TO NX%
2400       XSI(I%,J%)=0                            'initial vorticity zero
2410       P(I%,J%)=I%-1                           'initial stream function
2420    NEXT J%
2430 NEXT I%
2440 '
2450 FOR J%=JL% TO JR%                             'set value of P on plate to
2460    FOR I%=1 TO WID%                           ' zero
2470       P(I%,J%)=0
2480       IS%(I%,J%)=1                            'the plate is a boundary
2490    NEXT I%
2500 NEXT J%
2510 '
2520 FOR I%=1 TO NY%                               'set up array IS% which flags
2530    IS%(I%,1)=1                                ' boundaries
2540    IS%(I%,NX%)=1                              'right and left edges are
```

```
2550 NEXT I%                                        ' boundaries
2560 '
2570 FOR J%=1 TO NX%                                'top and bottom are boundaries
2580     IS%(1,J%)=1
2590     IS%(NY%,J%)=1
2600 NEXT J%
2610 '
2620 CLS
2630 RETURN
2640 '
3000 '****************************************************************
3010 'subroutine to relax the stream function
3020 'input   variables: P(I%,J%), XSI(I%,J%)
3030 'output  variables: P(I%,J%)
3040 'global variables: IS%(I%,J%), NX%, NXMAX%, NY%, NYMAX%, WP, WP1
3050 'local   variables: I%, J%, TEMP
3060 '****************************************************************
3070 FOR I%=1 TO NYMAX%                             'loop over the lattice
3080     FOR J%=1 TO NXMAX%
3090         IF IS%(I%,J%)=1 THEN GOTO 3410         'don't relax boundary points
3100         IF I%=NY%-1 THEN GOTO 3260             'top edge is special
3110         IF J%<>2 THEN GOTO 3160                'front edge is special
3120             TEMP=P(I%-1,2)+P(I%+1,2)+P(I%,3)-XSI(I%,J%)
3130             P(I%,2)=WP*TEMP/3+WP1*P(I%,2)
3140             P(I%,1)=P(I%,2)
3150             GOTO 3410
3160         IF J%<>NX%-1 THEN GOTO 3220            'rear edge is special
3170             TEMP=P(I%-1,J%)+P(I%+1,J%)+P(I%,J%-1)-XSI(I%,J%)
3180             P(I%,J%)=WP*.333333*TEMP+WP1*P(I%,J%)
3190             P(I%,NX%)=P(I%,J%)
3200             GOTO 3410
3210 '                                             'an interior point
3220             TEMP=P(I%-1,J%)+P(I%+1,J%)+P(I%,J%-1)+P(I%,J%+1)-XSI(I%,J%)
3230             P(I%,J%)=WP*.25*TEMP+WP1*P(I%,J%)
3240             GOTO 3410
3250 '                                             'just under lattice top
3260         IF J%<>2 THEN GOTO 3310                'left corner is special
3270             TEMP=P(I%-1,2)+P(I%,3)-XSI(I%,2)+1
3280             P(I%,2)=WP*.5*TEMP+WP1*P(I%,2)
3290             P(I%,1)=P(I%,2): P(NY%,1)=P(I%,1)+1
3300             GOTO 3390
3310         IF J%<>NX%-1 THEN GOTO 3360            'right corner is special
3320             TEMP=P(I%-1,J%)+P(I%,J%-1)-XSI(I%,J%)+1
3330             P(I%,J%)=WP*.5*TEMP+WP1*P(I%,J%)
3340             P(I%,NX%)=P(I%,J%): P(NY%,NX%)=P(I%,NX%)+1
3350             GOTO 3390
3360         TEMP=P(I%-1,J%)+P(I%,J%-1)+P(I%,J%+1)-XSI(I%,J%)+1
3370         P(I%,J%)=WP*TEMP/3+WP1*P(I%,J%)
3380 '
3390         P(NY%,J%)=P(I%,J%)+1                    'top edge = row under it + 1
3400 '
3410     NEXT J%
3420 NEXT I%
3430 RETURN
3440 '
4000 '****************************************************************
4010 'subroutine to relax the vorticity
4020 'input   variables: P(I%,J%), XSI(I%,J%)
```

```
4030 'output variables: XSI(I%,J%)
4040 'global variables: JL%, JR%, NX%, NXMAX%, NY%, NYMAX%, RE4, WID%, WX, WX1
4050 'local  variables: I%, J%, TEMP
4060 '****************************************************************
4070 'calculate boundary conditions for XSI; front and back, then top
4080 FOR I%=1 TO WID%
4090     XSI(I%,JL%)=2*P(I%,JL%-1)                    'front-face b.c. for XSI
4100     XSI(I%,JR%)=2*P(I%,JR%+1)                    'back -face b.c. for XSI
4110 NEXT I%
4120 FOR J%=JL%+1 TO JR%-1                            'top -face  b.c. for XSI
4130     XSI(WID%,J%)=2*P(WID%+1,J%)
4140 NEXT J%
4150 '
4160 FOR I%=1 TO NYMAX%                               'relaxation loop over the lattice
4170     FOR J%=1 TO NXMAX%
4180         IF IS%(I%,J%)=1 THEN GOTO 4310           'don't relax a boundary point
4190             IF J%<>NX%-1 THEN GOTO 4270          'right edge is special
4200                 TEMP=XSI(I%+1,J%)+XSI(I%-1,J%)+XSI(I%,J%-1)
4210                 TEMP=TEMP+RE4*XSI(I%,J%-1)*(P(I%+1,J%)-P(I%-1,J%))
4220                 TEMP=TEMP+RE4*(XSI(I%+1,J%)-XSI(I%-1,J%))*(P(I%,J%+1)-P(I%,J%-1))
4230                 XSI(I%,J%)=WX*TEMP/(3+RE4*(P(I%+1,J%)-P(I%-1,J%)))+WX1*XSI(I%,J%)
4240                 XSI(I%,NX%)=XSI(I%,J%)
4250                 GOTO 4310
4260 '
4270                 TEMP=XSI(I%+1,J%)+XSI(I%-1,J%)+XSI(I%,J%+1)+XSI(I%,J%-1)
4280                 TEMP=TEMP-RE4*(P(I%+1,J%)-P(I%-1,J%))*(XSI(I%,J%+1)-XSI(I%,J%-1))
4290                 TEMP=TEMP+RE4*(P(I%,J%+1)-P(I%,J%-1))*(XSI(I%+1,J%)-XSI(I%-1,J%))
4300                 XSI(I%,J%)=WX*.25*TEMP+WX1*XSI(I%,J%)
4310     NEXT J%
4320 NEXT I%
4330 RETURN
4340 '
5000 '****************************************************************
5010 'subroutine to display the stream function and vorticity
5020 'input  variables: P(I%,J%), VORT%, XSI(I%,J%)
5030 'output variables: none
5040 'global variables: ASKII%(I%), BORDER%(I%), GRAPHICS%, INTERIOR%(I%,J%),
5050 '                  MONO%, NEGASKII%(I%), NX%, NY%, PLACE%(I%,J%)
5060 'local  variables: BUF%, I%, J%, M%, TINT%, Z%
5070 '****************************************************************
5080 IF VORT% THEN GOTO 5340                          'display XSI only
5090 '
5100 COLOR 7: LOCATE 25,1,0
5110 PRINT USING "iteration ### R=##.#";IT%,RE;
5120 LOCATE 25,53,0
5130 PRINT "Press e to end iterations";
5140 '
5150 IF GRAPHICS% THEN DEF SEG=&HB800 ELSE DEF SEG=&HB000    'display P
5160 FOR I%=1 TO NY%
5170     FOR J%=1 TO NX%
5180         Z%=CINT(P(I%,J%))                        'convert P to ASCII
5190         IF Z%>35 THEN Z%=35                       ' character; keep P
5200         IF Z%<-26 THEN Z%=-26                     ' between -26 and 35
5210         IF Z%=>0 THEN Z%=ASKII%(Z%) ELSE Z%=NEGASKII%(ABS(Z%))
5220 '
5230         M%=ABS(P(I%,J%)) MOD 10                   'find tint for P
5240         IF IS%(I%,J%)=1 THEN TINT%=BORDER%(M%) ELSE TINT%=INTERIOR%(M%)
5250 '
```

```
5260        BUF%=PLACE%(I%,J%)
5270        POKE BUF%+1,TINT%
5280        POKE BUF%,Z%
5290     NEXT J%
5300 NEXT I%
5310 IF NOT (GRAPHICS% AND MONO%) THEN RETURN
5320 '
5330 DEF SEG=&HB000                                    'display XSI
5340 FOR I%=1 TO NY%
5350     FOR J%=1 TO NX%
5360        Z%=CINT(10*XSI(I%,J%))                     'convert XSI to ASCII
5370        IF Z%>35 THEN Z%=35                         ' character; keep
5380        IF Z%<-26 THEN Z%=-26                       ' 10*XSI between -26,35
5390        IF Z%>=0 THEN Z%=ASKII%(Z%) ELSE Z%=NEGASKII%(ABS(Z%))
5400 '
5410        M%=ABS(10*XSI(I%,J%)) MOD 10               'find tint for XSI
5420        IF IS%(I%,J%)=1 THEN TINT%=BORDER%(M%) ELSE TINT%=INTERIOR%(M%)
5430        IF GRAPHICS% AND VORT% THEN GOTO 5460       'want tint for mono
5440          IF IS%(I%,J%)=1 THEN TINT%=10 ELSE TINT%=2
5450 '
5460        BUF%=PLACE%(I%,J%)
5470        POKE BUF%+1, TINT%
5480        POKE BUF%,Z%
5490     NEXT J%
5500 NEXT I%
5510 IF NOT VORT% THEN RETURN
5520 LOCATE 25,31:  PRINT "Type c to continue";
5530 IF INKEY$<>"c" THEN GOTO 5530
5540 RETURN
5550 '
6000 '*******************************************************************
6010 'subroutine to calculate and display the forces
6020 'input   variables: P(I%,J%), XSI(I%,J%)
6030 'output variables: none
6040 'global variables: JL%, JR%, RE, WID%
6050 'local   variables: FP, PV, I%, J%, P
6060 '*******************************************************************
6070 FV=XSI(WID%,JL%)/2                        'calculate viscous force
6080 FOR J%=JL%+1 TO JR%                        'integral over top surface
6090     FV=FV+XSI(WID%,J%)
6100 NEXT J%
6110 FV=FV-XSI(WID%,JR%)/2
6120 FV=FV/RE/(WID%-1)
6130 '
6140 P=0: FP=0                                  'calculate pressure force
6150 FOR I%=2 TO WID%                           'integrate up front face
6160     P=P-(XSI(I%,JL%)-XSI(I%,JL%-1)+XSI(I%-1,JL%)-XSI(I%-1,JL%-1))/2/RE
6170     FP=FP+P
6180 NEXT I%
6190 FP=FP-P/2
6200 '
6210 FOR J%=JL%+1 TO JR%                        'integrate along top surface
6220     P=P+(XSI(WID%+1,J%)-XSI(WID%,J%)+XSI(WID%+1,J%-1)-XSI(WID%,J%-1))/2/RE
6230 NEXT J%
6240 FP=FP-P/2
6250 '
6260 FOR I%=WID%-1 TO 1 STEP -1                 'integrate down back surface
6270     P=P+(XSI(I%,JR%+1)-XSI(I%,JR%)+XSI(I%+1,JR%+1)-XSI(I%+1,JR%))/2/RE
```

```
6280    FP=FP-P
6290 NEXT I%
6300 FP=FP/(WID%-1)                              'normalize pressure force
6310 '
6320 COLOR 7,0: LOCATE 25,22,0                   'print results
6330 PRINT USING "Fv=+##.###^^^^ Fp=+##.###^^^^";FV,FP;
6340 RETURN
6350 '
7000 '**************************************************************************
7010 'subroutine to display the options for continuing
7020 'input  variables: P(I%,J%), XSI(I%,J%)
7030 'output variables: AGAIN%, NX%, NXMAX%, NY%, NYMAX%, RE, RE4, VORT%,
7040 '                  WP, WP1, WX, WX1
7050 'global variables: NX%, NXMAX%, NY%, NYMAX%, RE, WP, WX
7060 'local  variables: C%, S$
7070 '**************************************************************************
7080 AGAIN%=0                                    'reset flag
7090 CLS: COLOR 7,0
7100 '
7110 LOCATE 1,21
7120 PRINT "Options for continuing the calculation"
7130 LOCATE 3,10:  PRINT "1) Resume iterating"
7140 LOCATE 5,10:  PRINT "2) Re-display stream function"
7150 LOCATE 7,10:  PRINT "3) Change sub-lattice which is being swept"
7160 LOCATE 9,10:  PRINT "4) Change relaxation parameters"
7170 LOCATE 11,10: PRINT "5) Change Reynolds number"
7180 LOCATE 13,10: PRINT "6) Save current solution on disk"
7190 LOCATE 15,10: PRINT "7) Display vorticity on this screen"
7200 LOCATE 17,10: PRINT "8) Start program again from the beginning"
7210 LOCATE 19,10: PRINT "9) End program"
7220 LOCATE 21,25: PRINT SPACES$(50)
7230 LOCATE 21,25: BEEP: INPUT "Enter the number of your choice";C%
7240 IF C%>=1 AND C%<=9 THEN GOTO 7260 ELSE GOTO 7220
7250 '
7260 IF C%<>1 THEN GOTO 7280                     'resume iterations
7270    CLS: RETURN
7280 IF C%<>2 THEN GOTO 7340                     'display stream functn
7290    CLS
7300    GOSUB 5000: GOSUB 6000
7310    LOCATE 25,53: BEEP: PRINT "Press any key to continue";
7320    IF INKEY$="" THEN GOTO 7320
7330    GOTO 7080
7340 IF C%<>3 THEN GOTO 7440                     'change sublattice
7350    LOCATE 22,15
7360    PRINT USING "The current lattice size is Nx=##, Ny=##";NX%,NY%
7370    LOCATE 23,15
7380    PRINT USING "The current sublattice is Nxmax=##, Nymax=##";NXMAX%,NYMAX%
7390    LOCATE 24,15: BEEP
7400    INPUT;"Enter the new values of Nxmax,Nymax";NXMAX%,NYMAX%
7410    IF NXMAX%<0 OR NXMAX%>NX% THEN NXMAX%=NX%
7420    IF NYMAX%<0 OR NYMAX%>NY% THEN NYMAX%=NY%
7430    GOTO 7080
7440 IF C%<>4 THEN GOTO 7530                     'change relaxation
7450    LOCATE 22,15
7460    PRINT USING "The current relaxation parameter for PSI is #.###";WP
7470    LOCATE 23,15: BEEP: INPUT;"Enter the new parameter for PSI";WP
7480    LOCATE 24,15
7490    PRINT USING "The current relaxation parameter for XSI is #.###";WX;
```

```
7500    LOCATE 25,15: BEEP: INPUT;"Enter the new parameter for XSI";WX
7510    WX1=1-WX: WP1=1-WP
7520    GOTO 7080
7530 IF C%<>5 THEN GOTO 7610                        'change Reynolds number
7540    LOCATE 22,15
7550    PRINT USING "The current lattice Reynolds number is ##.###";RE
7560    LOCATE 23,15: BEEP
7570    INPUT "Enter the new Reynolds number";RE
7580    IF RE<0 THEN RE=1
7590    RE4=RE/4
7600    GOTO 7080
7610 IF C%<>6 THEN GOTO 7720                        'save data
7620    LOCATE 22,15: BEEP
7630    INPUT "Enter name of file in which to store solution";S$
7640    OPEN S$ FOR OUTPUT AS #1
7650    WRITE #1,NX%,NY%
7660    FOR I%=1 TO NY%
7670       FOR J%=1 TO NX%
7680          WRITE #1,P(I%,J%),XSI(I%,J%)
7690       NEXT J%
7700    NEXT I%
7710    CLOSE #1: GOTO 7080
7720 IF C%<>7 THEN GOTO 7790                        'display XSI
7730    CLS
7740    VORT%=-1                                     'set flag
7750    IF GRAPHICS% THEN DEF SEG=&HB800 ELSE DEF SEG=&HB000
7760    GOSUB 5000
7770    VORT%=0
7780    GOTO 7080
7790 IF C%<>8 THEN END                              'end program
7800    AGAIN%=-1                                    ' or begin again
7810    CLS
7820    RETURN
7830 '
8000 '******************************************************************
8010 'subroutine to display the header screen
8020 'input   variables: none
8030 'output variables: GRAPHICS%,MONO%
8040 'global variables: none
8050 'local   variables: G$,M$,ROW%
8060 '******************************************************************
8070 SCREEN 0: CLS: KEY OFF                        'program begins in text mode
8080 '
8090 LOCATE 1,30: COLOR 15                          'display book title
8100 PRINT "COMPUTATIONAL PHYSICS";
8110 LOCATE 2,40: COLOR 7
8120 PRINT "by";
8130 LOCATE 3,33: PRINT "Steven E. Koonin";
8140 LOCATE 4,14
8150 PRINT "Copyright 1985, Benjamin/Cummings Publishing Company";
8160 '
8170 LOCATE 5,10                                    'draw the box
8180 PRINT CHR$(201)+STRING$(59,205)+CHR$(187);
8190 FOR ROW%=6 TO 19
8200    LOCATE ROW%,10: PRINT CHR$(186);
8210    LOCATE ROW%,70: PRINT CHR$(186);
8220 NEXT ROW%
8230 LOCATE 20,10
```

```
8240 PRINT CHR$(200)+STRING$(59,205)+CHR$(188);
8250 '
8260 COLOR 15                                    'print title, etc.
8270 LOCATE  7,36:  PRINT "PROJECT 6";
8280 COLOR 7
8290 LOCATE 10,23:  PRINT "2-D viscous incompressible flow about"
8300 LOCATE 11,32:  PRINT "a rectangular block"
8310 LOCATE 13,36:  PRINT "**********"
8320 LOCATE 15,25:  PRINT "Type 'e' to end while running."
8330 LOCATE 16,22:  PRINT "Type ctrl-break to stop at a prompt."
8340 '
8350 LOCATE 19,16: BEEP                          'get screen configuration
8360 INPUT "Does your computer have graphics capability (y/n)";G$
8370 IF LEFT$(G$,1)="y" OR LEFT$(G$,1)="n" GOTO 8400
8380    LOCATE 19,12:  PRINT SPACE$(58):  BEEP
8390    LOCATE 18,35:  PRINT "Try again,":  GOTO 8350
8400 IF LEFT$(G$,1)="y" GOTO 8430
8410    GRAPHICS%=0:  MONO%=-1
8420    RETURN
8430 GRAPHICS%=-1
8440 LOCATE 18,15:  PRINT SPACE$(55)
8450 LOCATE 19,15:  PRINT SPACE$(55)
8460 LOCATE 19,15: BEEP
8470 INPUT "Do you also have a separate screen for text (y/n)";M$
8480 IF LEFT$(M$,1)="y" OR LEFT$(M$,1)="n" GOTO 8500  'error trapping
8490    LOCATE 18,35:  PRINT "Try again,":  GOTO 8450
8500 IF LEFT$(M$,1)="y" THEN MONO%=-1 ELSE MONO%=0
8510 RETURN
8520 '
9000 '*******************************************************************
9010 'subroutine to print message about POKEing, set up arrays for POKEing
9020 'input  variables: none
9030 'output variables: ASKII%(I%), BORDER%(I%), INTERIOR%(I%), NEGASKII%(I%)
9040 'global variables: none
9050 'local  variables: I%
9060 '*******************************************************************
9070 CLS
9080 '
9090 LOCATE 1,34,0
9100 PRINT "C A U T I O N"
9110 LOCATE 3,9
9120 PRINT "For efficient screen display this program uses the POKE command";
9130 LOCATE 5,16
9140 PRINT "which places data directly into the screen buffers.";
9150 LOCATE 7,15
9160 PRINT "The monochrome display buffer is assumed to begin";
9170 LOCATE 9,14
9180 PRINT "at location &HB000 and the graphics buffer at &HB800.";
9190 LOCATE 11,20
9200 PRINT "These are appropriate for IBM hardware.";
9210 LOCATE 13,20
9220 PRINT "If your hardware is not IBM, check your";
9230 LOCATE 15,14
9240 PRINT "manuals for the locations of your display buffers and";
9250 LOCATE 17,12
9260 PRINT "modify the subroutine beginning at line 5000 accordingly";
9270 LOCATE 21,27
9280 PRINT "Press any key to continue";: BEEP
```

```
9290 IF INKEY$="" THEN GOTO 9290
9300 LOCATE ,,0
9310 CLS
9320 '
9330 FOR I%=0 TO 9                              'set up arrays to convert
9340    ASKII%(I%)=I%+48                        ' P and XSI to ASCII
9350 NEXT I%                                    'numbers 0-9
9360 FOR I%=10 TO 35
9370    ASKII%(I%)=I%+55                        'letters A-Z
9380    NEGASKII%(I%-9)=I%+87                    'letters a-z
9390 NEXT I%
9400 '
9410 IF NOT GRAPHICS% THEN GOTO 9500
9420    RESTORE 9470                            'set up arrays with colors
9430    FOR I%=0 TO 9                           ' for interior and border
9440       READ INTERIOR%(I%)                   ' points
9450       BORDER%(I%)=INTERIOR%(I%)+96
9460    NEXT I%
9470    DATA 4,5,1,3,2,12,13,9,11,10
9480    RETURN
9490 '
9500 FOR I%=0 TO 9                              'monochrome screen uses either
9510    INTERIOR%(I%)=2:  BORDER%(I%)=10        ' regular or boldface only
9520 NEXT I%
9530 RETURN
9540 '
10000 '**********************************************************************
10010 'subroutine to display information about the display
10020 'input   variables:   none
10030 'output variables:    none
10040 'global variables:    FIRST%
10050 'local  variables:    none
10060 '**********************************************************************
10070 CLS
10080 '
10090 LOCATE 1,10,0
10100 PRINT "The program will display the streamlines (and the vorticity"
10110 LOCATE 3,11,0
10120 PRINT "if you have two monitors).  If the vorticity is shown, it"
10130 LOCATE 5,23,0
10140 PRINT "is shown on the text only screen."
10150 LOCATE 8,15,0
10160 PRINT "The lines of constant vorticity and the streamlines"
10170 LOCATE 10,17,0
10180 PRINT "are indicated by letters and numbers as follows:"
10190 LOCATE 12,25,0
10200 PRINT "0-9,A-Z  corresponding to 0-35"
10210 LOCATE 13,23,0
10220 PRINT "and a-z  corresponding to -1 to -26"
10230 LOCATE 15,5,0
10240 PRINT"(The values of the vorticity displayed are ten "
10250 LOCATE 15,52,0
10260 PRINT "times the actual value)"
10270 LOCATE 18,7,0
10280 PRINT"The colors are to give greater contrast to neighboring streamlines."
10290 LOCATE 20,9,0
10300 PRINT "The boundary and plate are distinguished by a yellow background"
10310 LOCATE 22,15,0
```

```
10320 PRINT "(graphics monitor) or boldface (text only screen)."
10330 '
10340 LOCATE 25,28,0
10350 PRINT "Type any key to continue";
10360 IF INKEY$="" GOTO 10360
10370 '
10380 CLS
10390 FIRST%=0
10400 RETURN
10410 '.
11000 '*********************************************************************
11010 'subroutine to switch from mono to graphics screen
11020 'input   variables: none
11030 'output  variables: none
11040 'global  variables: none
11050 'local   variables: none
11060 '*********************************************************************
11070 DEF SEG=0
11080 POKE &H410, (PEEK(&H410) AND &HCF) OR &H10
11090 SCREEN 1,0,0,0: SCREEN 0: WIDTH 80: LOCATE ,,1,6,7
11100 RETURN
11110 '
12000 '*********************************************************************
12010 'subroutine to switch from graphics to mono screen
12020 'input   variables: none
12030 'output  variables: none
12040 'global  variables: none
12050 'local   variables: none
12060 '*********************************************************************
12070 DEF SEG=0
12080 POKE &H410, (PEEK(&H410) OR &H30)
12090 SCREEN 0:  WIDTH 80:  LOCATE ,,1,12,13
12100 RETURN
12110 '
```

C.7 Project VII

This program solves the non-linear reaction-diffusion equations of the Brusselator (Eqs. (VII.3)) in two dimensions with no-flux boundary conditions. The concentrations of species A and B, the diffusion constants, the size of the lattice, and the time step are defined in subroutine 1000. Initial conditions for the X and Y concentrations are taken to be random fluctuations about their equilibrium values (lines 1280-1370). Time evolution (loop 160-210) is then by the alternating-direction method (7.19), carried out in subroutine 4000. Note that some of the coefficients required for the inversions of the tri-diagonal matrices are calculated only once at the beginning and stored (subroutine 2000, called from line 1260). At each time step, the X and Y concentrations are displayed as character arrays on the monitors (subroutine 3000).

A typical result from this program can be obtained by choosing A and B concentrations of 2 and 5, respectively, diffusion constants Dx=Dy=0.001, lattice sizes Nx=Ny=20, a time step of 0.07, and initial fluctuations of 0.1.

```
10  '*****************************************************************
20  'Project 7: The Brusselator in two dimensions
21  'COMPUATATIONAL PHYSICS by Steven E. Koonin
22  'Copyright 1985, Benjamin/Cummings Publishing Company
30  '*****************************************************************
40  DIM X(24,79),Y(24,79)                 'arrays for concentrations of X and Y
50  DIM ALPHAXX(24),BETAXX(24),GAMMAXX(24)'coefficients for x-recursion of X
60  DIM ALPHAYX(24),BETAYX(24),GAMMAYX(24)'coefficients for x-recursion of Y
70  DIM ALPHAXY(79),BETAXY(79),GAMMAXY(79)'coefficients for y-recursion of X
80  DIM ALPHAYY(79),BETAYY(79),GAMMAYY(79)'coefficients for y-recursion of Y
90  DIM PLACE%(24,79),ASKI%(62)           'array for buffer location,ASCII code
100 '
110 GOSUB 6000                            'display the header screen
120 GOSUB 1000                            'initialize the calculation
130 TIME=0
140 WHILE INKEY$<>"e"
150     TIME=TIME+DT                      'increment time
160     GOSUB 4000                        'take a time step
170     GOSUB 3000                        'display X and Y
180     IF OVRFLOW% THEN OVRFLOW%=0:  GOTO 120
190 WEND
200 '
210 GOSUB 5000                            'display options for continuing
220 '
1000 '*****************************************************************
1010 'subroutine to initialize the calculation
1020 'input  variables: none
1030 'output variables: ADT, BDT, BP1DT, CXP, CXZ, CYP, CYZ, DT, NX%, NY%
1040 '                  X(I%,J%), Y(I%,J%)
1050 'global variables: none
1060 'local  variables: A, B, D1, D2, FLUCT, H, I%, J%
1070 '*****************************************************************
1080 IF MONO% THEN GOSUB 8000:  CLS       'clear text screen
1090 IF GRAPHICS% THEN GOSUB 7000:  CLS   'clear graphic screen
```

```
1100 '
1110 LOCATE 10,10                                          'input parameters
1120 INPUT;"Enter concentrations of species A,B";A,B
1130 LOCATE 12,10
1140 INPUT;"Enter diffusion coefficients Dx,Dy";D1,D2
1150 LOCATE 14,10
1160 INPUT;"Enter lattice sizes Nx(<=24),Ny(<=79)";NX%,NY%
1170 IF NX%>24 THEN NX%=24                                 'keep lattice small
1180 IF NY%>79 THEN NY%=79
1190 LOCATE 16,10
1200 INPUT;"Enter time step";DT
1210 '
1220 H=1/(NX%-1)                                           'useful constants
1230 BDT=B*DT: ADT=A*DT: BP1DT=1-(B+1)*DT
1240 CXP=-D1*DT/H^2: CXZ=1+2*D1*DT/H^2
1250 CYP=-D2*DT/H^2: CYZ=1+2*D2*DT/H^2
1260 GOSUB 2000                                            'recursion coeffs.
1270 '                                                     ' and buffer loc.
1280 LOCATE 18,10                                          'initial conditions
1290 INPUT;"Enter size of initial fluctuations (>0,<1)";FLUCT
1300 IF FLUCT<0 THEN FLUCT=0
1310 IF FLUCT>1 THEN FLUCT=1
1320 FOR I%=1 TO NX%                                       'sweep the lattice
1330    FOR J%=1 TO NY%
1340       X(I%,J%)=A*(1-FLUCT+FLUCT*RND)
1350       Y(I%,J%)=B/A*(1-FLUCT+FLUCT*RND)
1360    NEXT J%
1370 NEXT I%
1380 '
1390 CLS
1400 GOSUB 3000                                            'display X,Y
1410 RETURN
1420 '
2000 '****************************************************************
2010 'subroutine to find the ALPHA and GAMMA coefficients
2020 'input   variables: none
2030 'output variables: ALPHAXX, ALPHAXY, ALPHAYX, ALPHAYY, GAMMXX, GAMMAXY,
2040 '                   GAMMAYX, GAMMAYY, PLACE%(I%,J%)
2050 'global variables: CXP, CXZ, CYP, CYZ, NX%, NY%
2060 'local   variables: BOTTOM%, I%, IC%, IR%, J%, LEDGE%
2070 '****************************************************************
2080 '                                                     'x-direction
2090 ALPHAXX(NX%)=1: ALPHAYX(NX%)=1                        'no-flux boundary conditions
2100 FOR I%=NX% TO 1 STEP -1
2110    GAMMAXX(I%)=1/(CXZ+CXP*ALPHAXX(I%))
2120    ALPHAXX(I%-1)=-CXP*GAMMAXX(I%)
2130    GAMMAYX(I%)=1/(CYZ+CYP*ALPHAYX(I%))
2140    ALPHAYX(I%-1)=-CYP*GAMMAYX(I%)
2150 NEXT I%
2160 '                                                     'y-direction
2170 ALPHAXY(NY%)=1: ALPHAYY(NY%)=1                        'no-flux boundary conditions
2180 FOR J%=NY% TO 1 STEP -1
2190    GAMMAXY(J%)=1/(CXZ+CXP*ALPHAXY(J%))
2200    ALPHAXY(J%-1)=-CXP*GAMMAXY(J%)
2210    GAMMAYY(J%)=1/(CYZ+CYP*ALPHAYY(J%))
2220    ALPHAYY(J%-1)=-CYP*GAMMAYY(J%)
2230 NEXT J%
2240 '
```

```
2250 LEDGE%=39-NY%/2:   IF NY% MOD 2=1 THEN LEDGE%=LEDGE%+2        'center lattice
2260 BOTTOM%=12+NX%/2
2270 FOR I%=1 TO NX%                              'establish array for buffer loc
2280    IR%=BOTTOM%-I%+1
2290    FOR J%=1 TO NY%
2300       IC%=LEDGE%+J%-1
2310       PLACE%(I%,J%)=160*(IR%-1)+2*(IC%-1)
2320    NEXT J%
2330 NEXT I%
2340 RETURN
2350 '
3000 '********************************************************************
3010 'subroutine to display X and Y
3020 'input   variables: OVRFLOW%, TIME, UNDRFLOW%, X(I%,J%), Y(I%,J%)
3030 'output  variables: OVRFLOW%, UNDRFLOW%
3040 'global  variables: ASKI%(I%), GRAPHICS%, MONO%, NX%, NY%, PLACE%(I%,J%)
3050 'local   variables: BUF%, CH%, I%, J%, TINT%
3060 '********************************************************************
3070 IF NOT (OVRFLOW% OR UNDRFLOW%) THEN GOTO 3160        'display error message
3080    LOCATE 12+NX%/2+1,16:  PRINT SPACE$(64);
3090    LOCATE ,16
3100    IF UNDRFLOW% THEN PRINT "Concentration is negative, type c to continue";
3110    IF OVRFLOW% THEN PRINT "Overflow in concentration, type c to continue";
3120    IF INKEY$<>"c" THEN GOTO 3120
3130    UNDRFLOW%=0:  OVRFLOW%=-1
3140    RETURN
3150 '
3160 LOCATE 12+NX%/2+1,1
3170 PRINT USING " Time=###.##        ";TIME;
3180 LOCATE ,22,0
3190 PRINT "This screen is the X concentration.       Press e to end";
3200 '
3210 IF GRAPHICS% THEN DEF SEG=&HB800 ELSE DEF SEG=&HB000        'display X
3220 FOR I%=1 TO NX%
3230    FOR J%=1 TO NY%
3240       CH%=5*X(I%,J%): IF CH%>61 THEN CH%=61
3250       IF CH%<0 THEN CH%=0
3260       IF GRAPHICS% THEN GOTO 3280
3270          TINT%=7: GOTO 3290
3280       TINT%=((CH%+15)MOD 15) + 1 : IF TINT%=8 THEN TINT%=1
3290       BUF%=PLACE%(I%,J%):   CH%=ASKI%(CH%)
3300       POKE BUF%,CH% : POKE BUF%+1,TINT%
3310    NEXT J%
3320 NEXT I%
3330 IF NOT (GRAPHICS% AND MONO%) THEN RETURN
3340 '
3350 DEF SEG=&HB000                                          'display Y
3360 FOR I%=1 TO NX%
3370    FOR J%=1 TO NY%
3380       CH%=5*Y(I%,J%): IF CH%>61 THEN CH%=61
3390       IF CH%<0 THEN CH%=0
3400       BUF%=PLACE%(I%,J%): CH%=ASKI%(CH%)
3410       POKE BUF%,CH% : POKE BUF%+1,7
3420    NEXT J%
3430 NEXT I%
3440 '
4000 '********************************************************************
4010 'subroutine to take a time step
```

```
4020 'input   variables: X(I%,J%), Y(I%,J%)
4030 'output variables: OVRFLOW%, UNDRFLOW%, X(I%,J%), Y(I%,J%)
4040 'global variables: ALPHA and GAMMA arrays, ADT, BDT, BP1DT, CXP, CYP, DT
4050 '                   NX%, NY%
4060 'local  variables: BETA arrays, I%, J%, SX, SY, X2Y
4070 '****************************************************************************
4080 FOR J%=1 TO NY%                          'for each value of y, do an x-sweep
4090 '                                        'backward sweep for BETA coefficients
4100    BETAXX(NX%)=0: BETAYX(NX%)=0          'no-flux boundary conditions
4110    FOR I%=NX% TO 1 STEP -1
4120       X2Y=X(I%,J%)^2*Y(I%,J%)*DT
4130       SX=X(I%,J%)*BP1DT+ADT+X2Y
4140       SY=Y(I%,J%)+BDT*X(I%,J%)-X2Y
4150       BETAXX(I%-1)=(SX-CXP*BETAXX(I%))*GAMMAXX(I%)
4160       BETAYX(I%-1)=(SY-CYP*BETAYX(I%))*GAMMAYX(I%)
4170    NEXT I%
4180 '                                        'forward x sweep
4190    X(1,J%)=BETAXX(0)/(1-ALPHAXX(0))      'starting values for no-flux b.c.
4200    Y(1,J%)=BETAYX(0)/(1-ALPHAYX(0))
4210    FOR I%=1 TO NX%-1
4220       X(I%+1,J%)=ALPHAXX(I%)*X(I%,J%)+BETAXX(I%)
4230       Y(I%+1,J%)=ALPHAYX(I%)*Y(I%,J%)+BETAYX(I%)
4240    NEXT I%
4250 NEXT J%
4260 '
4270 FOR I%=1 TO NX%                          'for each value of x, do a y sweep
4280 '                                        'backward sweep for BETA coefficients
4290    BETAXY(NY%)=0: BETAYY(NY%)=0          'no-flux boundary conditions
4300    FOR J%=NY% TO 1 STEP -1
4310       BETAXY(J%-1)=(X(I%,J%)-CXP*BETAXY(J%))*GAMMAXY(J%)
4320       BETAYY(J%-1)=(Y(I%,J%)-CYP*BETAYY(J%))*GAMMAYY(J%)
4330    NEXT J%
4340 '                                        'forward y sweep
4350    X(I%,1)=BETAXY(0)/(1-ALPHAXY(0))      'starting values for no-flux b.c.
4360    Y(I%,1)=BETAYY(0)/(1-ALPHAYY(0))
4370    FOR J%=1 TO NY%-1
4380       X(I%,J%+1)=ALPHAXY(J%)*X(I%,J%)+BETAXY(J%)
4390       Y(I%,J%+1)=ALPHAYY(J%)*Y(I%,J%)+BETAYY(J%)
4400 '                                        'check for reasonable values
4410       IF Y(I%,J%+1)>20 OR X(I%,J%+1)>20 THEN OVRFLOW%=-1 ELSE OVRFLOW%=0
4420       IF Y(I%,J%+1)<-1 OR X(I%,J%+1)<-1 THEN UNDRFLOW%=-1 ELSE UNDRFLOW%=0
4430       IF (UNDRFLOW% OR OVRFLOW%) THEN RETURN
4440    NEXT J%
4450 NEXT I%
4460 RETURN
4470 '
5000 '****************************************************************************
5010 'subroutine to list options for continuing
5020 'input   variables: DT, Y(I%,J%)
5030 'output variables: ADT, BDT, BP1DT, CXP, CXZ, CYP, CYZ, DT
5040 'global variables: A, ASKI%(I%), B, H, GRAPHICS%, NX%, NY%, PLACE%(I%,J%)
5050 'local  variables: A%, BUF%, CH%, I%, J%, TINT%
5060 '****************************************************************************
5070 CLS
5080 LOCATE 1,21
5090 PRINT "Options for continuing the calculation"
5100 LOCATE 3,10:  PRINT "1)  Resume iterating
5110 LOCATE 5,10:  PRINT "2)  Display Y-concentration on this screen";
```

```
5120 LOCATE 7,10:  PRINT "3)  Change time step";
5130 LOCATE 9,10:  PRINT "4)  Begin again with new parameters";
5140 LOCATE 11,10: PRINT "5)  End program";
5150 '
5160 LOCATE 13,25:  PRINT SPACE$(50)
5170 LOCATE 13,25:  BEEP:  INPUT "Enter the number of your choice";A%
5180 IF A%>=1 AND A%<=5  THEN GOTO 5200 ELSE GOTO 5160
5190 '
5200 IF A%<>1 THEN GOTO 5230
5210   CLS:  GOSUB 3000
5220   RETURN 140
5230 IF A%<>2 THEN GOTO 5410
5240   CLS
5250   IF GRAPHICS% THEN DEF SEG=&HB800 ELSE DEF SEG=&HB000          'display Y
5260   FOR I%=1 TO NX%
5270    FOR J%=1 TO NY%
5280       CH%=5*Y(I%,J%):  IF CH%>61 THEN CH%=61
5290       IF CH%<0 THEN CH%=0
5300       IF GRAPHICS% THEN GOTO 5320
5310         TINT%=7:  GOTO 5330
5320       TINT%=((CH%+15)MOD 15) + 1 : IF TINT%=8 THEN TINT%=1
5330       BUF%=PLACE%(I%,J%):  CH%=ASKI%(CH%)
5340       POKE BUF%,CH% : POKE BUF%+1,TINT%
5350    NEXT J%
5360   NEXT I%
5370   LOCATE 12+NX%/2+1,20
5380   PRINT "Y-concentration          Type c to continue";
5390   IF INKEY$<>"c" THEN GOTO 5390
5400   GOTO 5070
5410 IF A%<>3 THEN GOTO 5500
5420   LOCATE 17,10
5430   PRINT USING "The current time step is #.###";DT
5440   LOCATE 18,10:  INPUT "Enter new value of DT";DT:
5450   BDT=B*DT: ADT=A*DT: BP1DT=1-(B+1)*DT
5460   CXP=-D1*DT/H^2: CXZ=1+2*D1*DT/H^2
5470   CYP=-D2*DT/H^2: CYZ=1+2*D2*DT/H^2
5480   GOSUB 2000                                        'recursion coeffs.
5490   CLS:  GOSUB 3000:  RETURN 140
5500 IF A%<>4 THEN END
5510   RETURN 120
5520 '
6000 '*******************************************************************
6010 'subroutine to display the header screen
6020 'input  variables: none
6030 'output variables: ASKI%(I%), GRAPHICS%, MONO%
6040 'global variables: none
6050 'local  variables: G$,M$,ROW%
6060 '*******************************************************************
6070 SCREEN 0: CLS: KEY OFF                    'program begins in text mode
6080 '
6090 LOCATE 1,30: COLOR 15                     'display book title
6100 PRINT "COMPUTATIONAL PHYSICS";
6110 LOCATE 2,40: COLOR 7
6120 PRINT "by";
6130 LOCATE 3,33: PRINT "Steven E. Koonin";
6140 LOCATE 4,14
6150 PRINT "Copyright 1985, Benjamin/Cummings Publishing Company";
6160 '
```

```
6170 LOCATE 5,10                                    'draw the box
6180 PRINT CHR$(201)+STRING$(59,205)+CHR$(187);
6190 FOR ROW%=6 TO 19
6200     LOCATE ROW%,10: PRINT CHR$(186);
6210     LOCATE ROW%,70: PRINT CHR$(186);
6220 NEXT ROW%
6230 LOCATE 20,10
6240 PRINT CHR$(200)+STRING$(59,205)+CHR$(188);
6250 '
6260 COLOR 15                                       'print title, etc.
6270 LOCATE  7,36:  PRINT "PROJECT 7";
6280 COLOR 7
6290 LOCATE 10,25:  PRINT "The Brusselator in two dimensions"
6300 LOCATE 13,36:  PRINT "**********"
6310 LOCATE 15,25:  PRINT "Type 'e' to end while running."
6320 LOCATE 16,22:  PRINT "Type ctrl-break to stop at a prompt."
6330 '
6340 LOCATE 19,16: BEEP                             'get screen configuration
6350 INPUT "Does your computer have graphics capability (y/n)";G$
6360 IF LEFT$(G$,1)="y" OR LEFT$(G$,1)="n" GOTO 6390
6370   LOCATE 19,12:  PRINT SPACE$(58):  BEEP
6380   LOCATE 18,35:  PRINT "Try again,":  GOTO 6340
6390 IF LEFT$(G$,1)="y" GOTO 6420
6400   GRAPHICS%=0:  MONO%=-1
6410   GOTO 6510
6420 GRAPHICS%=-1
6430 LOCATE 18,15:  PRINT SPACE$(55)
6440 LOCATE 19,15:  PRINT SPACE$(55)
6450 LOCATE 19,15: BEEP
6460 INPUT "Do you also have a separate screen for text (y/n)";M$
6470 IF LEFT$(M$,1)="y" OR LEFT$(M$,1)="n" GOTO 6490  'error trapping
6480   LOCATE 18,35:  PRINT "Try again,":  GOTO 6440
6490 IF LEFT$(M$,1)="y" THEN MONO%=-1 ELSE MONO%=0
6500 '                                              'fill array with ASCII code
6510 FOR I%=0 TO 9                                  'numbers 0-9
6520     ASKI%(I%)=I%+48
6530 NEXT I%
6540 FOR I%=10 TO 35                                'letters A-Z
6550     ASKI%(I%)=I%+55
6560 NEXT I%
6570 FOR I%=36 TO 61                                'letters a-z
6580     ASKI%(I%)=I%+61
6590 NEXT I%
6600 '
6610 RETURN
6620 '
7000 '*******************************************************************
7010 'subroutine to switch from mono to graphics screen
7020 'input  variables: none
7030 'output variables: none
7040 'global variables: none
7050 'local  variables: none
7060 '*******************************************************************
7070 DEF SEG=0
7080 POKE &H410, (PEEK(&H410) AND &HCF) OR &H10
7090 SCREEN 1,0,0,0:  SCREEN 0:  WIDTH 80:  LOCATE ,,1,6,7
7100 RETURN
7110 '
```

```
8000 '**********************************************************************
8010 'subroutine to switch from graphics to mono screen
8020 'input   variables: none
8030 'output variables: none
8040 'global variables: none
8050 'local   variables: none
8060 '**********************************************************************
8070 DEF SEG=0
8080 POKE &H410,   (PEEK(&H410) OR &H30)
8090 SCREEN 0:   WIDTH 80: LOCATE ,,1,12,13
8100 RETURN
8110 '
```

C.8 Project VIII

This program uses variational or Path Integral Monte Carlo methods to solve the two-center two-electron problem of the H_2 molecule using the trial wavefunction specified by Eqs. (VIII.6,8). Subroutine 1000 defines the interproton separation, the wavefunction parameter β, and which method of calculation is to be used; Eq. (VIII.8) for the wavefunction parameter a is also solved here (lines 1220-1250). The main calculations are done in subroutine 3000: thermalization in loop 3410-3440 and data-taking in loop 3660-4120. The Metropolis steps for the variational calculation are taken in subroutine 5000, while time steps for the PIMC calculations are taken in subroutine 9000. Both methods use subroutines 6000 (to find the local energy (VIII.5) of a given configuration), 7000 (to calculate the wavefunction and various distances), and 14000 (to generate a starting configuration). The PIMC calculation also uses subroutine 10000 (to calculate the drift vector of Eq. (VIII.14)) and subroutine 12000 (to generate the initial ensemble). For either method, observations, taken very FREQ% steps (line 3720), are divided into groups to estimate the step-to-step correlations in the energy. As the steps proceed, estimates for the electronic eigenvalue and intermolecular potential (VIII.1) are displayed. (Note that the latter is relative to the energy of two hydrogen atoms at infinite separation.) If requested, the energy auto-correlation function for 200 steps is calculated explicitly and displayed in subroutine 8000.

A representative calculation from this program can be had by choosing an interproton separation of 0 (so that the neutral Helium atom is being described), a wavefunction parameter BETA=0.25 Å^{-1}, a variational calculation (rather than PIMC), a Metropolis step size of 0.40 Å, 20 thermalization steps, and a calculation of the energy (rather than the correlation function) using 10 groups of 5 samples taken with a sampling frequency of 6. To do a PIMC calculation for the same situation, select many of the same parameters, together with an ensemble size of 10 and a time step of 0.01. The results generated by calculations like these can be improved by accumulating more observations to get better statistics and by extrapolating to a vanishing time step.

```
10 '**********************************************************************
20 'Project 8: Monte Carlo solution of the H2 molecule
21 'COMPUTATIONAL PHYSICS by Steven E. Koonin
22 'Copyright 1985, Benjamin/Cummings Publishing Company
30 '**********************************************************************
40 DIM ENSEMBLE(6,50)                       'storage for the ensemble
50 DIM WEIGHT(50)                           'weight of each ensemble member
60 DIM CONFIG(6),CSAVE(6)                   'storage for individual configs.
70 DIM DRIFT(6)                             'the drift vector
```

```
80 DIM ECORR(200)                              'energies for the correlation fnct.
90 HBM=7.6359: E2=14.409                        'physical constants
100 ABOHR=HBM/E2
110 '
120 GOSUB 15000                                 'display header screen
130 GOSUB 1000                                  'input physical parameters
140 GOSUB 2000                                  'define functions needed
150 IF VARY%<>0 THEN GOSUB 3000                 'do the calculation
160 '
170 GOTO 130                                    'start another calculation
180 '
1000 '***********************************************************************
1010 'subroutine to input the physical parameters
1020 'input  variables: none
1030 'output variables: A,ALPHA,BETA,S,S2,VARY%
1040 'global variables: ABOHR
1050 'local  variables: AOLD
1060 '***********************************************************************
1070 CLS
1080 '
1090 LOCATE 9,20
1100 PRINT USING "The current interproton separation is ##.####";S
1110 LOCATE 10,22: BEEP
1120 INPUT "Enter new interproton separation (<0 to stop)";S
1130 IF S<0 THEN STOP
1140 S2=S/2                                     'half proton separation
1150 '
1160 LOCATE 11,20
1170 PRINT USING "The current wavefunction parameter beta is #.####";BETA
1180 LOCATE 12,22: BEEP
1190 INPUT "Enter new wavefunction parameter beta";BETA
1200 '
1210 ALPHA=2*ABOHR                              'parameter alpha
1220 A=ABOHR: AOLD=0                            'iteration for parameter A
1230 WHILE ABS(A-AOLD)>.000001
1240     AOLD=A: A=ABOHR/(1+EXP(-S/AOLD))
1250 WEND
1260 '
1270 LOCATE 14,20: BEEP
1280 INPUT "Enter 0 to start again, 1 for variational, or 2 for PIMC";VARY%
1290 RETURN
1300 '
2000 '***********************************************************************
2010 'subroutine to define the functions needed
2020 'input  variables: none
2030 'output variables: FNCHI,FNCHIP,FNCHIPP,FND2CHI,FNDIST,FNF,FNFP,FNFPP,
2040 '                   FND2F
2050 'global variables: A,ALPHA,BETA
2060 'local  variables: none
2070 '***********************************************************************
2080 DEF FNCHI(R)=EXP(-R/A)                     'electron-nucleus function
2090 DEF FNCHIP(R)=-FNCHI(R)/A                  'its first  derivative
2100 DEF FNCHIPP(R)=FNCHI(R)/A^2                'its second derivative
2110 DEF FND2CHI(R)=FNCHIPP(R)+2*FNCHIP(R)/R    'its laplacian
2120 '
2130 DEF FNF(R)=EXP(R/(ALPHA*(1+BETA*R)))       'electron-electron function
2140 DEF FNFP(R)=FNF(R)/(ALPHA*(1+BETA*R)^2)    'its first and second derivs
2150 DEF FNFPP(R)=FNFP(R)^2/FNF(R)-2*BETA*FNF(R)/ALPHA/(1+BETA*R)^3
```

```
2160 DEF FND2F(R)=FNFPP(R)+2*FNFP(R)/R          'its laplacian
2170 '
2180 DEF FNDIST(X,Y,Z)=SQR(X^2+Y^2+Z^2)        'length of a vector
2190 '
2200 RETURN
2210 '
3000 '************************************************************************
3010 'subroutine to do the variational or PIMC calculation
3020 'input   variables: none
3030 'output variables: DELTA,DT,HBMDT,NENSEM%,SQDT
3040 'global variables: A,ABOHR,ACCEPT%,BETA,CORR%,ECORR,EPSILON,E2,FREQ%,HBM,
3050 '                  NGROUP%,S,SIZE%,VARY%
3060 'local   variables: AVGE,GROUPE,GROUPE2,IGROUP%,ISTEP%,MORE%,NTHERM%,SIGE,
3070 '                  SIGE1,SIGE2,SUMA,SUME,SUME2,SUMSIG,U
3080 '************************************************************************
3090 CLS
3100 '
3110 IF VARY%=1 THEN GOTO 3290
3120    LOCATE 8,10
3130    PRINT USING "The current ensemble size is ##";NENSEM%
3140    LOCATE ,12: BEEP
3150    INPUT "Enter new ensemble size (<=50, <=0 to stop)";NENSEM%
3160    IF NENSEM%<=0 THEN RETURN
3170    IF NENSEM%>50 THEN NENSEM%=50
3180 '
3190    GOSUB 12000                              'generate initial ensemble
3200 '
3210    LOCATE ,10
3220    PRINT USING "The current time step is #.#####";DT
3230    LOCATE ,12: BEEP
3240    INPUT "Enter new time step (<=0 to stop)";DT
3250    IF DT<=0 THEN GOTO 3090
3260    HBMDT=HBM*DT: SQDT=SQR(HBMDT)            'useful constants
3270    GOTO 3350
3280 '
3290 LOCATE 8,10
3300 PRINT USING "The current Metropolis step for sampling PHI is #.####";DELTA
3310 LOCATE ,12: BEEP
3320 INPUT "Enter new Metropolis step size (0 to stop)";DELTA
3330 IF DELTA<=0 THEN RETURN
3340 '
3350 LOCATE ,10                                 'thermalization
3360 PRINT USING "The current number of thermalization steps is ###";NTHERM%
3370 LOCATE ,12: BEEP                           'thermalization
3380 INPUT "Enter number of thermalization steps";NTHERM%
3390 IF NTHERM%<=0 THEN GOTO 3460
3400    IF VARY%=1 THEN GOSUB 14000             'generate starting configuration
3410    FOR ISTEP%=1 TO NTHERM%                 'do the thermalization steps
3420       IF VARY%=1 THEN GOSUB 5000           'take a Metropolis step
3430       IF VARY%=2 THEN GOSUB 9000           'take a time step
3440    NEXT ISTEP%
3450 '
3460 GOSUB 13000                                'get sampling parameters
3470 '
3480 MORE%=NGROUP%                              'number of groups to run
3490 SUME=0: SUME2=0: SUMSIG=0                  'zero sums
3500 '
3510 CLS                                        'print header information
```

```
3520 LOCATE 9,10
3530 IF VARY%=1 THEN GOTO 3570
3540    PRINT USING "PIMC calculation with time step=##.####";DT;
3550    PRINT USING " Ensemble size=##";NENSEM%
3560    GOTO 3580
3570 PRINT USING "Variational calculation with Metropolis step=##.####";DELTA
3580 LOCATE 11,10: PRINT USING "Interproton separation=##.####";S
3590 LOCATE 12,10: PRINT USING "Wavefunction parameter beta=##.####";BETA
3600 LOCATE 13,10: PRINT USING "Wavefunction parameter a  =##.####";A
3610 LOCATE 14,10
3620 PRINT USING "Sampling frequency=####";FREQ%
3630 '
3640 IF VARY%=1 THEN SUMA=0                          'zero acceptance count
3650 '
3660 FOR IGROUP%=NGROUP%-MORE%+1 TO NGROUP%          'loop over groups
3670    GROUPE=0: GROUPE2=0: ACCEPT%=0               'zero group sums
3680    FOR ISTEP%=1 TO FREQ%*SIZE%                  'loop over steps
3690       IF INKEY$="e" THEN GOTO 3090              'check for interrupt
3700       IF VARY%=1 THEN GOSUB 5000                'take a Metropolis step
3710       IF VARY%=2 THEN GOSUB 9000                'take a time step
3720       IF ISTEP% MOD FREQ%<>0 THEN GOTO 3820     'sometimes measure energy
3730          IF VARY%=1 THEN GOSUB 6000             'calculate epsilon
3740          GROUPE=GROUPE+EPSILON                  'update group sums
3750          GROUPE2=GROUPE2+EPSILON^2
3760 '
3770          LOCATE 15,1                            'display this sample
3780          PRINT USING "Sample ## of ## ";ISTEP%/FREQ%,SIZE%
3790          LOCATE 15,17
3800          PRINT USING "in group ###   E=+###.####";IGROUP%,EPSILON
3810 '
3820    NEXT ISTEP%
3830    SUME=SUME+GROUPE                             'update total sums
3840    SUME2=SUME2+GROUPE2
3850    IF VARY%=1 THEN SUMA=SUMA+ACCEPT%
3860 '
3870    GROUPE=GROUPE/SIZE%                          'group average
3880    GROUPE2=GROUPE2/SIZE%-GROUPE^2
3890    IF GROUPE2<0 THEN GROUPE2=0
3900    SIGE=SQR(GROUPE2/SIZE%)                      'group std. deviation
3910    SUMSIG=SUMSIG+GROUPE2/SIZE%                  'update sum of variances
3920 '
3930    LOCATE 16,1                                  'display current totals
3940    PRINT USING "Group ### of ###";IGROUP%,NGROUP%
3950 '
3960    LOCATE 16,18
3970    PRINT USING "Eigenvalue=+###.#### +- ##.####";GROUPE,SIGE
3980 '
3990    AVGE=SUME/(IGROUP%*SIZE%)
4000    SIGE1=SQR((SUME2/(IGROUP%*SIZE%)-AVGE^2)/IGROUP%/SIZE%)
4010    SIGE2=SQR(SUMSIG/IGROUP%^2)
4020    IF S>.01 THEN U=AVGE+E2/S+E2/ABOHR ELSE U=0 'net molecular potential
4030    LOCATE 17,1
4040    PRINT USING "Grand average E=+###.#### +- ##.####";AVGE,SIGE1
4050    LOCATE 17,38
4060    PRINT USING "+- ##.####   U=+##.####";SIGE2,U
4070 '
4080    IF VARY%<>1 THEN GOTO 4110
4090       LOCATE 17,61
```

```
4100     PRINT USING " Acceptance=#.####";SUMA/(IGROUP%*SIZE%*FREQ%)
4110    IF CORR% THEN ECORR(IGROUP%)=EPSILON          'store E if correlation
4120 NEXT IGROUP%
4130 '
4140 IF NOT CORR% THEN GOTO 4180
4150    GOSUB 8000                                    'calculate correlation
4160    GOTO 3090                                     'start again
4170 '
4180 LOCATE ,10: BEEP                                 'ask for more groups
4190 INPUT;"How many more groups (0 to stop)";MORE%
4200 LOCATE ,10: PRINT SPACE$(65)
4210 NGROUP%=NGROUP%+MORE%
4220 IF MORE%=0 THEN GOTO 3090 ELSE GOTO 3660         'start again or do more
4230 '
5000 '**************************************************************
5010 'subroutine to take a Metropolis step
5020 'input   variables: CONFIG
5030 'output variables: CONFIG
5040 'global variables: ACCEPT%,DELTA,PHI,W
5050 'local   variables: CSAVE,I%,WTRY
5060 '**************************************************************
5070 FOR I%=1 TO 6
5080    CSAVE(I%)=CONFIG(I%)                          'save current configuration
5090    CONFIG(I%)=CONFIG(I%)+DELTA*(RND-.5)          'generate trial configuration
5100 NEXT I%
5110 '
5120 GOSUB 7000                                       'calculate the trial PHI
5130 WTRY=PHI^2
5140 '
5150 IF WTRY<W*RND THEN GOTO 5200                     'reject if trial PHI too small
5160    W=WTRY                                        'update the weight
5170    ACCEPT%=ACCEPT%+1                             'increment the acceptance count
5180    GOTO 5240
5190 '
5200 FOR I%=1 TO 6                                    'restore the old configuration
5210    CONFIG(I%)=CSAVE(I%)
5220 NEXT I%
5230 '
5240 RETURN
5250 '
6000 '**************************************************************
6010 'subroutine to calculate epsilon for a given configuration
6020 'input   variables: CONFIG
6030 'output variables: EPSILON
6040 'global variables: CHI1L,CHI1R,CHI2L,CHI2R,E2,F,HBM,R1L,R1R,R12,R2L,R2R,S,
6050 '                  X1,X2,Y1,Y2,Z1,Z2
6060 'local   variables: CROSS,R1LDOTR12,R1RDOTR12,R2LDOTR12,R2RDOTR12,SR12Z,
6070 '                   TEMP,TPOP,VPOP
6080 '**************************************************************
6090 GOSUB 7000                                       'calculate PHI and distances
6100 R1DOTR12=X1*(X1-X2)+Y1*(Y1-Y2)+Z1*(Z1-Z2)        'calculate useful dot prods.
6110 SR12Z=S*(Z1-Z2)
6120 R1LDOTR12=R1DOTR12+SR12Z/2
6130 R1RDOTR12=R1DOTR12-SR12Z/2
6140 R2LDOTR12=R1LDOTR12-R12^2
6150 R2RDOTR12=R1RDOTR12-R12^2
6160 '                                                'kinetic energy
6170 TPOP=2*FND2F(R12)/F                              'correlation contribution
```

```
6180 TEMP=FND2CHI(R1R)+FND2CHI(R1L)                    'one-electron contributions
6190 TPOP=TPOP+TEMP/(CHI1R+CHI1L)
6200 TEMP=FND2CHI(R2R)+FND2CHI(R2L)
6210 TPOP=TPOP+TEMP/(CHI2R+CHI2L)
6220 TEMP=FNCHIP(R1L)*R1LDOTR12/R1L                    'cross terms
6230 CROSS=(TEMP+FNCHIP(R1R)*R1RDOTR12/R1R)/(CHI1L+CHI1R)
6240 TEMP=-FNCHIP(R2R)*R2RDOTR12/R2R
6250 TEMP=TEMP-FNCHIP(R2L)*R2LDOTR12/R2L
6260 CROSS=CROSS+TEMP/(CHI2L+CHI2R)
6270 TPOP=TPOP+2*FNFP(R12)/F*CROSS/R12
6280 TPOP=-.5*HBM*TPOP                                 'final factors
6290 '
6300 VPOP=-E2*(1/R1L+1/R1R+1/R2L+1/R2R-1/R12)          'potential energy
6310 '
6320 EPSILON=TPOP+VPOP                                 'total energy
6330 RETURN
6340 '
7000 '*********************************************************************
7010 'subroutine to calculate PHI and the distances for a given configuration
7020 'input   variables: CONFIG
7030 'output variables: CHI1L,CHI1R,CHI2L,CHI2R,F,PHI,R1L,R1R,R12,R2L,R2R,X1,
7040 '                  X2,Y1,Y2,Z1,Z2
7050 'global variables: S2
7060 'local  variables: none
7070 '*********************************************************************
7080 X1=CONFIG(1): Y1=CONFIG(2): Z1=CONFIG(3)          'electron coordinates
7090 X2=CONFIG(4): Y2=CONFIG(5): Z2=CONFIG(6)
7100 R1L=FNDIST(X1,Y1,Z1+S2)                           'electron-nucleus distances
7110 R1R=FNDIST(X1,Y1,Z1-S2)
7120 R2L=FNDIST(X2,Y2,Z2+S2)
7130 R2R=FNDIST(X2,Y2,Z2-S2)
7140 R12=FNDIST(X1-X2,Y1-Y2,Z1-Z2)                     'inter-electron distance
7150 F=FNF(R12)                                        'electron-electron function
7160 CHI1R=FNCHI(R1R): CHI1L=FNCHI(R1L)                'electron-nucleus functions
7170 CHI2R=FNCHI(R2R): CHI2L=FNCHI(R2L)
7180 PHI=(CHI1R+CHI1L)*(CHI2R+CHI2L)*F                 'total wavefunction
7190 RETURN
7200 '
8000 '*********************************************************************
8010 'subroutine to calculate and print the energy auto-correlation function
8020 'input   variables: ECORR
8030 'output variables: none
8040 'global variables: none
8050 'local  variables: CORR,I%,K%,SUMEI,SUMEIK,SUME2I,SUME2IK,SUMEIEK
8060 '*********************************************************************
8070 CLS                                               'print header
8080 LOCATE 1,24: PRINT "Energy auto-correlation function";
8090 '
8100 FOR K%=0 TO 40                                    'loop over interval
8110    SUMEI=0: SUMEIK=0                               'zero sums
8120    SUME2I=0: SUME2IK=0: SUMEIEK=0
8130    FOR I%=1 TO 200-K%                              'calculate sums
8140       SUMEI=SUMEI+ECORR(I%)
8150       SUMEIK=SUMEIK+ECORR(I%+K%)
8160       SUME2I=SUME2I+ECORR(I%)^2
8170       SUME2IK=SUME2IK+ECORR(I%+K%)^2
8180       SUMEIEK=SUMEIEK+ECORR(I%)*ECORR(I%+K%)
8190    NEXT I%
```

```
8200    CORR=(200-K%)*SUMEIEK-SUMEI*SUMEIK                          'calc. correlation
8210    CORR=CORR/SQR((200-K%)*SUME2I-SUMEI^2)
8220    CORR=CORR/SQR((200-K%)*SUME2IK-SUMEIK^2)
8230 '
8240    IF K%<20 THEN LOCATE K%+2,1 ELSE LOCATE K%-18,40          'display results
8250    PRINT USING "k=## correlation=+##.####";K%,CORR
8260 NEXT K%
8270 '
8280 LOCATE 23,28: BEEP                                           'prompt and wait
8290 PRINT "Press any key to continue";
8300 IF INKEY$="" THEN GOTO 8300
8310 RETURN
8320 '
9000 '*************************************************************************
9010 'subroutine to step the ensemble for 1 time step
9020 'input   variables: ENSEMBLE,WEIGHT
9030 'output variables: EPSILON
9040 'global variables: CONFIG,DRIFT,DT,ETA,SQDT
9050 'local   variables: EBAR,IENSEM%,K%,NORM,WBAR
9060 '*************************************************************************
9070 EBAR=0: WBAR=0                                               'zero E and weight sum
9080 FOR IENSEM%=1 TO NENSEM%                                     'loop over ensemble
9090    FOR K%=1 TO 6                                             'get a configuration
9100       CONFIG(K%)=ENSEMBLE(K%,IENSEM%)
9110    NEXT K%
9120    GOSUB 10000                                               'calculate the drift
9130    FOR K%=1 TO 6                                             'move each coordinate
9140       GOSUB 11000                                            'get gaussian rnd. numb
9150       CONFIG(K%)=CONFIG(K%)+DRIFT(K%)+ETA*SQDT
9160    NEXT K%
9170 '
9180    GOSUB 6000                                                'calculate new epsilon
9190 '
9200    WEIGHT(IENSEM%)=WEIGHT(IENSEM%)*EXP(-EPSILON*DT)'modify the weight
9210    EBAR=EBAR+WEIGHT(IENSEM%)*EPSILON                         'contribution to energy
9220    WBAR=WBAR+WEIGHT(IENSEM%)
9230 '
9240    FOR K%=1 TO 6                                             'store new config.
9250       ENSEMBLE(K%,IENSEM%)=CONFIG(K%)
9260    NEXT K%
9270 NEXT IENSEM%
9280 '
9290 EPSILON=EBAR/WBAR                                            'weighted avg. energy
9300 '
9310 NORM=NENSEM%/WBAR                                            'renormalize weights
9320 FOR IENSEM%=1 TO NENSEM%
9330    WEIGHT(IENSEM%)=NORM*WEIGHT(IENSEM%)
9340 NEXT IENSEM%
9350 RETURN
9360 '
10000 '************************************************************************
10010 'subroutine to calculate the drift vector
10020 'input   variables: CONFIG
10030 'output variables: DRIFT
10040 'global variables: CHI1L,CHI1R,CHI2L,CHI2R,F,HBMDT,R1L,R1R,R12,R2L,R2R,
10050 '                  X1,X2,Y1,Y2,Z1,Z2
10060 'local   variables: FACTA,FACTB,FACTE
10070 '************************************************************************
```

```
10080 GOSUB 7000                                        'calculate the trial function
10090 '                                                 'calculate useful factors
10100 FACTA=HBMDT*(FNCHIP(R1L)/R1L+FNCHIP(R1R)/R1R)/(CHI1L+CHI1R)
10110 FACTB=HBMDT*(FNCHIP(R1L)/R1L-FNCHIP(R1R)/R1R)/(CHI1L+CHI1R)
10120 FACTE=HBMDT*FNFP(R12)/F/R12
10130 '
10140 DRIFT(1)=FACTA*X1+FACTE*(X1-X2)                   'drift for electron 1
10150 DRIFT(2)=FACTA*Y1+FACTE*(Y1-Y2)
10160 DRIFT(3)=FACTA*Z1+FACTB*S2+FACTE*(Z1-Z2)
10170 '                                                 'calculate useful factors
10180 FACTA=HBMDT*(FNCHIP(R2L)/R2L+FNCHIP(R2R)/R2R)/(CHI2L+CHI2R)
10190 FACTB=HBMDT*(FNCHIP(R2L)/R2L-FNCHIP(R2R)/R2R)/(CHI2L+CHI2R)
10200 '
10210 DRIFT(4)=FACTA*X2-FACTE*(X1-X2)                   'drift for electron 2
10220 DRIFT(5)=FACTA*Y2-FACTE*(Y1-Y2)
10230 DRIFT(6)=FACTA*Z2+FACTB*S2-FACTE*(Z1-Z2)
10240 '
10250 RETURN
10260 '
11000 '*****************************************************************
11010 'subroutine to generate a gaussian random number
11020 'input   variables: none
11030 'output  variables: ETA
11040 'global  variables: none
11050 'local   variables: IG%
11060 '*****************************************************************
11070 ETA=0                                             'sum 12 uniform random numbers
11080 FOR IG%=1 TO 12
11090     ETA=ETA+RND
11100 NEXT IG%
11110 ETA=ETA-6                                         'subtract 6
11120 RETURN
11130 '
12000 '*****************************************************************
12010 'subroutine to generate the ensemble at time t=0
12020 'input   variables: none
12030 'output  variables: ENSEMBLE,WEIGHT
12040 'global  variables: A,CONFIG,NENSEM%,
12050 'local   variables: DELTA,IENSEM%,ISTEP%,K%
12060 '*****************************************************************
12070 DELTA=1.5*A                                       'step for Metropolis of PHI^2
12080 GOSUB 14000                                       'generate initial configuration
12090 '
12100 FOR ISTEP%=1 TO 20                                'do 20 thermalization steps
12110     GOSUB 5000                                    'take a Metropolis step
12120 NEXT ISTEP%
12130 '
12140 FOR ISTEP%=1 TO 10*NENSEM%                        'now generate the ensemble
12150     GOSUB 5000                                    'take a Metropolis step
12160     IF ISTEP% MOD 10<>0 THEN GOTO 12210           'use every 10'th configuration
12170     IENSEM%=ISTEP%/10                             'store the configuration
12180     FOR K%=1 TO 6
12190         ENSEMBLE(K%,IENSEM%)=CONFIG(K%)
12200     NEXT K%
12210 NEXT ISTEP%
12220 '
12230 FOR IENSEM%=1 TO NENSEM%                          'set all weights to 1
12240     WEIGHT(IENSEM%)=1
```

```
12250 NEXT IENSEM%
12260 RETURN
12270 '
13000 '***********************************************************
13010 'subroutine to input the sampling parameters
13020 'input   variables: none
13030 'output variables: CORR%,FREQ%,NGROUP%,SIZE%
13040 'global variables: none
13050 'local   variables: none
13060 '***********************************************************
13070 LOCATE ,10: BEEP
13080 INPUT "Enter 0 to calculate energy, 1 to calculate correlation";CORR%
13090 IF CORR%=0 THEN GOTO 13140
13100   CORR%=-1                              'set correlation flag to TRUE
13110   FREQ%=1: NGROUP%=200: SIZE%=1         'sampling params for correlation
13120   GOTO 13270
13130 '
13140 LOCATE ,10
13150 PRINT USING "The current number of groups is ###";NGROUP%
13160 LOCATE ,12: BEEP                        'get sampling params for a run
13170 INPUT "Enter the new number of groups";NGROUP%
13180 LOCATE ,10
13190 PRINT USING "The current number of samples in each group is ###";SIZE%
13200 LOCATE ,12: BEEP
13210 INPUT "Enter the new number of samples in each group";SIZE%
13220 LOCATE ,10
13230 PRINT USING "The current sampling frequency is ###";FREQ%
13240 LOCATE ,12: BEEP
13250 INPUT "Enter the new sampling frequency";FREQ%
13260 '
13270 RETURN
13280 '
14000 '***********************************************************
14010 'subroutine to generate a starting configuration
14020 'input   variables: none
14030 'output variables: CONFIG,W
14040 'global variables: A,S2,PHI
14050 'local   variables: I%
14060 '***********************************************************
14070 FOR I%=1 TO 6                           'put electrons near origin
14080   CONFIG(I%)=(RND-.5)*A
14090 NEXT I%
14100 CONFIG(3)=CONFIG(3)+S2                  'center elec. 1 at right
14110 CONFIG(6)=CONFIG(6)-S2                  'center elec. 2 at left
14120 '
14130 GOSUB 7000                              'calculate PHI
14140 W=PHI*PHI                               'initialize weight
14150 RETURN
14160 '
15000 '***********************************************************
15010 'subroutine to display the header screen
15020 'input   variables: none
15030 'output variables: none
15040 'global variables: none
15050 'local   variables: ROW%
15060 '***********************************************************
15070 SCREEN 0: CLS: KEY OFF                  'program begins in text mode
15080 '
```

```
15090 LOCATE 1,30: COLOR 15                           'display book title
15100 PRINT "COMPUTATIONAL PHYSICS";
15110 LOCATE 2,40: COLOR 7
15120 PRINT "by";
15130 LOCATE 3,33: PRINT "Steven E. Koonin";
15140 LOCATE 4,14
15150 PRINT "Copyright 1985, Benjamin/Cummings Publishing Company";
15160 '
15170 LOCATE 5,10                                     'draw the box
15180 PRINT CHR$(201)+STRING$(59,205)+CHR$(187);
15190 FOR ROW%=6 TO 19
15200    LOCATE ROW%,10: PRINT CHR$(186);
15210    LOCATE ROW%,70: PRINT CHR$(186);
15220 NEXT ROW%
15230 LOCATE 20,10
15240 PRINT CHR$(200)+STRING$(59,205)+CHR$(188);
15250 '
15260 COLOR 15                                        'print title, etc.
15270 LOCATE  7,36:  PRINT "PROJECT 8";
15280 COLOR 7
15290 LOCATE 10,21:  PRINT "Monte Carlo solution of the H2 molecule";
15300 LOCATE 13,36:  PRINT "**********"
15310 LOCATE 15,25:  PRINT "Type 'e' to end while running."
15320 LOCATE 16,22:  PRINT "Type ctrl-break to stop at a prompt."
15330 '
15340 LOCATE 21,8
15350 PRINT "This code finds the Born-Oppenheimer potential of the H2 molecule"
15360 LOCATE 22,12
15370 PRINT "using variational and Path-Integral Monte Carlo methods to";
15380 LOCATE 23,11
15390 PRINT "solve the two-center two-electron problem.  All lengths are";
15400 LOCATE 24,16
15410 PRINT "expressed in Angstroms and all energies are in eV.";
15420 LOCATE 25,31,1:  BEEP
15430 PRINT "Type c to continue";
15440 IF INKEY$<>"c" GOTO 15440
15450 LOCATE ,,0
15460 '
15470 RETURN
15480 '
```

References

[Ab64]   *Handbook of Mathematical Functions*, eds. M. Abramowitz and I. A. Stegun (Dover, New York, 1964).

[Ab78]   R. Abraham and J. E. Marsden, *Foundations of Mechanics, Second Edition* (Benjamin-Cummings Publishing Corp., Reading, 1978) Chapter 8.

[Ac70]   F. S. Acton, *Numerical Methods that Work* (Harper and Row, New York, 1970).

[Ad84]   S. L. Adler and T. Piran, Rev. Mod. Phys. **56**, 1 (1984).

[Ar68]   V. I. Arnold and A. Avez, *Ergodic Problems in Classical Mechanics* (Benjamin Publishing Corp., New York, 1968).

[Ba64]   A. Baker, Phys. Rev. **134**, B240 (1964).

[Be68]   H. A. Bethe and R. W. Jackiw, *Intermediate Quantum Mechanics* (Benjamin Publishing Corp., New York, 1968).

[Be69]   P. R. Bevington, *Data Reduction and Error Analysis for the Physical Sciences* (McGraw-Hill, New York, 1969).

[Bo74]   R. A. Bonham and M. Fink, *High Energy Electron Scattering* (Van Nostrand-Reinhold, New York, 1974).

[Bo76]   J. A. Boa and D. S. Cohen, SIAM J. Appl. Math., **30**, 123 (1976).

[Bu81]   R. L. Burden, J. D. Faires, and A. C. Reynolds, *Numerical Analysis, Second Edition* (Prindle, Weber, and Schmidt, Boston, 1981).

[Br68]   D. M. Brink and G. R. Satchler, *Angular Momentum, Second edition* (Clarendon Press, Oxford, 19680.

[Ca80]   J. M. Cavedon, Thesis, Universite de Paris-Sud, 1980.

[Ce79]   D. M. Ceperley and M. H. Kalos in *Monte Carlo Methods in Statistical Physics*, ed. K. Binder (Springer-Verlag, New York/Berlin, 1979).

[Ce80]   D. M. Ceperley and B. J. Alder, Phys. Rev. Lett. **45**, 566 (1980).

[Ch57]   S. Chandrasekhar, *An Introduction to the Study of Stellar Structure* (Dover, New York, 1957).

[Ch84]   S. Chandrasekhar, Rev. Mod. Phys. **56**, 137 (1984).

[Ch84a] S. A. Chin, J. W. Negele, and S. E. Koonin, Ann. Phys, **157**, 140 (1984).

[Fl78]   H. Flocard, S. E. Koonin, and M. S. Weiss, Phys. Rev. C **17**, 1682 (1978).

[Fo63]   L. D. Fosdick in *Methods in Computational Physics, vol 1*, ed. B. Alder *et al.*, p. 245 (Academic Press, New York, 1963).

[Fo66]   T. deForest, Jr. and J. D. Walecka, Adv. Phys. **15**, 1 (1966).

[Fr73]   J. L. Friar and J. W. Negele, Nucl. Phys. **A212**, 93 (1973).

[Fr75]   J. L. Friar and J. W. Negele, Adv. Nucl. Phys. **8**, 219 (1975).

[Fr77]   B. Frois *et al.*, Phys. Rev. Lett. **38**, 152 (1977).

[Go49]   P. Gombas, *Die Statische Theorie des Atoms* (Springer, Vienna, 1949).

[Go67]   A. Goldberg, H.M. Schey, and J. L. Schwartz, Am. J. Phys. **35**, 177 (1967).

[Go80]   H. Goldstein, *Classical Mechanics, Second Edition* (Addison-Wesley Publishing Company, Reading, 1980).

[Ha64]   J. M. Hammersley and D. C. Handscomb, *The Monte Carlo Method* (Methuen, London, 1964).

[He50]   G. Herzberg, *Spectra of Diatomic Molecules* (D. Van Nostrand Company, Inc., New York, 1950).

[He64]   M. Hénon and C. Heiles, Astron. J. **69**, 73 (1964).

[He80]   R. H. G. Helleman in *Fundamental Problems in Statistical Mechanics, vol. 5*, ed. E. G. D. Cohen (North Holland Publishing, Amsterdam, 1980) pp. 165-233.

[He82]   M. Hénon, Physica **5D**, 412 (1982).

[Ho57]   R. Hofstadter, Ann. Rev. Nuc. Sci., **7**, 231 (1957).

[Hu63]   K. Huang, *Statistical Mechanics* (John Wiley and Sons, New York, 1963).

[Ka79]   K. K. Kan, J. J. Griffin, P. C. Lichtner, and M. Dworzecka, Nucl. Phys. **A332**, 109 (1979).

[Ka81]   M. H. Kalos, M. A. Lee, P. A. Whitlock. and G. V. Chester, Phys. Rev. B **24**, 115 (1981).

[Ka85]   M. H. Kalos and P. A. Whitlock, *The Basics of Monte Carlo Methods* (J. Wiley and Sons, New York, in press).

[Ke78]   B. W. Kernighan and P. J. Plauger, *The Elements of Programming Style, Second Edition* (McGraw-Hill Book Company, New York, 1978).

[Ki71]   L. J. Kieffer, At. Data, **2**, 293 (1971).

[Kn69]   D. Knuth *The Art of Computer Programming, Volume 2: Seminumerical Algorithms* (Addison-Wesley Publishing Company, Reading, 1969).

[La59]   L. D. Landau and E. M. Lifshitz, *Course of Theoretical Physics, Volume 6, Fluid Mechanics* (Pergamon Press, Oxford, 1959).

[Li65]   H. Lipkin, N. Meshkov, and A. J. Glick, Nucl. Phys. **62**, 188 (1965), and the two papers following.

[Mc73]   B. McCoy and T. T. Wu, *The Two–Dimensional Ising Model* (Harvard University Press, Cambridge, 1973).

[Mc80]   J. B. McGrory and B. H. Wildenthal, Ann. Rev. Nuc. Sci., **30**, 383 (1980).

[Me53]   N. Metropolis, A. Rosenbluth, M. Rosenbluth, A. Teller, and E. Teller, J. Chem. Phys. **21**, 1087 (1953).

[Me68]   A. Messiah, *Quantum Mechanics* (J. Wiley & Sons, Inc., New York, 1968]

[Ne66]   R. G. Newton, *Scattering Theory of Waves and Particles* (McGraw-Hill Book Co., New York, 1966).

[Ni77]   G. Nicolis and I. Prigogine, *Self–organization in Nonequilibrium Systems* (J. Wiley and Sons, New York, 1977), Chapter 7.

[Re82]   P. J. Reynolds, D. M. Ceperley, B. J. Alder, and W. A. Lester, Jr., J. Chem. Phys. **77**, 5593 (1982).

[Ri67]   R. D. Richtmeyer and K. W. Morton, *Difference Methods for Initial–value Problems, Second Edition* (Interscience, New York, 1967).

[Ri80]   S. A. Rice in *Quantum Dynamics of Molecules (1980)*, ed. R. G. Wolley (Plenum Publishing Corp., New York, 1980) pp. 257-356.

[Ro76]   P. J. Roache, *Computational Fluid Dynamics* (Hermosa, Albuquerque, 1976).

[Ru63]   H. Rutishauser, Comm. of the ACM, **6**, 67, Algorithm 150 (February, 1963).

[Sh80]   R. Shankar, Phys. Rev. Lett. **45**, 1088 (1980).

[Sh83]   S. L. Shapiro and S. A. Teukolsky, *Black Holes, White Dwarfs, and Neutron Stars* (J. Wiley & Sons, Inc., New York, 1983) Chapter 3.

[Sh84]   T. E. Shoup, *Applied Numerical Methods for the Microcomputer* (Prentice Hall, Inc., Englewood Cliffs, 1984).

[Si74]   I. Sick, Nucl. Phys. **A218**, 509 (1974).

[Va62]   R. Varga, *Matrix Iterative Analysis* (Prentice-Hall, Englewood Cliffs, 1962).

[Wa66]   E. L. Wachspress, *Iterative Solution of Elliptic Systems and Applications to the Neutron Diffusion Equations of Reactor Physics* (Prentice-Hall, Englewood Cliffs, 1966).

[Wa67]   T. G. Waech and R. B. Bernstein, J. Chem. Phys. **46**, 4905 (1967).

[Wa73]   S. J. Wallace, Ann. Phys. **78**, 190 (1973); Phys. Rev. D**8**, 1846 (1973); Phys. Rev. D**9**, 406 (1974).

[We71]   R. C. Weast, *Handbook of Chemistry and Physics*, 52nd edition, (The Chemical Rubber Company, Cleveland, 1971).

[We80]   M. Weissbluth, *Atoms and Molecules* (Academic Press, New York, 1980).

[Wh77]   R. R. Whitehead, A. Watt, B. J. Cole, and I. Morrison, Adv. Nucl. Phys. **9**, 123 (1977).

[Wi74]   A. T. Winfree, Sci. Amer. **220**, 82 (June, 1974).

[Wu62]   T.-Y. Wu and T. Ohmura, *Quantum Theory of Scattering* (Prentice-Hall, Englewood Cliffs, 1962).

[Ye54]   D. R. Yennie, D. G. Ravenhall, and D. N. Wilson, Phys. Rev. **95**, 500 (1954).

To help in the production of future editions of this book and related books, we would appreciate your candid responses to the following questions.

1. Please circle your answers to the following:
    A. I found the level of the mathematics in this text

        a) too elementary   b) about right   c) too advanced

    B. I found the level of the physics in this text

        a) too elementary   b) about right   c) too advanced

    C. I found the level of the programming in this text

        a) too elementary   b) about right   c) too advanced

2. Do you see a need for a similar book at a lower (freshman/sophomore) level? _____

3. What is your discipline or principal research interest? _____ If you are a student, what is your academic level? _____

4. In what other languages (e.g., FORTRAN, PASCAL, C) would you like to see the programs written? _____

5. For what other hardware or operating systems would you like to see this book available? _____

6. Please use the space below for any other comments you wish to communicate to the publisher and author.

Please give your complete name and address below so that we can inform you of future editions of this book or similar books.

_____
_____
_____
_____

For your convenience, simply detach and fold this postage-paid form.